The Nature of Physical Computation

OXFORD STUDIES IN PHILOSOPHY OF SCIENCE

General Editor:
　P. Kyle Stanford, University of California, Irvine

Founding Editor:
　Paul Humphreys, University of Virginia

Advisory Board
　Anouk Barberousse (European Editor)
　Robert W. Batterman
　Jeremy Butterfield
　Peter Galison
　Philip Kitcher
　James Woodward

Recently Published in the Series:

Mathematics and Scientific Representation
Christopher Pincock

Simulation and Similarity: Using Models to Understand the World
Michael Weisberg

Systematicity: The Nature of Science
Paul Hoyningen-Huene

Causation and Its Basis in Fundamental Physics
Douglas Kutach

Reconstructing Reality: Models, Mathematics, and Simulations
Margaret Morrison

The Ant Trap: Rebuilding the Foundations of the Social Sciences
Brian Epstein

Understanding Scientific Understanding
Henk de Regt

The Philosophy of Science: A Companion
Anouk Barberousse, Denis Bonnay, and Mikael Cozic

Calculated Surprises: The Philosophy of Computer Simulation
Johannes Lenhard

Chance in the World: A Skeptic's Guide to Objective Chance
Carl Hoefer

Brownian Motion and Molecular Reality: A Study in Theory-Mediated Measurement
George E. Smith and Raghav Seth

Branching Space-Times: Theory and Applications
Nuel Belnap, Thomas Müller, and Tomasz Placek

Causation with a Human Face: Normative Theory and Descriptive Psychology
James Woodward

The Nature of Physical Computation
Oron Shagrir

The Nature of Physical Computation

ORON SHAGRIR

OXFORD
UNIVERSITY PRESS

Oxford University Press is a department of the University of Oxford. It furthers
the University's objective of excellence in research, scholarship, and education
by publishing worldwide. Oxford is a registered trade mark of Oxford University
Press in the UK and certain other countries.

Published in the United States of America by Oxford University Press
198 Madison Avenue, New York, NY 10016, United States of America.

© Oxford University Press 2022

All rights reserved. No part of this publication may be reproduced, stored in
a retrieval system, or transmitted, in any form or by any means, without the
prior permission in writing of Oxford University Press, or as expressly permitted
by law, by license, or under terms agreed with the appropriate reproduction
rights organization. Inquiries concerning reproduction outside the scope of the
above should be sent to the Rights Department, Oxford University Press, at the
address above.

You must not circulate this work in any other form
and you must impose this same condition on any acquirer.

Library of Congress Cataloging-in-Publication Data
Names: Shagrir, Oron, 1961– author.
Title: The nature of physical computation / Oron Shagrir.
Description: New York, NY, United States of America : Oxford University Press, [2022] |
Series: Oxford studies in philosophy of science |
Includes bibliographical references and index.
Identifiers: LCCN 2021027461 (print) | LCCN 2021027462 (ebook) |
ISBN 9780197552384 (hardcover) | ISBN 9780197552407 (epub)
Subjects: LCSH: Computer science—Philosophy. | Semantic computing. | Computers—Philosophy.
Classification: LCC QA76.167.S53 2021 (print) | LCC QA76.167 (ebook) | DDC 006—dc23
LC record available at https://lccn.loc.gov/2021027461
LC ebook record available at https://lccn.loc.gov/2021027462

DOI: 10.1093/oso/9780197552384.001.0001

1 3 5 7 9 8 6 4 2

Printed by Integrated Books International, United States of America

Contents

Introduction	1
1. Desiderata of a Theory of Computation	6
1.1 Scope	6
1.2 Features	10
1.3 Summary	24
2. Turing's Computability	26
2.1 The 1936 Affair	27
2.2 Turing's Analysis	33
2.3 Who Is "the Computer"?	39
2.4 Effective Computability and Machine Computation	46
2.5 Summary	48
3. Preamble to Machine Computation	49
3.1 Gandy's Account of Machine Computation	49
3.2 Generic Computation	55
3.3 Algorithmic Computation	60
3.4 Physical Computation	70
3.5 Summary	87
4. Computation as Step-Satisfaction	88
4.1 Cummins's Account of Computation	89
4.2 Is Step-Satisfaction Necessary for Computation?	95
4.3 Neural Computation	103
4.4 Summary	118
5. Computation as Implementation	119
5.1 Triviality Results	120
5.2 Avoiding Triviality	129
5.3 From Implementation to Computation	137
5.4 Summary	144
6. Computation as Mechanism	145
6.1 An Outline of the Mechanistic Account	145
6.2 What Is "Mechanistic" in the Mechanistic Account?	148
6.3 Computational and Mechanistic Explanations	151
6.4 Rules, Medium-Independence, and Teleological Functions	167
6.5 Summary	174

Contents

7. The Semantic View of Computation — 175
   - 7.1 What Is a Semantic View of Computation? — 175
   - 7.2 Objections to the Semantic View — 189
   - 7.3 Summary — 200

8. An Argument for the Semantic View — 201
   - 8.1 Simultaneous Implementation — 201
   - 8.2 The Master Argument: From Simultaneous Implementation to the Semantic Individuation of Computational States — 207
   - 8.3 Objection 1: Computational Individuation Is More Basic — 214
   - 8.4 Objection 2: Externalism Without Content — 221
   - 8.5 Summary — 228

9. Computing as Modeling — 229
   - 9.1 What Is Modeling? — 229
   - 9.2 The Modeling Notion of Computation — 238
   - 9.3 Others Who Have Linked Computing to Modeling — 243
   - 9.4 The Methodological Role of Modeling — 249
   - 9.5 Computational Explanations — 253
   - 9.6 Summary — 263

   Conclusion — 264

Acknowledgments — 267
Bibliography — 271
Name Index — 301
Subject Index — 307

Introduction

From laptops to smartphones, computing systems are everywhere today. Even the brain is thought to be a kind of computing system. What does it mean, however, to say that a physical system computes? What is it about laptops, smartphones, and nervous systems such that they are deemed to compute—and why does it seldom occur to us to describe stomachs, hurricanes, or rocks in this way? Answering these questions turns out to be a notoriously difficult task, and scholars have put forward very different accounts of physical computation. Some have even described this situation as a foundational crisis. While I am not sure I would go that far, it is certainly true that clarifying the nature of computation is key to laying the conceptual foundations of the computational sciences, including computer science and engineering, as well as the cognitive and neural sciences. Not surprisingly, philosophers have increasingly focused their attention on the nature of computation. They ask whether computation is objective, to what extent it is pervasive or even trivial, what the precise relations are between computation and representations, and more. In recent years, many philosophers have settled on the mechanistic account (discussed in Chapter 6).

In this book, I offer an extended argument for a variant of the semantic view of computation. This view states that semantic properties are involved in the nature of computing systems. Thus, laptops, smartphones, and nervous systems compute because they have certain semantic properties. Stomachs, hurricanes, and rocks, for instance, which do not have semantic properties, do not compute. The variant of the semantic view defended here consists of three elements. One is *implementation*: a physical computing system implements a formalism of some kind, such as an abstract automaton (elaborated on in Chapter 5). Another element is *representation*: some of the system's physical magnitudes represent objects and properties in a target domain, such as in the individual's physical environment (discussed in Chapter 7). The third and final element is *mirroring*: the computing system preserves certain relations in the target (or "represented") domain (the topic of Chapter 9). These three elements are related in that the input-output function of the implemented formalism underlies the mirroring relation between the computing system and the target domain (explained in Chapter 9). I call this characterization *the modeling account of computation*.

Alongside my positive thesis for the semantic view, I also argue against three premises that have stood in the way of an adequate account of physical computation. The first of these—*the logical dogma*—is that there is a strong linkage between the mathematical theories we find in logic and computer science (e.g., computability theory, automata theory, proof theory) and physical computation. Scholars have described physical computing systems as *syntactic engines* (Stich 1991) and *automatic formal systems* (Haugeland 1981b) as being "illuminated by the idea of a Turing machine" (Crane 2016: 70), while they have described computing processes as executions of *programs* (Cummins 1988), *algorithms* (Copeland 1996), and *effective procedures* (Crane 2016), or as implementations of *automata*. I claim that while these notions are perfectly sound in their original contexts, they lead us astray when applied in the context of physical computing systems. When it comes to characterizing physical computation, we should be careful about using these notions—or entirely refrain from using them. To clarify, I do not deny that physical computing systems can be seen as implementing some formalism. Rather, I contend that the implemented formalism need not be tied to the type of formalism found, for example, in mathematical theories of computability. Much of the first part of the book is devoted to refuting this dogma.

A second premise—*the architectural dogma*—is that the difference between computing and non-computing physical systems has to do (at least in part) with a distinct *abstract causal structure*. This is not meant to imply that non-computing physical systems lack such a structure, but rather to say that they lack the *right kind* of structure. Another way to put it is that computing systems possess the right kind of architectural or functional profile. This distinct profile is often associated with discrete, digital, or stepwise architectures, which are usually markedly different from more continuous and dynamic ones. Physical computing systems have thus been said to have "specific architectural" properties (Newell and Simon 1976: 117) and are defined as *physical symbol systems* (Newell and Simon 1976). Physical computation has been characterized as "the generation of output strings of digits from input strings of digits in accordance with a general rule" (Piccinini 2008b: 34),[1] and computing processes have been described as *step-satisfaction* (Cummins 1988) or as *syntactic processes* (Fodor 1980; Stich 1983)—meaning (roughly) that they occur "in a languagelike medium" (Fodor 1994: 9) or in classical architectures (Fodor and Pylyshyn 1988).

Architectural accounts, as I call them, share the view that possessing the right kind of architectural profile is a necessary condition for computing; therefore, systems that lack this select profile do not compute. These accounts diverge, however, on the question of whether the select architecture is also a *sufficient* condition of computation. Some say yes, while others identify other (necessary)

---

[1] More recently, however, Piccinini (2015) confines this characterization to digital computation only (as discussed in Chapter 6).

features. Some architectural accounts even impose a semantic criterion on computation in addition to the select architecture—namely, that computation operates on semantically evaluable entities (see, e.g., Fodor 1980; Pylyshyn 1984; Fodor and Pylyshyn 1988). I argue (mainly in Chapter 4) against the necessity of architectural profiles. These profiles play a minimal role, if any, in distinguishing computing from non-computing physical systems: either the proposed architectural profile excludes paradigmatic cases of computing or it encompasses too much by applying to virtually every physical system.

It is worth mentioning that I do not deny that computational vehicles are identified with architectural profiles (etc.). In fact, I think that computations are, in some sense, abstract and medium-independent (as will be discussed in Chapters 5 and 6). What I do reject is the notion that computation favors certain kinds of architectural profiles (e.g., digital ones) over others (e.g., more continuous ones). Additionally, I do not deny that architectural profiles are relevant to characterizing various types of computation (e.g., digital computation). Instead, I claim that they are much less relevant to characterizing computing systems. More generally, I deny that the features that are relevant to the distinction between different kinds of computing systems are also relevant to the distinction between computing and non-computing systems (see also Sprevak 2018; Lee 2021).

The third premise is that an account of computation must be *substantive* in some sense, at least if the account is meant to provide solid foundations for the computational sciences. While I believe that my account advances a substantive notion of computation, I think that the initial stipulation of a substantivity premise—that is, starting from the premise that computation is substantive—often leads to overly strong requirements for an account of computation. I will caution against three such requirements that are sometimes associated with a substantivity premise (I discuss a fourth requirement, about the explanatory role of computation, in Chapter 2).[2] The first pertains to the objectivity of computation. Some scholars have argued that computation must be objective. But, as I note in Chapter 2, we can resist a strong form of objectivity about computation without compromising the idea that scientists discover the computational properties of physical systems. A second requirement is that computation should be naturalized. This naturalistic constraint underpins some philosophical theories about the nature of the mind—the computational theory of mind, computational functionalism, and computationalism—and is yet another reason to insist that computation is non-semantic (see Chapter 7). However, I will suggest that there

---

[2] Coelho Mollo (forthcoming), for example, who is sympathetic to the substantivity premise, says: "Given its central place in the computational-representational basic framework of the cognitive sciences, philosophers aim to produce naturalistic theories that yield a robust, objective, non-trivial notion of computation in physical systems." He also recommends avoiding pancomputationalism.

is very little evidence that computation plays this naturalizing role—and even if it does, the naturalistic assumption is consistent with the semantic view. Lastly, many accounts of computation seek to avoid *pancomputationalism*—namely, the claim that every physical system performs computations. There are several forms of pancomputationalism, some with more devastating consequences than others (see Piccinini and Anderson 2018). I suggest that we can live with modest forms of pancomputationalism. The account that I propose is consistent with *very limited pancomputationalism* (my term)—namely, that every physical system *could*, under certain circumstances, perform some computation (Chapter 7).

Some authors have called for a more pluralistic approach to computation (e.g., Chalmers 2012; Lee 2021). I am sympathetic to this call. Given that rapid developments in theory and technology have significantly altered the meaning of the term *computation* over the past century or so, it is inevitable that one will encounter different meanings and uses of the word. In fact, over the years, the concept of computation has indeed undergone dramatic changes. Nonetheless, pluralism about computation cannot mean that everything goes. We have good reason to prefer one conception of computation over another if the latter is incoherent or applicable only in a very narrow domain. My argument for the semantic view proceeds along these lines. While this view does not cover every single meaning or use of the term *computation* that has emerged over the years, I argue that it is more effective than non-semantic accounts in distinguishing computing from non-computing physical systems. Indeed, the semantic view is far more applicable than assumed by its opponents: it is hospitable to new paradigms of computing, and it is especially suited to the use of computation in the contemporary cognitive and neural sciences.

The book consists of three parts, each of which is made up of three chapters. Part I provides general background. Chapter 1 deals with the desiderata of an account of physical computation; Chapter 2 addresses Turing's analysis of human computability; and Chapter 3 distinguishes between different kinds of machine computation. Although varying in scope, these chapters have a common theme—namely, that the linkage between the mathematical theory of computability and the notion of physical computation is weak (see also Copeland et al. 2016).

Part II reviews existing accounts of physical computation. While aiming to cover most major accounts, I analyze three influential accounts in greater depth: Robert Cummins's *step-satisfaction account* (Chapter 4), David Chalmers's *implementing-an-automaton account* (Chapter 5), and Gualtiero Piccinini's *mechanistic account* (Chapter 6). I focus on these accounts for several reasons. First, they explicitly analyze *physical* computation. Second, they have been extensively discussed in the literature on physical computation. Third, these accounts are good representatives of non-semantic accounts: Chalmers's and Piccinini's are explicitly non-semantic, while Cummins's account is non-semantic at least regarding physical functions. Last, and most importantly, I argue that while none of these accounts is satisfactory, each highlights certain key features of physical computation that I eventually adopt in my positive account. While I reject Cummins's characterization of computation in terms of step-satisfaction, my account relies on his notion of *simulation representation* (also known as *input-output representation*). I agree with Chalmers that *medium-independence* (which he characterizes in terms of *organizational invariance*) is necessary for computing, but I also agree with Piccinini that it is insufficient for computing (although for different reasons than those cited by Piccinini). I therefore agree with Piccinini that medium-independence must be supplemented with another feature that defines computation, but, unlike Piccinini, I doubt that a teleological function is the missing element. This paves the way for the semantic view of computation.

Part III focuses on the semantic account. I first explain its primary claim—that semantic properties are involved in the nature of computation—and distinguish it from other, closely related views. I also address various arguments for and against the semantic view (Chapter 7). I then present and defend what I describe as the *master argument* for the semantic view (Chapter 8). This argument has been challenged by Piccinini (2008a, 2015), Coelho Mollo (2018), Dewhurst (2018a), and others, and I respond to their objections. In Chapter 9, I introduce and defend the mirroring aspect of my account and argue that it is at least central to current computational approaches in the cognitive and neural sciences. With this, I complete the modeling characterization of computation.

In summary, the book defends a variant of the semantic view of computation. In the first part of the book, I set the stage for an adequate account; in the second, I highlight some difficulties with extant non-semantic accounts; and in the final part, I articulate and defend the semantic view, and advance a specific (modeling) account.

1
Desiderata of a Theory of Computation

In this chapter, I outline the various demands that arise in a philosophical account of computation. This task is important for two reasons. One is that different lists of demands may lead to different accounts of computation. This can explain why certain accounts of computing that are successful in one domain fail to apply in other domains. The second reason is that some lists of demands set the threshold too high, and thus eventually lead to dead ends. We need a list of desiderata that sets the stage for a doable project, one that is not overly ambitious and therefore ends up with too little. In this chapter, I discuss the two lists of desiderata put forward by Brian Cantwell Smith (1996, 2002) for a general theory of computation and by Gualtiero Piccinini (2007, 2015) for an account of physical (concrete) computation. These authors differ in approach: while Smith focuses on the theory's *scope*, Piccinini formulates a set of *features* that should be included in the theory. I discuss each of these approaches in turn.[1]

1.1 Scope

Smith (2002: 24ff.) states that a comprehensive theory of computing must meet three criteria. The *empirical criterion* is to do justice to real-world examples of computing (such as calculators and desktops). The *conceptual criterion* is to acknowledge related concepts such as *interpretation*, *representation*, and *semantics*. The *cognitive criterion* is to provide solid grounds for the computational theory of mind and for cognitive science.[2]

The account I propose in this book aims to meet these criteria, at least to some degree. The first aim is to account for physical computation (a goal parallel to Smith's *empirical criterion*)—namely, to relate both to real-world examples of computing, such as laptops and smartphones, and to more recent technologies such as neural, quantum, DNA, membrane, and other styles of computing. The

---

[1] See also Fresco (2008) for a discussion of Smith and Piccinini.
[2] In his *On the Origin of Objects*, Smith (1996: 4–13, esp. 5) mentions only two, in which the "conceptual" refers to what he later labels as "cognitive."

The Nature of Physical Computation. Oron Shagrir, Oxford University Press. © Oxford University Press 2022.
DOI: 10.1093/oso/9780197552384.003.0002

second aim is to pay special attention to the claim that cognitive and/or neural systems compute. This aim is similar to Smith's *cognitive criterion*. As for Smith's *conceptual criterion*, I agree that an account of computation should acknowledge closely related concepts such as *interpretation*, *representation*, and *semantics*. This, of course, does not mean providing accounts for these notions, which is even harder than accounting for computation—but rather explaining how these notions relate to computation.

That said, we must restrict the scope of the account. Setting an overly wide scope by trying to account for too much may lead to despairing conclusions (Fresco 2008). Smith himself famously summed up his project with the gloomy remarks that "computation is not a subject matter" (1996: 73); that "there will never be a satisfying and intellectually productive 'theory of computation' of the sort I initially set out to find" (1996: 74); and that "we will never have a theory of computing because there is nothing there to have a theory of" (2010: 38).

One restriction is that we do not have to account for every use of the term *computation*, or every real-world example of it. Neither will I distinguish between the derivatives of computation—that is, computing, computational, etc.—nor, by the same token, will I make too much of the differences between computing entities, such as agents, systems, processes, states, events, and so forth. I will also refer to calculation and its derivatives as synonymous with computation. This is not to say that there are no differences between these terms, but I will draw attention to such differences only when they are important.

A second restriction pertains to various types of theses that state that the physical universe is best modeled as a giant computer or as a network of computational processes of one sort or another (such as a deterministic cellular automaton). The most renowned thesis in this genre was put forward by the computer pioneer Konrad Zuse (1967). Zuse's thesis states that the physical universe is fundamentally a cellular automaton. Whether this thesis is true is an open question, though many believe it to be false.[3] In any event, my account of physical computation does not address these theses. The account is not about the fundamentals of the physical universe, but about the fundamentals of physical computing systems.

A third restriction concerns the use of computer models, simulations, and other tools in the sciences. Here we must distinguish between two meanings, or uses, of the term *computation* in the computational sciences. In one use, *computation* refers to the extensive use of computer models, simulations, and methods in the study of systems and functions. Computational astrophysics, for example,

---

[3] See Wheeler (1990); Schmidhuber (2000); Wolfram (2002); Dodig-Crnkovic and Müller (2011); Dodig-Crnkovic (2017); Copeland, Sprevak, and Shagrir (2017); Copeland, Shagrir, and Sprevak (2018); and Piccinini and Anderson (2018).

refers to the use of these computing tools in studying the heavens. In another use, *computation* refers to the view that the studied system itself computes. This is often the case in the cognitive and brain sciences (and computer science as well), where the nervous system itself is often described as computing. In some sciences, we find only the former use of *computation*. In computational astrophysics, for example, no one describes the modeled systems—the heavens, planetary systems, atmospheres, and so on—as computing systems. In other sciences, we find both uses of computation. Take, for example, Stern and Travis's introduction to the *Science* 2006 special issue on computational neuroscience, in which they define computational neuroscience as the employment of computer models and simulations to study the brain:

> Computational neuroscience is now a mature field of research. In areas ranging from molecules to the highest brain functions, scientists use mathematical models and computer simulations to study and predict the behavior of the nervous system. Modeling has become so powerful these days that there is no longer a one-way flow of scientific information. There is considerable intellectual exchange between modelers and experimentalists. The results produced in the simulation lab often lead to testable predictions and thus challenge other researchers to design new experiments or reanalyze their data as they try to confirm or falsify the hypotheses put forward. (Stern and Travis 2006: 75)

This excerpt is in line with the former meaning of computation—namely, the use of computational tools (such as models and simulations) in neuroscience. Immediately thereafter, however, Stern and Travis assert that the modeled nervous system itself computes, reflecting the latter meaning of *computation*: "Understanding the dynamics and computations of single neurons and their role within larger neural networks is at the center of neuroscience. How do single-cell properties contribute to information processing and, ultimately, behavior?" (2006: 75).[4]

In the present work, I focus on the latter meaning of computation—namely, that a given modeled system performs computations. The aim is to account for the (alleged) fact that laptops, smartphones, brains, and perhaps other systems compute. Thus, henceforward, whenever I use the term *computation*, I am referring to a computing system (unless explicitly stated otherwise). In particular, I will treat computational models, computational descriptions, and computational explanations as models, descriptions, and explanations that refer to

---

[4] A similar dual position can be found in Churchland and Sejnowski (1992); Koch (1999); O'Reilly and Munakata (2000); and Dayan and Abbott (2001).

computing systems.⁵ This is not to say that there are no interesting relationships between the two definitions of the term *computation*, but rather that the account presented here will not dwell upon the use of computer models and simulations in science (i.e., the former definition of computation), which is a separate area of study in its own right.⁶

Yet another restriction pertains to Smith's *cognitive criterion*—namely, providing "intelligible foundation for the computational theory of mind . . . that underlies traditional artificial intelligence and cognitive science" (2002: 24). While my aim is certainly to explicate the concept of computation in the cognitive (and other relevant) sciences, this should be distinguished from other goals that are not necessarily part of an account of physical computation (or at least are not my goals here). This is for three reasons. First, I am not seeking to provide arguments in favor of the computational enterprise in cognitive (or any other related) science, but rather to understand the conceptual framework that underlies the computational enterprise. This understanding may help us to assess the prospects of computational approaches in cognitive, neural, and other sciences. But, ultimately, whether or not these approaches prove to be useful (and if so, to what extent) is largely an empirical issue.⁷ Second, I will not be dealing with the question "What *kinds* of computations are carried out by a system?" One may get the impression that I am seeking to argue in favor of a certain type of neural computation, as opposed to more *classical* approaches (to use Fodor and Pylyshyn's label). However, this is not the case. Rather, my aim is to make sense of the notion of computation that appears in non-classical approaches, not to endorse the theoretical and empirical virtues of those approaches. Whether or not the cognitive system is classical is a question to be settled by further scientific investigation—not by a philosophical account of computation.

Third, my aim is not to provide solid grounds for *philosophical* theories about the mind. Smith says that a theory of computation "must provide a tenable foundation for the computational theory of mind" (1996: 5). But I want to distinguish the claim that the mind/brain computes from certain *philosophical pictures* about the computing mind/brain—in particular, from the influential *computational theory of mind*,⁸ which maintains certain assumptions and agendas that may

---

⁵ But see Piccinini (2015), who distinguishes between computational explanations that are always about computing systems and computational descriptions that might not be.

⁶ For a discussion of the philosophy of computer models and simulation, see, e.g., Humphreys (2004); Frigg and Reiss (2009); Winsberg (2010); and Weisberg (2013).

⁷ Among those who advance arguments against, or alternatives to, computational approaches to cognition are Dreyfus (1972); Searle (1980; 1992); van Gelder (1995); Chemero (2009); Hutto and Myin (2012, 2017); and Hutto et al. (2018). See also Orlandi (2018), who argues that certain theories in perception are only compatible with some accounts of computation.

⁸ According to the latter view, "the mind literally is a digital computer" (Horst 2015); more specifically, "thinking is a computational process involving the manipulation of semantically interpretable strings of symbols, which are processed according to algorithms" (Schneider 2011: 13). In particular,

or may not reflect the theoretical framework and empirical practices of cognitive science. My aim, therefore, is not to provide "a tenable foundation for computational theory of mind"; if anything, my account undermines some of this theory's underlying premises about computation (see my reply to Objection 5 in Chapter 7). Instead, I seek to account for the actual usage of the term *computation* in the cognitive and neural sciences.

While all this is, I admit, still very loose and tentative, the scope will become clearer when we look at the desired features of such an account.

1.2 Features

Piccinini (2015: 11–15) lists six desired features of an account of computing: *objectivity, explanation, the right things compute, the wrong things don't compute, miscomputation is explained*, and *taxonomy*. I will discuss these features in a somewhat wider perspective, labeling the desiderata a bit differently. The *meaning* desideratum, as I will call it, is to explain what it means to say that a physical system computes (Section 1.2.1). The *ontological* desideratum is to explain the objectivity status of computing systems (Section 1.2.2). The *utility* desideratum is to elucidate the role (such as an explanatory role) of computational descriptions (Section 1.2.3). While this book is mainly concerned with fulfilling the first desideratum, I will also say something about the others.

1.2.1 Meaning

When we say that certain systems, modules, processes, or mechanisms *compute*, we mean that they are similar in certain respects to each other. Even more importantly, we want to emphasize that they are different in some respects from other, non-computing systems. Thus, the *meaning* desideratum boils down to *classification conditions* that correctly classify cases of computation as well as non-computation. Piccinini formulates this demand in terms of two criteria:

> *The right things compute.* A good account of computing mechanisms should entail that paradigmatic examples of computing mechanisms, such as digital computers, calculators, both universal and non-universal Turing machines, and finite state automata, compute. (2015: 12)

---

mental operations, as computing processes, are causally sensitive to the syntactic, non-semantic, and non-intentional structure of the symbol (Schneider 2011: 12).

The wrong things don't compute. A good account of computing mechanisms should entail that all paradigmatic examples of non-computing mechanisms and systems, such as planetary systems, hurricanes, and digestive systems, don't perform computations. (2015: 12)

As Piccinini implies, it is unrealistic to have a precise formulation of necessary and sufficient conditions that will clearly classify every system into one of the two classes. There are disputable and borderline cases, such as lookup tables. We would be extremely pleased if our conditions were to correctly classify "paradigmatic examples" of computing and non-computing cases.

Now, what you include in the class of computing systems—and, even more importantly, in the *contrast class* of non-computing systems—pretty much determines the account of computing you end up with. Changing the context, that is, the systems included in each class, can lead to very different accounts of computing. To illustrate the point about the relationships between the inclusive (*things-that-compute*) and contrast (*things-that-don't-compute*) classes that you start with, on the one hand, and the account of computation you end up with, on the other, we must digress a little and compare two characterizations of computation.

Gödel characterizes computation procedures as being "mechanical," which he describes as "purely formal, i.e., refer only to the outward structure of the formulas, not to their meaning, so that they could be applied by someone who knew nothing about mathematics, or by a machine" (1933: 45). Jack Copeland provides a somewhat similar characterization of a mechanical computation procedure, saying that it is one that "demands no insight or ingenuity on the part of the human being carrying it out" (Copeland 2015).[9] In contrast, Sejnowski, Koch, and Churchland claim that "mechanical and causal explanations of chemical and electrical signals in the brain are different from computational explanations. The chief difference is that a computational explanation refers to the information content of the physical signals" (1988: 1300). These two characterizations are strikingly different. Gödel views computation as mechanical procedures that are blind to content, while Sejnowski, Koch, and Churchland argue that computational explanations refer to informational content, while mechanical ones do not. Leaving aside the validity of these characterizations, it is worth noting that they arrive at very different, and indeed contrasting, characterizations (assuming, of course, that computational explanations and computational procedures are related). I would like to suggest that the characterizations are different partly because they are made in very different contexts.

---

[9] Copeland and Gödel also refer to certain finiteness constraints; these are discussed in detail in Chapter 2.

Gödel thought about computation in the context of logic and mathematics, and more specifically in the context of formal systems. He contrasted modes, methods, and procedures that are part of mathematical thinking. One class includes the effective computational procedures, or, as Gödel often calls them, *mechanical procedures*. The contrast class of non-computational or non-mechanical procedures includes other modes of mathematical understanding, thinking, and creativity, which are sometimes referred to as "intuition" or "ingenuity."[10] Following his incompleteness results, Gödel's concern was the extent to which non-computational methods can be expressed by computational ones—or, in other words, whether mathematical thinking can be formalized.[11] In this context of logic and mathematics, it is natural to view computation—that is, mechanical procedures—in terms of blindness to the content of the formulas. When performing computations, the mathematician attends to the "outward structure of the formulas, not to their meaning." When intuition or ingenuity is involved, the mathematician might also take into account the content and the meaning of mathematical expressions.

Like Gödel, Copeland (2015) places the notion of a computation (mechanical) procedure in the context of logic and mathematics. He writes, "The Church-Turing thesis concerns the notion of an *effective* or *mechanical* method in logic and mathematics. 'Effective' and its synonym 'mechanical' are terms of art in these disciplines." For Copeland, the procedure demands "no insight or ingenuity on the part of the human being carrying it out," which underscores that the relevant context here is related to "the human being." The agent that carries out the computation procedure is an (idealized) human being. Gödel also alludes to a human computer when referring to "someone who knew nothing about mathematics." This does not mean that only a human can carry out a mechanical procedure. In fact, Gödel explicitly raises the possibility of computation "by a machine."[12] But this possibility only points to the default, which is human calculation; the benchmark for computability is that which can be calculated by a human, though he or she can be replaced by a machine. The contrast class of non-computing includes modes of thinking that *do* demand "insight or ingenuity." If we extend the *non-computing* class beyond the mathematical domain, we could perhaps add to the non-computing class other phenomena, methods,

---

[10] These rubrics are taken from Turing's (1939) characterization of mathematical thinking.

[11] Gödel's answer to this question is a cautious no. This answer is a consequence of his incompleteness results, which indicate that no formal system captures (in the sense of derivation) all mathematical truths, and of his inclination toward rational optimism, which is the view that the mathematician can in principle prove any mathematical truth. Gödel discusses these issues at greater length in his Gibbs lecture (Gödel 1951).

[12] In a 1963 note added to his 1931 paper on incompleteness, Gödel writes that the "characteristic property [of a formal system] is that reasoning in them, in principle, can be completely replaced by mechanical devices" (p. 195 n. 70).

and processes—such as imagining, hallucinating, dreaming, and feeling—all of which are non-mechanical in the sense that they are sensitive to meaning.

Crucially, the term *computation*, as applied in the context of physical systems, is no longer contrasted with *personal-level* phenomena, such as insight, ingenuity, intuition, and perhaps dreaming and hallucinating. We contrast physical computing systems with other physical systems—such as planetary systems, hurricanes, and digestive systems—that do not compute (to use Piccinini's examples). Even when we say that neural or cognitive processes "compute," we are not ascribing computation to personal-level processes, but rather arguing that the subpersonal, non-conscious processes *underlying* personal-level phenomena compute. For example, we mean that the non-conscious processes underpinning what we see, attend to, recognize, learn, create, and perhaps even dream and intuit are computing processes. The claim, in its strongest form, is that these subpersonal computing mechanisms underpin *all* personal-level phenomena, computing and non-computing alike. Not surprisingly, the *contrast class* of the non-computing cases undergoes a similar change. When we claim that the subpersonal mechanisms are *computing*, we are contrasting them not with the personal-level cases of insight, ingenuity, and intuition, but with processes that are "merely" electrical, chemical, and biological. This is the context of *physical computation*. We want to distinguish physical processes that are described as *computing* from those that are not described as such.

Sejnowski, Koch, and Churchland's characterization should be understood in the context of *physical computation*. They aim to identify the properties that distinguish computational descriptions (and explanations) from other "mechanical and causal" descriptions that merely refer to the chemical and electrical signals in the brain. They appear to claim that these computational properties are relevant to cognition—even to the personal-level phenomena that Gödel classified as non-computing. This claim is perfectly consistent with the assertion that mathematical intuition and ingenuity, when considered at the personal level, is non-computing. The claim, once again, is that the computational properties of the subpersonal neural processes underlying personal-level phenomena are relevant to, and perhaps even constitute, mathematical intuition and ingenuity.

Let us set aside the question of whether or not Sejnowski, Koch, and Churchland's characterization of computation is correct; it is undoubtedly controversial. The point is that their characterization and that of Gödel make sense in the context in which they were made, but less so in other contexts. Gödel's characterization of computing procedures—as referring to the "outward structure of the formulas, not to their meaning"—seems quite reasonable in the context of mathematical (and perhaps other personal-level) thinking, where we want to distinguish computation, which is mechanical in this sense, from intuition, insight, and ingenuity, which seemingly are not. In the context of physical computation, however, it is not as helpful to characterize computation as mechanical, as "mechanical" does not immediately differentiate computation from other electrical, chemical, and biological descriptions. After all, the latter processes are mechanical—that is, blind to meaning—too (indeed, Sejnowski, Koch, and Churchland refer to the *non-computing* descriptions as "mechanical").

Sejnowski, Koch, and Churchland's suggestion—that computational descriptions allude to the informational content of the cells—is plausible in the context of physical computation, where we want to distinguish computation from other, non-computational descriptions. While one may not endorse this characterization of computing, it is at least the sort of characterization that we should take seriously, given that non-computational descriptions—such as electrical, chemical, and biological descriptions—do *not* seem to refer to informational content. However, Sejnowski, Koch, and Churchland's characterization makes little sense in the context of personal-level thinking. It is very odd to characterize computation, in this context, in terms of descriptions that refer to informational content, as it is nonsense to say that non-computing personal-level processes (e.g., intuition, ingenuity, and dreaming) do not refer to informational content.

The upshot of this digression is that our eventual account of computation crucially depends on the classes of *the-right-things-compute* and *the-wrong-things-don't-compute* with which we began. Different classes can lead us to very different answers to the question of meaning. When we compare human personal-level computation with other (non-computing) human capacities, we end up with one answer. When we compare computational with non-computational properties of a physical system or a process, we may end up with a very different answer.

What, then, should be included in the classes of *the-right-things-compute* and *the-wrong-things-don't-compute*? Given that we are aiming to account for physical computation, I would modify Piccinini's paradigmatic examples of computing systems. As for the class of computing systems, we would surely want to include digital computers and calculators in this class. I also think that analog computers are paradigmatic *the-right-things-compute* cases: although they have proven less useful than digital computers, they do have a long and interesting history and, more importantly, a great deal of relevance to current computational work in cognitive science—as the nervous system is no more digital than it is analog.[13] Conversely, I think we had better not include the *abstract* automata and Turing machines as obvious paradigmatic cases of computing systems, since they are not physical or concrete systems (though we should surely say something about them).

---

[13] See Piccinini and Bahar (2013) and Maley (2018).

My view is that we should also include the nervous system in the class of computing systems. One might argue that including the nervous system among the computing systems from the start prejudges the question of whether the nervous system is computational, which is an empirical question. I would still insist on its inclusion, for three reasons. First, the view that nervous systems compute is very widespread. True, we are more confident that laptops compute (one might even take this to be an *analytic* supposition), but the view that the brain computes is deeply entrenched in the cognitive and neural sciences.[14] Second, the aim of an account of physical computation is not to vindicate the claim that brains compute, but rather to explicate the widespread assertion that they do. Indeed, it might turn out one day that this assertion is false and that the brain does not compute—but this also applies to other cases of computing and non-computing systems. It might turn out that hurricanes compute. Yet we understand hurricanes as paradigmatic cases of non-computing systems, as we aim to explicate the widespread (albeit perhaps false) assertion that hurricanes do not compute. Lastly, we cannot ignore the numerous attempts, both in philosophy and in the sciences, to establish a link between laptops and brains—namely, to point out a number of interesting commonalities between brains and computing systems such as laptops. Consequently, those who develop accounts of physical computation aspire to highlight these commonalities. We might find out, while developing these accounts, that there are actually no such commonalities; in that case, we might then decide to exclude brains from the class of computing systems. That said, it makes sense to start with the understanding that motivated our accounts of physical computation in the first place—namely, the promise that laptops and brains both belong to the class of computing systems.[15]

What about the contrast class—that of *the-wrong-things-don't-compute*? Piccinini cites planetary systems, hurricanes, and digestive systems as paradigmatic cases of non-computing systems; one could add to the list rocks, toasters, and perhaps other systems. I do not think that Piccinini's examples are controversial. We could agree that moving-in-orbits, storming, and digesting are non-computing processes. What is more controversial is whether these systems have computational properties at all. If they do, it may well mean that these systems compute under other descriptions (other than the descriptions *moving-in-orbits*, *storming*, and *digesting*). This controversy is directly related to the thesis of *pancomputationalism*—namely, the claim that every physical system computes (I

---

[14] Christof Koch, e.g., says: "The brain computes! This is accepted as a truism by the majority of neuroscientists engaged in discovering the principles employed in the design and operation of nervous systems" (1999: 1). See also the statements by Stern and Travis earlier in this chapter.

[15] One might think about this process, in which we start from a class of paradigmatic examples ("intuitions") but remain open to revising this class as we attempt to formulate a coherent account, in terms of reflective equilibrium (Daniels 2020).

discuss this thesis in detail in Chapter 5). Piccinini appears to hold the view that planetary systems, hurricanes, and stomachs do not have computational properties at all.[16] Chalmers (2011) is more open to the possibility that stomachs (etc.) compute. He thinks that the important difference is that stomachs do not digest in virtue of their computational properties, whereas "with cognition... the claim is that it is *in virtue* of implementing some computation that a system is cognitive" (332–333).

I suppose that those who take the first approach—that digestive systems (stomachs) lack computational properties—tend to characterize the contrast class in *across-system terms*. Thus, Piccinini contrasts computing systems with other physical systems, such as planetary systems, hurricanes, and digestive systems, which we do not refer to as computing systems. Those who adopt the second approach—that digestive systems possess computational properties—might prefer to characterize the contrast class in *within-system terms*. Sejnowski, Koch, and Churchland, for example, contrast computational explanations with "mechanical and causal explanations of chemical and electrical signals in the brain." Thus, they take the contrast class of computing mechanisms to be non-computing mechanisms of the same computing systems. They assert that the computational properties (or explanations) of the (neural) mechanisms extend beyond their chemical and electrical properties. But I would not make too much of this distinction. One can follow the across-system convention without subscribing to the notion that digestive systems lack computational properties; in that case, the computational properties would not be part of the description (and explanation) of digestive, *qua* digestive, processes (this is perhaps closer to how Chalmers describes the contrast class).

To recap: First, an account of physical computation should identify paradigmatic examples of non-computing systems. Most scholars, I think, would agree with Piccinini that digestive systems (etc.), *qua* digestive, are non-computing. Second, the question as to whether or not the paradigmatic examples of non-computing systems have computational properties at all is more controversial. An account of computation need not decide about this issue in advance—both approaches make sense, at least until one develops arguments for and against them. Third, there is the question of whether to characterize the contrast class in within-system or across-system terms. Here too, I do not think that the characterization makes too much of a difference, as long as the computation account is clear about the distinction between computing and non-computing.

---

[16] More recently, Piccinini has put forward a more nuanced position about this, saying that "we accept that there is a sense in which a physical system may perform computations even though it... does not have the *function* to compute" (Piccinini and Anderson 2018: 24).

Piccinini advances two more desiderata—*explaining miscomputation* and *taxonomy*— that are relevant to the identity conditions of computation. I deal with these two desiderata only briefly and focus instead on the classification criteria discussed previously. *Explaining miscomputation* is formulated as follows:

Explaining miscomputation . . . A good account of computing mechanisms should explain how it's possible for a physical system to miscompute. (2015: 14)

When we define a computational capacity (or norm) of a system, we also want to indicate when the system fails to exercise that capacity. The *miscomputation* desideratum allows us to explain this failure. Piccinini provides one account of miscomputation; others have provided somewhat different ones (Fresco and Primiero 2013; Dewhurst 2014; Tucker 2018; Colombo 2021). Although I agree that computation is normative, I will not address this (important) desideratum here.

The *taxonomy* desideratum is articulated as follows:

Taxonomy. Different classes of computing mechanisms have different capacities.... Any account of computing systems whose conceptual resources explain or shed light on those differences is preferable to an account that is blind to those differences. (2015: 14)

This desideratum seeks to explicate the criteria for individuating *types* of computation. These criteria should pertain not only to systems or mechanisms as a whole, but also to types of events, states, and interactions that may be part of the system or mechanism. Although a systematic account of different kinds of computing systems, such as digital versus analog computation, is certainly important, I do not attempt to provide one. I do discuss, however, the distinction between the criteria for distinguishing computing from non-computing systems (*the-right-things-compute* and *the-wrong-things-don't-compute*) and the criteria for classifying different kinds of computation. These two sets of criteria are not identical (see also Lee 2021 and Sprevak 2018). I argue that while architectural profile is relevant to the individuation of computational types (kinds), it is irrelevant to the individuation of computation as such—that is, distinguishing computing from non-computing systems (Chapter 4). Some scholars claim that while semantic properties are relevant to the individuation of computation, they are irrelevant to the individuation of computational types (see Chapter 7). I argue, to the contrary, that semantic properties are relevant both to the individuation of computation and to the individuation of computational types.

1.2.2 Ontology

A second desideratum is to clarify the extent to which computation is objective. Some scholars have insisted that the distinction between computing and non-computing is a matter of fact. Piccinini, for example, puts this demand in the form of the following desideratum:

> *Objectivity.* An account with objectivity is such that whether a system performs a particular computation is a matter of fact. (2015: 11)

In my view, concerns about objectivity are overrated. There is no reason to impose very strong objectivity constraints on an account of physical computation (see also Fresco 2015). What is meant here by objectivity, or "a matter of fact"? One option is to contrast objectivity with observer-dependence. On this understanding, if computers are objective, then they are in the company of other ("natural") kinds such as electrons, neurons, and proteins, which are presumably observer-independent. When scientists study these systems, they appeal to "empirical facts about these systems" (Piccinini 2007: 503). If they are not objective—that is, if they are observer-dependent—then computers have more in common with toasters, chairs, and credit cards, whose identities at least partly depend on the (supposedly intentional) properties of those who observe, use, or design them.[17]

Why should it be a matter of concern that computation is observer-dependent? The answer is mainly that such a denial of objectivity contradicts standard practices and assumptions in the computational sciences, where scientists arguably *discover* observer-independent facts about the systems they study. I address this concern by drawing a distinction between two subclasses of computing systems. One subclass includes computing systems whose computational properties are all objective; we can call these *objective computing systems*. The nervous system might be included in this class. The other subclass is that of *conventional computing systems*. Some of the computational properties of these systems are not objective. We might want to include in this class smartphones, laptops, and some artifacts. Other artifacts, such as robots, may be objective computing systems; I leave that to the reader to decide. This distinction, between objective and conventional computing systems, is not outlandish. Consider the closely related notion of *representation*. Most of us would agree that there are many things whose representational power is a matter of interpretation

---

[17] This sense of observer-relativity is introduced, e.g., by Searle, who says that "there is a distinction between those features that we might call intrinsic to nature and those features that exist relative to the intentionality of observers, users, etc." (1995: 9). I discuss other, more nuanced views in my reply to Objection 4 in Chapter 7.

("derivative") and not a matter of fact. Examples might include words, maps, and even data structures in my laptop. This does not mean that all representations are derivative. Dretske (1988), for one, suggests that there are natural ("objective") systems of representations alongside the conventional ("non-objective") ones; brains are perhaps natural systems of representations. The same distinction between objective and conventional might apply within the overall category of computing systems.

Let us start with the former class, that of objective computing systems. In reference to minds and brains, Piccinini writes that "psychologists and neuroscientists are in the business of discovering which computations are performed by minds and brains. When they disagree, they address their opponents by mustering empirical evidence about the systems they study" (Piccinini 2007: 503). The denial of *objectivity*, however, does not imply that the computational properties of minds and brains are not objective. There may be conventional computing systems whose computational properties (or at least some of them) are not a matter of fact: these properties are derivative, or a matter of interpretation. However, this does not imply that the computational properties of minds and brains are non-objective. It may well be that minds and brains (and perhaps other systems) are objective computing systems. If that is the case, scientists have good reason to search for, and discover, their (objective) computational properties.

If I am right about this, we should distinguish between two claims about objectivity:

Strong objectivity (SO): Every computational property of every (computing) physical system is objective.

Partial objectivity 1 (PO1): Every computational property of some (computing) physical systems (e.g., brains) is objective.

Clearly, satisfying PO1 is enough to meet the desideratum that the computational properties of minds and brains are objective. There is no reason to adopt SO—that is, to assume that the computational properties of all computing systems are objective.

Let us turn to the class of conventional systems, which may include laptops, smartphones, and other machines. One could insist that the computational properties of these systems are also a "matter of fact." As Piccinini puts it:

Computer scientists and engineers appeal to empirical facts about the systems they study to determine which computations are performed by which mechanisms. They apply computational descriptions to concrete mechanisms in a way entirely analogous to other bona fide scientific descriptions. (2007: 503)

If this is the case, the distinction between SO and PO1 is not helpful, since, arguably, there are no conventional computing systems whatsoever. Even smartphones and laptops are objective computing systems.

I agree with Piccinini that computer scientists and engineers discover computational properties of smartphones and laptops, and that these properties are objective. However, this does not yet imply that *all* the computational properties of these systems must be objective. Assume, for the sake of argument, that a system computes if (a) it implements a finite automaton, and (b) it operates on representations. It follows that being computational depends on two features (properties): *implementation* and *representation*. Assume, further, that satisfying the former feature (implementation) is always a matter of fact, but satisfying the latter, semantic condition is a matter of fact in some cases, but conventional in others (as described previously). If one does not like this example, one can always replace implementation and representation with other features, such as *medium-independence* and *teleological function*. The identity of the features is not important for the point that I am making. The point is that a conventional computing system can possess some objective computational properties (implementation) alongside non-objective computational properties (some representations).[18]

Arguably, smartphones, laptops, and some other artifacts are conventional computing systems, in that their representational capabilities are observer-dependent. They depend on our interpretation of their states as representing chess pieces, next month's salaries, and so on. However, some of their computational features are objective: they implement, as a matter of fact, certain finite-state automata. If this is indeed the case, then there is something to *discover* about conventional systems—that is, the objective features of implementation. Scientists aim to do so and may well study and discover the finite automata implemented by the conventional systems.

If this is true, we should distinguish between SO (strong objectivity) and a weaker (partial) objectivity condition:

Partial objectivity 2 (PO2): Some computational properties of every (computing) physical systems (e.g., laptops) are objective.

Clearly, PO2 is consistent with PO1. Thus, the conjunctive claim (PO1 & PO2)—that all the computational properties of minds and brains are objective, as are some of the computational properties of laptops (etc.)—is a consistent one.

---

[18] This category, of conventional systems with objective properties, is not distinctive to computing systems. Dretske's *conventional representational systems Type II* have natural (objective) signs (indicators) whose function (representational power) to indicate is conventional (Dretske 1988).

The upshot is this: Even if there are good reasons to think that computation is observer-independent, there is no need to adopt a strong objectivity (SO) constraint for an account of physical computation. We can be satisfied with a weaker desideratum, namely, PO1 & PO2. My account meets this weaker desideratum.

Another way to understand objectivity is by contrasting it with "free interpretation" (Piccinini 2015: 11). As an example of free interpretation, one may mention the triviality results (discussed in Chapter 5), which enable us to apply any computational description to any physical system. On this understanding, computers are objective in the sense that there are some strict constraints on the way we can assign computational descriptions to physical systems. Assuming that there are also strict constraints on the way we apply toaster-descriptions (etc.) to physical systems, computers might belong to the class of toasters, credit cards, and chairs, and yet be considered objective. More importantly, these constraints leave little room for the scientist (or an observer who is not necessarily a designer or user) to decide whether something is a computer and, if so, what it computes.[19]

I believe that my account meets this objectivity desideratum. I will suggest that there are strict constraints on whether a physical system does or does not compute. However, I am hesitant to impose any objectivity desideratum from the start. One might still insist that some computational descriptions are observer-dependent[20] or loosely constrained by empirical facts about the systems,[21] and yet that they serve as a useful tool in the computational sciences. I see no compelling reason to rule out these more instrumental and less committal approaches in advance. It might turn out that computational descriptions are useful, even if they are not objective at all.

1.2.3 Utility

The utility desideratum's purpose is to explicate the *relevance* and *role* of computational properties (descriptions). The utility desideratum goes beyond the meaning desideratum. The *meaning* question is *what* we mean when we describe a system as a computer, and it arises irrespective of whether or not we actually

---

[19] Coelho Mollo (forthcoming) draws a useful distinction in this context between *thin* perspectives, which are relatively unconstrained, and *thick* perspectives, which are more constrained.

[20] See Schweizer (2019b), who argues that "computational descriptions of physical systems are not founded upon deep ontological distinctions, but rather upon interest-relative human conventions. Hence physical computation is a 'conventional' rather than a 'natural' kind" (p. 27).

[21] See Cao (2018), who argues that "neural computation and neural representation are, in practice, thinner, more liberal, and more observer-relative notions than the types of computation and representation often assumed in theoretical psychology or computational cognitive science" (293).

apply computational descriptions (or refer to computational properties). The *utility* question complements the meaning question by asking *why* we apply computational descriptions (or refer to computational properties) in certain contexts.

The utility question arises regardless of whether or not computational properties are objective. Being objective does not make properties useful. The mass and odor of retinal nerve cells are presumably objective properties, but we do not refer to these properties when describing the contribution of these cells to the visual task of edge detection. Nor do we assume, like Chalmers (2011), that every concrete physical system, including stomachs, has objective computational properties. The alleged fact that these properties are objective does not necessarily make them useful in the context of digestion. The utility question, then, is: Why do we refer to the computational properties of neural processes, but not to the computational properties of digestive processes? Why refer to computational properties of nerve cells and not, say, to their mass and odor?

Being conventional does not render properties useless. Words (i.e., the physical marks on my paper) are representational entities that are (presumably) conventional but are very useful for purposes of communication. The computational properties of laptops and smartphones, even if partly or wholly conventional, are surely useful. Even if every computational property is conventional, we do not actually apply a computational description to every system. In reality, we apply computational descriptions to very few systems. The utility question, then, is why we should apply computational descriptions to some systems and not to others. In other words, what is the utility of computational properties in certain contexts and not in others?

The utility of computation can be more, less, or not at all substantial. We might use the term *computation* in a certain metaphorical sense, or perhaps to attract more attention to certain fields, increase the likelihood of winning grants, and so forth.[22] But utility may play a more substantial role in the study of certain systems. Presumably, when we describe the nervous system as computing, we assume that its computational properties play a substantial role, such as in cognizing. Chalmers (2011), for example, alludes to central conceptions in the philosophy of mind, where computational properties play a formative role in cognition.

Piccinini highlights the *explanatory role* of (physical) computation by invoking the following desideratum:

---

[22] See Boden (2006) and Miłkowski (2013: chap. 2) for a discussion of the "computer metaphor" of the mind.

Explanation ... A good account of computing mechanisms should say how appeals to program execution, and more generally to computation, explain the behavior of computing systems. (2015: 12)

Like Piccinini, I also refer to the explanatory role of computation, although my answer differs significantly from those of Piccinini and others. The question I wish to raise at this point, however, is whether or not an account of computation *should* assign a substantive role for computation. While I do think that computation has a substantive explanatory role, I want to suggest that this substantivity is not a desideratum of an account of computation.

William Ramsey (2007) argues for the substantivity of the notion of *mental representation*. He says that any account of representation should meet "the job description challenge." A successful job description of representation should "enable us to distinguish the representational from the non-representational and should provide us with conditions that delineate the sort of job representations perform, *qua* representations, in a physical system" (p. 27). He further argues that a successful account should indicate how the possession of content is "relevant to what it does in the cognitive system" (p. 27), and "how it is used" (p. 27). Admittedly, Ramsey does not come up with a crisp formulation of these conditions, but states that the posited notion must perform "important explanatory work in a given account of cognition" (p. 3).[23] Although his emphasis is on representation rather than computation, it would be helpful to fine-tune the utility requirement by comparing it with Ramsey's job description challenge.

I agree that an account of a theoretical concept—be it representation, computation, or any other—must clarify its contribution to the scientific investigation. An account of computation should clarify the contribution of computational properties to certain functions, and in particular to cognitive functions. This is precisely what the utility requirement is. But there may be some differences between Ramsey's requirement and mine. One possible difference has to do with the demand that the notion will play the role "*qua* representation" (p. 27) or be "distinctively representational" (p. 31), respectively, with computation. A possible way to understand this *qua*-phrase is that the account should clearly "distinguish the representational from the non-representational" (p. 27) and delineate the sort of job done by *these* representations in a physical system. I agree with this requirement. The account should provide classificatory criteria for distinguishing between computing and non-computing systems—which is the *meaning* requirement. It should also outline the sort of job done by the systems classified as computing systems—which is the *utility* requirement. Perhaps a further way to understand the *qua*-phrase, however, is in terms of the demand

[23] For further discussion, see Ramsey (2007: 27–34).

that the theoretical notion is somehow related to our pre-theoretical conception.[24] I would qualify this demand, however: while I do think that an account of computation should relate the theoretical notion to *some* features of the pre-theoretical conception, I also insist that it is wrong to decide in advance which pre-conceptual features should be reflected in the theoretical notion. As we saw earlier, invoking the pre-theoretical notion of "mechanical," as used in logic and mathematics, can lead us astray when characterizing physical computation.

A second possible difference has to do with Ramsey's requirement that the utility of the notion must be substantive—namely, that it must, for example, perform real explanatory work.[25] I do not impose such a demand in advance. Our goal is not to require an explanatory role, but to clarify whether the notion of computation has one. As philosophers of science, our job is to uncover the role, substantial or otherwise, of computation by examining how the notion functions in scientific theories. I especially resist the requirement to substantiate certain philosophical agendas, such as naturalism (see Chapter 7).[26] Whether or not these agendas are part of the scientific investigation is something to be explored rather than presumed.

It is true that some will find it disappointing if it turns out that computational properties are idle, and that their descriptions do not have a distinct explanatory role. It is also true that if two accounts of computing adequately distinguish computing from non-computing systems, and one assigns a substantial utility to computation and the other does not, we might favor the former over the latter. My point is just that we do not have to commit to substantivity in advance, as it may not be present at all.

1.3 Summary

I have attempted to delineate the scope of an account of physical computation by calling for certain limitations on the proposed inquiry. I have also (re-)formulated the list of desiderata for such an account. A key (meaning) desideratum is to formulate the *classification criteria* that distinguish computing from non-computing physical systems, and ones that distinguish between types of computation. A second desideratum is to clarify to what extent computation is objective. I proposed that the familiar concerns about the (non-)objectivity of

---

[24] See Ramsey (2007: 10–14).
[25] Opponents of this view might include Schweizer (2016, 2019a), who advocates for a sort of computational instrumentalism, and Cao (2018), who proposes that the explanatory virtues of computation in neuroscience might not go beyond those of computational modeling.
[26] Ramsey writes that the agenda is to clarify "how representation can be part of a naturalistic, mechanistic explanation" (2007: 26–27).

computation are assuaged by a weaker objectivity condition, and that we should be open to the possibility that computational properties are not objective at all (although I will not advocate that latter position). A third desideratum addresses the utility of computation. The aim is to specify the role, that is, the explanatory role, of computational descriptions (or properties) when applied to physical systems. Here too, we do not demand in advance that this role be substantive.

2
Turing's Computability

Many philosophical accounts of computing subscribe, in one way or another, to the notion that "to compute" is to follow or to execute an *effective procedure* or an *algorithm*. I use the terms *effective computation, effective calculation*, and *algorithmic computation* interchangeably in reference to any computation (calculation) performed by means of an effective procedure or an algorithm of this sort (in Chapter 3, however, I consider drawing a distinction between effective procedures and algorithms). Similarly, a *function* (of positive numbers) is deemed effectively computable (calculable) if, as Church puts it, "there exists an algorithm for the calculation of its values" (1936a: 102). In the following chapters, my aim is to cut through the tight relationship between algorithms and physical computation. The first step, made in this chapter, is to separate the notions of algorithmic computation, as studied by Church, Turing, and the other founders of computability, and that of machine computation (at this point I will use the more general term *machine*; however, I will gradually disambiguate it to distinguish between physical systems and other kinds of machines).

This chapter focuses on Turing's analysis, which reduces effective computability to *Turing machine computability* (Turing 1936: sec. 9). Turing's analysis is of interest for several reasons. First, Turing provided a precise characterization of what is *effectively computable* (in terms of Turing machine computability). Second, while there were others who offered a precise characterization of effective computability around the same time, Turing's characterization stands out in that it involves an analysis of the *process* of computing. Third, Turing introduced the notion of an *automatic machine* (now known as a *Turing machine*).[1] This notion lies at the heart of computability theory and automata theory even today. Turing also introduced the notion of a *universal Turing machine*: a Turing machine that can simulate the operations of any particular Turing machine, and can thereby compute any function that is computable by any Turing machine. This notion has inspired the development of general-purpose digital electronic computers that now dominate virtually every activity in daily life.[2]

---

[1] Turing referred to it simply as *a-machine*. The term *Turing machine* was coined by Church (1937a) in his review of Turing (1936). The current format of the Turing machine is mainly due to Post (1947).

[2] See Copeland (2012) for a detailed discussion of the impact of Turing's ideas on the developments of the modern computer.

Turing's ingenious work richly deserves the appreciation it has garnered. However, I wish to draw attention to a major limitation of his analysis. As others have already noted, Turing analyzed a *restrictive* class of computations—namely, the calculations that can be carried out by a *human computer*, an idealized human who calculates the values of a function by using (perhaps) a pencil and paper (e.g., Kleene 1952; Gandy 1988; Sieg 1994; Copeland 1997). While this analysis is of immense theoretical and practical importance, it cannot be taken to be the basis of machine computation.

This chapter proceeds as follows. First, I place Turing's 1936 paper in its historical context (Section 2.1), then give an outline of his analysis (Section 2.2). Next, I digress to engage in a discussion of the nature of human computation (Section 2.3). Finally, I discuss the scope of Turing's analysis with respect to machine computation (Section 2.4), arguing that the constraints Turing imposes on effective computation are too restrictive to be applied to computation more generally.

Some guidance for the reader: The material in Sections 2.1 and 2.2 certainly deserves a more fully developed review. I use the notes to expand on certain pertinent points and to suggest further reading. I recommend Sieg (2009), who provides an extensive discussion of much of the material presented in Sections 2.1 and 2.2, and Copeland (2004b), who provides a detailed guide to the material on Turing presented in Section 2.2. Much of what I say in these sections relies on their work. Those who are less interested in the historical and technical background of computability can skip these sections, reading only the summarizing paragraphs of each section. Readers who are less interested in the discussion pertaining to the nature of human computation are advised to skip Sections 2.3.1–2.3.3.

2.1 The 1936 Affair

Four pioneering papers were published in 1936, each of which provides a precise mathematical characterization of effective computability. Alonzo Church (1936a) characterized the effectively computable functions (over the positives) in terms of *lambda-definability*—an undertaking he began in the early 1930s (Church 1933), and which was carried on by Stephen Kleene and Barkley Rosser. Kleene (1936) characterized the general recursive functions, based on the expansion of *primitive recursiveness* by Herbrand (1931) and Gödel (1934).[3] Emil

---

[3] Church (1936a) also refers to this characterization. Subsequently, Kleene (1938) expanded the definition to partial functions. For a historical discussion, see Kleene (1981), as well as Adams (2011), who also discusses some early history.

Post (1936) in New York described "finite combinatory processes" carried out by a "problem solver or worker" (p. 289). Meanwhile, young Alan Turing in Cambridge provided a somewhat similar characterization, but offered the precise characterization in terms of Turing machines. Although Turing was referring to the computability of real numbers, he remarked that "it is almost equally easy to define and investigate computable functions" (p. 58) of countable domains.[4] All these precise characterizations were quickly proven to be extensionally equivalent, as they all define the same class of functions.[5]

Church and Turing—and to some degree Post—formulated versions of what is now known as the *Church-Turing thesis* (CTT). Church's classic formulation was as follows:

> We now define the notion, already discussed, of an effectively calculable function of positive integers by identifying it with the notion of a recursive function of positive integers (or of a λ-definable function of positive integers). (Church 1936a: 100)

Kleene coined the term *thesis* and formulated the thesis as follows:

> Thesis I. Every effectively calculable function (effectively decidable predicate) is general recursive. (Kleene 1943: 60)[6]

In this book, we will adhere to Kleene's formulation. The statement is called a "thesis" because it links a pre-theoretical notion—that of effective (algorithmic) calculability—with the precise notion of *general recursiveness*, or Turing machine computability.[7] Arguably, due to the pre-theoretical notion, such a statement is not subject to mathematical proof.[8] But we will leave aside questions of provability and focus on what is meant by "effective computation." To address this, we should first explicate the motives that prompted the attempts to characterize computability, which culminated in the so-called 1936 Affair.

---

[4] Many scholars have noted that the Turing machine operates on strings of symbols (e.g., digits), and therefore its applicability to functions of natural numbers requires interpretation (see the discussion in my reply to Objection 2 in Chapter 7). I will not discuss here the interesting debate on acceptable and "deviant" interpretations (see, e.g., Shapiro 1982; Rescorla 2007; Boker and Dershowitz 2009; Copeland and Proudfoot 2010).

[5] See Kleene (1981) for the historical details.

[6] It should be noted that, in Kleene's characterization, the "easy part" of the thesis is omitted. Later, Kleene (1952) discusses Church's thesis and Turing's thesis, but today this formulation is known as the Church-Turing thesis.

[7] Some have described the notion as "intuitive" or "vague," rather than "pre-theoretic"—see the discussion in Shapiro (2013).

[8] The claim that CTT is not provable has been challenged, e.g., by Mendelson (1990), but see also the discussion in Folina (1998, 2006) and Shapiro (1981, 1993, 2013); for further discussion, see Copeland and Shagrir (2019) and Boker and Dershowitz (forthcoming).

The use of algorithms in solving mathematical problems goes back at least as far as Euclid. Its use as a method of proof is found in Descartes, and its association with formal proof is noted by Leibniz. But algorithms came to the fore of modern mathematics only toward the end of the nineteenth century and the early twentieth. This development had two related sources. The first was the publication of various foundational works in mathematics that explored the concept of the algorithm, starting with Frege and culminating in Hilbert's finitistic program in the 1920s. Although there are substantive differences between these enterprises, they share one core idea, which is to provide mathematics with epistemological foundations by means of *logical calculus*.[9] The term *calculus* refers to a system of logical axioms and inference rules whose constructions are "effective"—namely, that there is an algorithm (effective procedure) that one can use to decide whether or not a string of symbols is a (formal) proof or derivation. In the 1920s, Hilbert also required that there be a meta-theoretic consistency proof of the axiomatic system, notably of the one (supposedly) embedding number theory.[10] The demand was that the meta-theoretic proofs exploit only finitistic means, of which effective procedures were considered to be a major, if not the only, resource. As is well known, the prospects of Hilbert's program were dashed with the publication of Gödel's incompleteness results in 1931.[11]

The other source was the growing number of decision problems that attracted mathematicians.[12] The most pertinent problem was the *Entscheidungsproblem* ("decision problem"), which concerned the decidability of logical systems. It was raised by Hilbert and his students during the 1920s; Hilbert and Ackermann (1928) specifically introduced the *Entscheidungsproblem* with respect to the restricted functional calculus (first-order predicate logic), describing it as the "most fundamental problem of mathematical logic."[13] The problem asked whether there was an algorithm for deciding whether or not a formula in the calculus is derivable;[14] such an algorithm, it was hoped, would provide a decision

---

[9] See Sieg (2009) for further historical exposition of the early works on computability.

[10] See, e.g., Hilbert (1926); for a survey and discussion, see Sieg (2013a) and Zach (2019).

[11] When he published his paper, Gödel commented that his results did not undermine the prospects of Hilbert's program (1931: 195)—but apparently changed his mind about this not long after.

[12] Thus, in his review of Turing (1936), Church (1937a) refers to:

the notion of effectiveness as it appears in certain mathematical problems (various forms of the *Entscheidungsproblem*, various problems to find complete sets of invariants in topology, group theory, etc., and in general any problem which concerns the discovery of an algorithm). (p. 43)

[13] Hilbert and Ackermann, *Grundzüge der Theoretischen Logik* (1928). See the translation in Gandy (1988: 58). The problem appears to have been formulated by Behmann as early as 1921 (Gandy 1988: 57; Mancosu 1999: 320–321).

[14] Establishing the completeness of first-order logic (Gödel 1929), the problem applies equally well to the validity of formulas.

procedure for the provability of any mathematical sentence.[15] Turing (1936) and Church (1936a, 1936b) aimed at this problem in 1936, and both proved, independently of each other, that there was no such algorithm for the derivability (or validity) of any formula in first-order predicate logic.

Before Church and Turing, however, there was no agreement on the precise definition of effective computability, and therefore of decidability. Indeed, there was not even a unified terminology. In formulating his famous *10th problem*—of deciding any Diophantine equation with rational coefficients—Hilbert refers to "*a process according to which it can be determined by a finite number of operations whether the equation is solvable in rational integers*" (1902: 458). Church, as we have seen, used the term *effective calculation*, as well as *effectively calculable function* (1933, 1936a). Gödel, who used the terms *mechanical procedure* and *finite procedure*, defined the procedure's properties in the context of formal systems.[16] The property of *being mechanical* was spelled out in Gödel's 1933 address to the Mathematical Association of America, "The Present Situation in the Foundations of Mathematics." Gödel opened the address with a rough characterization of formal systems:

> The outstanding feature of the rules of inference being that they are purely formal, i.e., refer only to the outward structure of the formulas, not to their meaning, so that they could be applied by someone who knew nothing about mathematics, or by a machine (p. 45).

We discussed this property in Chapter 1. The property of *being finite* is stressed in Gödel's Princeton 1934 address, where he characterized a "formal mathematical system" as follows:

> We require that the rules of inference, and the definitions of meaningful formulas and axioms, be constructive; that is, for each rule of inference there shall be a finite procedure for determining whether a given formula B is an immediate consequence (by that rule) of given formulas $A_1, \ldots, A_n$, and there shall be a finite procedure for determining whether a given formula A is a meaningful formula or an axiom. (p. 346)

---

[15] A nice illustration is provided by von Neumann (1927):

> The very day on which the undecidability does not obtain any more, mathematics as we now understand it would cease to exist; it would be replaced by an absolutely mechanical prescription [*eine absolut mechanische Vorschrift*] by means of which anyone could decide the provability or unprovability of any given sentence. Thus we have to take the position: it is generally undecidable, whether a given well-formed formula is provable or not. (Translation in Sieg 2009: 526)

[16] Gödel often vacillated between the terms *mechanical procedure* and *finite procedure* when referring to formal systems (see discussion in Shagrir 2006a). But in his 1972 note—in which he states that it is clear that he does not take the two to be synonymous—he talks about finite but non-mechanical procedures.

At this point, in 1934, Gödel understood the finite procedures to be the (primitive) recursive ones but did not assert that the latter was a precise definition of the former.[17]

Given the centrality of effective computability in logic and mathematics, it is perhaps surprising that the attempts to characterize it were not as extensive prior to the 1930s. In any event, this situation changed with the publication of the incompleteness results (Gödel 1931), which shattered the collective optimism about certain decidability questions, and, consequently, called for a more accurate definition of the notion of *effective computability*. As Martin Davis explains, "A *positive solution* to a decision problem consists of giving an algorithm for solving it; a *negative solution* consists of showing that no algorithm for solving the problem exists" (1958: xvi). When the problem has a positive solution, we "simply" have to find a pertinent algorithm for it—namely, an algorithm that solves the problem (we then might have to explain, or even prove, why there is a match between the proposed algorithm and the problem). When the problem has a negative solution, however, we must show that *no* algorithm solves it, and this already calls for a precise characterization of the notion of algorithm.

Two issues about undecidability are often mentioned in the context of the search for the precise characterization. One is the generality of the incompleteness results. The results were proved with respect to the system *P* and its extensions. The system *P* is "essentially the system obtained when the logic of *PM* [*Principia Mathematica*] is superposed upon the Peano axioms" (Gödel 1931: 151). The extensions are the "ω-consistent systems that result from *P* when recursively definable classes of axioms are added" (p. 185 n. 53). At the time, it was still an open question whether the undecidability results apply to every formal system; the question, in other words, was whether the recursively definable classes encompassed the relevant, effectively definable classes. But to determine the answer to this question, one would need a precise definition of the latter. Gödel himself wrote the following on this point:

---

[17] In fact, in a letter to Martin Davis (February 15, 1965), Gödel denies that the 1934 paper anticipated the Church-Turing thesis:

> It is not true that footnote 3 is a statement of Church's thesis. The conjecture stated there only refers to the equivalence of "finite (computation) procedure" and "recursive procedure." However, I was, at the time of these lectures, not at all convinced that my concept of recursion comprises all possible recursions; and, in fact, the equivalence between my definition and Kleene's [1936] is not quite trivial. (Quoted in Davis 1982: 8)

In a letter to Kleene (dated November 29, 1935), Church, who had apparently met Gödel early in 1934, reports Gödel's note to Davis:

> In regard to Gödel and the notions of recursiveness and effective calculability, the history is the following. In discussion with him the notion of lambda-definability, it developed that there was no good definition of effective calculability. My proposal that lambda-definability be taken as a definition of it he regarded as thoroughly unsatisfactory. (Quoted in Davis 1982: 9)

When I first published my paper about undecidable propositions the result could not be pronounced in this generality, because for the notions of mechanical procedure and of formal system no mathematically satisfactory definition had been given at that time. This gap has since been filled by Herbrand, Church and Turing. (193?: 166)[18]

Subsequently in that paper, and many times after, Gödel attributed the most convincing characterization of mechanical procedure to Turing. In a 1963 note added to the 1931 paper, Gödel wrote that Turing's characterization established the generality of the incompleteness results:

> In consequence of later advances, in particular of the fact that due to A. M. Turing's work a precise and unquestionably adequate definition of the general notion of formal system can now be given, a completely general version of Theorems VI and XI [the incompleteness results] is now possible. That is, it can be proved rigorously that in *every* consistent formal system that contains a certain amount of finitary number theory there exist undecidable arithmetic propositions and that, moreover, the consistency of any such system cannot be proved in the system. (1963: 195)

In a paragraph in the 1964 postscript to his 1934 paper, Gödel adds that "Turing's work gives an analysis of the concept of 'mechanical procedure' (alias 'algorithm' or 'computation procedure' or 'finite combinatorial procedure'). This concept is shown to be equivalent with that of a 'Turing machine'" (1964: 369–370).

The other problem is the *Entscheidungsproblem*. Given incompleteness, it was very likely to be undecidable.[19] But, again: to demonstrate unsolvability, one must show that no effective (algorithmic) procedure, of which there are infinitely many, solves the problem. One must therefore come up with a precise definition of *effective computability*. As previously noted, Church and Turing aimed to do just that. Church eventually proved the unsolvability of the *Entscheidungsproblem* by means of the notions of lambda-definability and recursiveness. Turing came up with a different approach: he reduced the concept of an algorithmic procedure to that of a Turing machine, then proved that no Turing machine could solve the *Entscheidungsproblem*.

In summary: We have briefly reviewed the fascinating events within the world of logic and mathematics that led to the precise mathematical characterization of effective computability. In this historical context, the characterization

---

[18] Davis dates the article to 1938. See his introduction to this paper in Gödel (1995).
[19] See Gandy (1988) for discussion.

of computability is tightly linked to the development of rigorous definitions for the notions of formal system and decision procedure. These were required in order to establish central undecidability results—in particular, the generality of Gödel's incompleteness results and the insolvability of Hilbert and Ackermann's *Entscheidungsproblem*.

2.2 Turing's Analysis

The Church-Turing thesis asserts that every effectively computable function is a general recursive function (or Turing machine computable). But what are the grounds for this thesis? In his book, Kleene (1952) lists four kinds of justification. The first two are the arguments of *confluence* and of *non-refutation*, which are prevalent in current textbooks in computability, logic, and automata theory. The confluence argument states that many characterizations of computation that differ in their goals, approaches, and details nonetheless encompass the same class of computable functions. As we have seen, the confluence of four such characterizations appeared in 1936, and many more characterizations have followed.[20] The non-refutation argument states that the thesis, though refutable, has not been refuted despite the many efforts and attempts to find a counterexample.[21] Both arguments are of an inductive nature: the more examples you have (either of yet another precise characterization of computability, or of yet another computable function that turns out to be recursive), the more the thesis is confirmed. Indeed, these arguments strengthen the impression that the thesis is not subject to mathematical proof.

The other two arguments are more direct, in that they deal, in one way or another, with the process of computing.[22] One argument, put forward by Church (1936a: 100–102), is known as the *step-by-step argument*.[23] Using Gödel's notion of *representability* (Gödel 1931; Kleene 1936), Church characterizes an effectively computable function as one that is calculable in logic. As he puts it:

---

[20] Boolos and Jeffrey write: "Indeed, given any other plausible, precise characterization of computability that has been offered, one can prove by careful, laborious reasoning that our notion is equivalent to it in the sense that any function which is computable according to one characterization is also computable according to the other" (1989: 20).

[21] Boolos and Jeffrey write: "It [CTT] is refutable by finding a counterexample; and the more experience of computation we have without finding a counterexample, the better confirmed the thesis becomes" (1989: 20).

[22] Quinon (2021) further suggests that Turing's and Church's arguments are examples of Carnapian explication.

[23] The label is Gandy's (1988). The argument is analyzed in detail by Sieg (1997). The argument has two versions; here I mention the second one. A variant of this version also appears in Turing 1936 (sec. 9, II), and is elaborated and discussed in Kripke (2013).

Let us call a function F of one positive integer *calculable within* the logic if there exists an expression f in the logic such that $\{f\}(\mu) = \nu$ is a theorem when and only when $F(m) = n$ is true, μ and ν being the expressions which stand for the positive integers m and n. (1936a: 101)

The rationale behind this characterization is the tight relationship between effective computation and logical *derivation*. A function is (intentionally) effectively computable only if there is a derivation of the corresponding ("representing") logical formula ("expression"), when we replace the number values m and n with the corresponding constants m and n. This characterization highlights the close kinship that existed between formal systems and effective computability at the time. On the one hand, a formal system is characterized in terms of effective computability; on the other, effective computability is defined in terms of formal derivability. A variant of this characterization appears in Turing (1936), Hilbert and Bernays (1939), Church (1941: 41), and Gödel (1946).[24] Given this characterization, Church proceeds to show that if each step of the derivation is general recursive, then the defined function is recursive as well. What is left open, however, is the assumption that these basic steps must be recursive.[25] As Sieg points out, this argument is "semicircular in the sense that he [Church] assumed without good reason that the necessarily elementary calculation steps have to be recursive" (2006: 193).

Our focus here, however, is Turing's argument for the thesis, known as *Turing's analysis*. The argument is explicated by Kleene (1952), who refers to the conclusion as *Turing's thesis* (p. 376). For many years, this argument was not well known, so logic and computer science textbooks, even today, often ignore it.[26] One notable exception was Gödel (193?, 1951, 1964; Wang 1974), who said that Turing's analysis produces a "correct and unique" definition of "the concept of mechanical" in terms of "the sharp concept of 'performable by a Turing machine'"—and that it is "absolutely impossible that anybody who understands the question and knows Turing's definition should decide for a different concept"

---

[24] In a footnote added to his 1946 address (for the Davis anthology), Gödel *defined* a computable function f in a formal system S "if there is in S a computable term representing f" (1946: 84).

[25] Church views the argument not as a conclusive proof, but as a "positive justification ... for the selection of a formal definition to correspond to an intuitive notion" (1936a: 100).

[26] Turing's argument is mentioned in the early days of automata theory—e.g., by McCulloch and Pitts (1943); Shannon and McCarthy (1956) in their introduction to *Automata Studies*; and Minsky (1967: 108–111), who cites it almost in full in his *Finite and Infinite Machines*. However, even Minsky asserts that the "strongest argument in favor of Turing's thesis is the fact that ... satisfactory definitions of 'effective procedure' have turned out to be equivalent" (p. 111). Apart from Minsky, I know of no other mention of Turing's argument in logic and computer science textbooks. The two arguments typically given for the Church-Turing thesis are the confluence (equivalence) of definitions and the lack of counterexamples.

(Wang 1974: 84).[27] But Gödel does not explain here, or elsewhere, why the analysis is so convincing.[28] Turing's analysis has been fully appreciated more recently by Gandy (1988), Sieg (1994), and Copeland (2004b). Gandy (1988) and Soare (1996) even use it to prove the Church-Turing thesis.[29] Here, I will only provide an outline.[30]

Turing begins his 1936 paper by stating that "the 'computable' numbers may be described briefly as the real numbers whose expressions as a decimal are calculable by finite means" (1936: 58). He proposes an explicit definition of computability with (Turing) machine computability, arguing that its "justification lies in the fact that the human memory is necessarily limited" (p. 59). Turing then compares "a man in the process of computing a real number to a machine which is only capable of a finite number of conditions [configurations]" (p. 59). After an informal exposition of the machine's operations, Turing comments that it is his "contention that these operations include all those which are used in the computation of a number" (p. 60).

In section 9, Turing returns to justify the identification of (effective) computability with Turing machine computability. He provides three arguments, remarking that "all arguments which can be given are bound to be, fundamentally, appeals to intuition, and for this reason rather unsatisfactory mathematically" (p. 74). The first argument, discussed here, is a "direct appeal to intuition" and "is only an elaboration of the ideas of [section] 1" (p. 75). It is presented in part I of section 9, and then a modification is added in part III. The second argument, in part II, is a variant of the step-by-step argument noted previously. The third argument, in section 10, consists of "examples of large classes of numbers which are computable" (p. 75).

Turing's analysis (i.e., the first argument) rests on two ingenious ideas. One is that in order to characterize the computable functions (or numbers, or relations) we should focus on the computational *processes*: "The real question at issue is 'What are the possible processes which can be carried out in computing

---

[27] In his Gibbs lecture, Gödel said:

There are several different ways of arriving at such a definition, which, however, all lead to exactly the same concept. The most satisfactory way, in my opinion, is that of reducing the concept of finite procedure to that of a machine with a finite number of parts, as has been done by the British mathematician Turing. (Gödel 1951: 304–305)

For further discussion, see my "Gödel on Turing on Computability" (Shagrir 2006a).

[28] See also Church, who describes Turing's identification of effectiveness with Turing machine computability as "evident immediately" (1937a: 43), "an adequate representation of the ordinary notion" (1937b: 43), and as having "more immediate intuitive appeal" (1941: 41), but does not say why it is more convincing than other arguments.

[29] The rediscovery of Turing's analysis is underscored by Martin Davis's comment (1982: 14 n. 15) that "this [Turing's] analysis is still very much worth reading. I regard my having failed to mention this analysis in my introduction to Turing's paper in Davis (1965) as an embarrassing omission."

[30] For a detailed exposition, see Copeland (2004b, 2006) and Sieg (1994, 2002, 2008, 2009).

a number?'" (p. 74). This idea is supported by the fact that the computable functions are the products of certain computational (i.e., finite and mechanical) processes; thus, if we want to exhaust the set of computational functions, we must be clear about the set of processes that lead to those functions.

Turing puts forward three underlying assumptions about these processes. One is that they are symbolic processes, in the sense of writing and erasing symbols: "Computing is normally done by writing certain symbols on paper" (p. 75). He adds that we can assume that the paper is one-dimensional (that is, like a tape) and divided into squares, although this is not essential for the argument. A second assumption is that the process is *step-wise*, in that it consists of a sequence of steps; each such step involves a change in symbolic configurations. The number of steps in a single run is finite but unbounded when the function is defined for a certain input; otherwise, it consists of infinitely many steps ("not halting"). A third assumption is that there is an agent carrying out this computation; arguably, this agent is a *human* computer. Turing says that "we may compare a man in the process of computing a real number to a machine which is only capable of a finite number of conditions" (p. 59). For now, I will use the more neutral term *the computer* for the computing agent and discuss its relationship with human computers in Section 2.3.

The first idea, then, is to focus on processes. The problem with this direction, however, is that it is unclear how one might characterize all possible processes, given that there are an infinite number of them. Turing's truly wonderful (second) idea is to formulate a finite set of restrictive constraints that apply to each step of each process. These constraints should have two properties. The first is that they must be general enough and thus their truth is almost self-evident. The second property is that a Turing machine can mimic the operations restricted by these constraints. We can think of these constraints as "axioms of computability."[31]

Overall, the analysis can be described as a two-premise argument (Shagrir 2002):

Premise 1: "The computer" operates under the restrictive conditions 1–5 (to be specified in what follows).

Premise 2 ("Turing's theorem"): Any function that can be computed by a(ny) computer that is restricted by conditions 1–5 is Turing machine computable.

Conclusion ("Turing's thesis"): Any function that can be computed by "the computer" is Turing machine computable.

---

[31] This is in line with Gödel's suggestion to Church (in Church's 1935 letter to Kleene) to "state a set of axioms which would embody the generally accepted properties of this notion, and to do something on that basis" (Davis 1982: 9). That Turing's analysis fulfills Gödel's desideratum is suggested by Sieg (2002: 400).

Turing (somewhat informally) enumerates several constraints that can be summed up by the following restrictive conditions:

1. "The behaviour of the computer at any moment is determined by the symbols which he is observing, and his 'state of mind' at that moment" (p. 75).

He then formulates boundedness conditions on each of the two determining factors—namely, the observed symbols and states of mind:

2. "There is a bound B to the number of symbols or squares which the computer can observe at one moment" (p. 75).
3. "The number of states of mind which need be taken into account is finite" (p. 75).

There are certain "simple operations" (behavior) that the computer may perform at any given moment (computation step): a change in the symbols written on the tape, a change of the observed squares, and a change in state of mind. Turing then applies boundedness conditions on the first two kinds of operations (assuming that the number of states of mind is bounded):

4. "We may suppose that in a simple operation not more than one symbol is altered" (p. 76).
5. "Each of the new observed squares is within L squares of an immediately previously observed square" (p. 76).

Let us examine the rationale behind these constraints a little more closely. Turing does not give reasons for the first condition: he apparently takes it to be a *sine qua non* of the concept of effective computation. The reason for the second condition, he says, is that the computer can observe only a bounded region, and that there is a lower bound on the size of the symbols: "If we were to allow an infinity of symbols, then there would be symbols differing to an arbitrarily small extent" (p. 75). Hence, presumably, "the computer" cannot tell one from another. As for the third condition, Turing says that "the reasons for this are of the same character as those which restrict the number of symbols. If we admitted an infinity of states of mind, some of them will be 'arbitrarily close' and will be confused" (pp. 75–76). Following this somewhat obscure statement, Turing stresses that the restriction regarding states of mind "is not one which seriously affects computation," and that we can always replace states of mind with "writing more symbols on the tape" (p. 76). In part III, Turing suggests avoiding states of mind altogether "by considering a more physical and definite counterpart" (p. 79) such as written symbolic configurations.

At first glance, the fourth condition might appear too restrictive, as we might be able to change more than one symbol at a time. But it is apparent that we can extend the condition as long as "one symbol" is replaced with another number that serves as a fixed limit on the number of symbols that can be altered at any one moment. Following Turing, we may assume, without loss of generality, "that the squares whose symbols are changed are always 'observed' squares" (p. 76). As for the fifth condition, Turing says that it is reasonable to suppose that the distance between current and previously observed squares "does not exceed a certain fixed amount" (p. 76).

It is often said that a Turing machine is a model of a computer, arguably a human computer.[32] We can now see that this statement is imprecise. A Turing machine is a *letter machine*: at each point it "observes" only one square on the tape. "The computer" might observe more than that (Figure 2.1). Other properties of a Turing machine are also more restrictive. The machine can perform precisely three kinds of bounded operations at each step: it can change the observed symbol alone; it can change the observed square (and symbol) that is either immediately to the left or immediately to the right of the current square; and it can change the state of the finite program. "The computer," however, might change something other than the observed symbol, shift eyes to non-adjacent squares, and so forth. A Turing machine is therefore *more restrictive* than "the computer": its restrictive conditions are tighter than conditions 1–5.

The aim of the second premise in the analysis is to demonstrate that the restrictive conditions 1–5 are nonetheless bounded by the Turing limit—namely, that the operations performed by *any* computing agent that satisfies conditions 1–5 can be reduced to a finite series of successive operations performed by a Turing machine. A change in a bounded region of symbols can be reduced to a (possible) change in each of the single symbols that make up this region, and a bounded change of the observed square can be reduced to a sequence of shifts of one observed square. Turing provides an outline of the reduction, and Kleene (1952) and others provide more detailed demonstrations. Based on Gandy's insights, Sieg (2002) formulates the conditions in *formal* terms as mathematical axioms (we will return to these points in Section 2.3.2).

To summarize the argument (of Turing's analysis): the first premise encapsulates a novel characterization of effective computation—namely (to put it in Gödel's terms), a characterization of computation as a mechanical and finite procedure. The mechanical part is described in terms of the operations

---

[32] Thus, see Gödel's quote cited earlier, in which he says that the concept of mechanical procedure "is shown to be equivalent with that of a 'Turing machine' " (1963: 369–370).

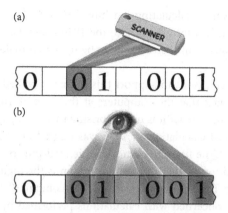

Figure 2.1 A human computer versus a Turing machine. (a) A Turing machine "observes" only one square on the tape at any given moment. (b) "The computer" can observe only a bounded number of squares at any given moment, but this might include more than a single square (from Shagrir 2016: 20).

on symbolic configurations; the operations refer to the form ("syntax") of the symbols, while making no direct reference to semantic properties. The finite part is captured in the terms of the conditions 1–5, which are explications of the finite means. The second premise states that every function that is effectively computable is also Turing machine computable. The conclusion ("Turing's thesis") is that the effectively computable functions—that is, those functions computable by "the computer"—are Turing machine computable.

2.3 Who Is "the Computer"?

The claim that Turing's analysis essentially applies to *human computers* was underscored by his student Robin Gandy.[33] In his 1980 paper on computability, Gandy wrote:

> Both Church and Turing had in mind calculation by an abstract human being using some mechanical aids (such as paper and pencil). The word "abstract" indicates that the argument makes no appeal to the existence of practical limits on time and space. (1980: 123–124)

In his historical 1988 paper, Gandy once again emphasized that Turing's "computability" relates to calculations by an ideal human, and that Turing "makes no

---

[33] See also Kleene (1952); Sieg (1994, 2002, 2008); and Copeland (2004b, 2006).

reference whatsoever to calculating machines" (Gandy 1988: 83). In Gandy's posthumously published introduction to the 1936 paper, he wrote that Turing "considers the actions of an abstract human being who is making a calculation" (2001: 11).

There are several reasons to support this human-oriented line of interpretation. One is the fact that the computers at the time of Turing's statements *were* humans, not machines: "It is not surprising that Turing does not mention machines. Numerical calculation in 1936 was carried out by human beings" (Gandy 2001: 12).[34] The first programmable, general-purpose computers were only manufactured in the 1940s. A second reason is the highly anthropomorphic language that Turing uses to describe "the computer." "Turing's analysis is quite explicitly concerned with calculations performed by a human being; there is no reference to machines other than those which he introduces to imitate the actions of a human computor" (Gandy 2001: 12). Third, the analysis essentially exploits the limitations of human computers, not of machines in general: "Turing's analysis of computation by a human being does not apply directly to mechanical devices" (Gandy 1980: 123), and "There are crucial steps in Turing's analysis where he appeals to the fact that the calculation is being carried out by a human being" (p. 124).

A decisive point in favor of the human-oriented interpretation, in my view, is that it places Turing's pioneering work on effective computability in its appropriate historical and philosophical context—namely, the role of effective computation in logic and mathematics (as discussed in Section 2.1). In this regard, the notion of *effective procedure* is tightly connected to *us*—the human computers. The notions of *decidability*, *formal derivability*, and *formal systems* are closely linked to what a human can or cannot do, at least in principle, when using an effective procedure. This does not mean that a machine cannot compute effectively, but rather that the benchmark of what counts as effectively computable is the human computer: something is effectively computable only if it *can* be computed by an idealized human being. To ignore the human connection is to miss a key and distinct aspect of the notion of *effective computation* in the context of logic and mathematics.

I would like to stress, however, that my main claim about Turing's analysis does not depend on whether Turing, Church, and other pioneers of computability were essentially referring to human calculation or not. Following Gandy, my claim (to be elaborated in Section 2.4) is that Turing's finiteness conditions are *too restrictive* with respect to machine computation. The human interpretation explains why Turing formulated finiteness conditions that do not apply to machine computation in general (the explanation being that Turing had human

---

[34] See, e.g., Grier (2005).

calculation rather than machine computation in mind). Nevertheless, the human interpretation itself invokes real issues about the nature of the human computer that have yet to be resolved. A full discussion of these issues is beyond the scope of the present work; for now, I will make a few pertinent comments that will later be expanded in the context of physical computation (the impatient reader can skip to Section 2.4).

2.3.1 Abstractness

There appears to be a disparity between *algorithms* (i.e., effective procedures), *Turing machines*, and *computable functions*—all of which are abstract, mathematical objects—and human computers, which are concrete or non-abstract.[35] This disparity is evident in the first premise of the analysis, where "the computer" equates the notion of *algorithm* with that of *calculation*. It also arises in the second premise, where the restrictions on computers are compared with those of Turing machines.

One way to reconcile this disparity was implicitly suggested by Gandy (1980), and more explicitly by Sieg (2008, 2009). They think about the restrictive conditions 1–5 as *mathematical* constraints or axioms that abstract from the limitations imposed on the human computer. The idea is that these mathematical axioms precisely capture the notion of *algorithm* (or *effective procedure*, or *effective/algorithmic computation*). This is because these conditions model a human computer. The human computer, one might say, is an implementation or concretization of the mathematical axioms (and by extension, human calculation is an implementation of a specific algorithm).[36]

One advantage of this approach is that it does not limit effective computation to human computers, but rather allows it to be executed by non-humans as well—or even by machines. This is simply because the mathematical axioms can be applied to (or model or implemented by) a variety of different systems. The systems in question may be tangible (e.g., human computers) or abstract (e.g., Turing machines); they may be human or non-human. In other words, the axioms define a particular class of (computing) systems—namely, those that satisfy the restrictions, irrespective of whether or not they are human. They can be seen

---

[35] An additional and notoriously hard question is how to draw the abstract/concrete distinction. I discuss some of these difficulties in the context of computing physical systems in Chapter 3; see Rosen (2020) for a more general discussion. A further question is whether physical computing systems have abstract properties. I think that they do, but I will remain neutral here about this ontological question (see also the relevant discussion about medium-independent properties in Chapters 5 and 6).

[36] Dresner (2010) argues that this abstraction relation is no different from the relation between measurement schemes (e.g., real-number scales) and physical objects.

as defining a certain kind of mathematical scheme (or set of abstract computers) that is applied to different systems (which can be seen as "implementing" them)—one of which is the human computer. However, the notion of algorithm is essentially tied to human calculation, because this mathematical scheme is abstracted from a human computer. Once abstracted, it can be applied to many other kinds of computing agents that are also said to compute effectively.

Another advantage of this approach is that it treats the second premise in Turing's analysis as a mathematical theorem. This premise links two different mathematical notions: the class of systems defined by the mathematical axioms 1–5, and that of the Turing machine. This is perhaps what Gandy meant when he said that Turing had proved a theorem (1980: 124) and called it "Turing's theorem" (Gandy 1988: 83; Sieg 1994).

2.3.2 Idealization: Competence and Performance

One may rightly argue that no real human can have unlimited time and space to complete the computation; in this sense, the restrictive conditions are perhaps too liberal.[37] But the human computer is an idealized entity.[38] The idealization can take one of two very different forms. One is an idealization in terms of the practical, real-world limitations of space, time, and material aids (e.g., pencils and paper). In principle, the human can use as much time and space as it takes to complete the computation. One might define this kind of idealization in terms of the *competence/performance distinction* (Chomsky 1965): *performance* is always limited by the amount of paper potentially available in the universe and by a given time span (e.g., the lifetime of a person, planet, or universe). *Competence*, however, goes beyond this: under ideal conditions, the human could, in principle, transcend these limitations. This kind of idealization appears to be required if computation is associated with *surveyability*—since there is no upper limit on the length of a formal proof, other than that it must be finite.[39]

The second sort of idealization concerns normativity. When the human follows an algorithm for addition, the assumption is that he or she is following it "properly"—calculation mistakes, inattention, forgetfulness, distractions,

---

[37] This assertion is associated with the claim that the "easy part" of CTT is false, as no human can have the real powers of a Turing machine.

[38] I contrast *idealized* with *abstract*—hence, idealization and abstraction. In particular, *idealized* might refer to real non-abstract systems. *Idealization* refers to the system (or one might say a different system; see Norton 2012) operating under ideal conditions. Thus, when Gandy talks about an abstract human being, I take it that he means an idealized human being.

[39] There are those who challenge the relationship between these idealizations and proofs. Some argue that very long computer-generated "proofs" are not available in practice to the human computer, as in the four-color problem (Tymoczko 1979; Teller 1980). More recently, some have pointed out that proofs must be polynomially bounded; otherwise, they cannot be surveyed ("in practice") by humans (Aaronson 2013).

and so forth are immaterial to the computation process. These mistakes are of a different kind from the previous ones. In the preceding cases, real humans will never be able to add very large numbers: they will die or run out of material aids beforehand. This is not the claim here. When asked to calculate "67 + 58," even in the actual world, the human computer usually replies "125." The problem is that occasionally the human—when tired, distracted, or the like—might sometimes reply "126." Idealization is therefore required to tell which reply is the correct one. Here too, one can describe the difference in terms of the competence/performance distinction. Competence is associated with the *correct* application of the (specific) set of instructions, whereas performance is associated with the *actual* application, which might involve all kinds of faults.[40]

2.3.3 Cognitive Versus Non-Cognitive

The restrictive conditions 1–5, as their name suggests, restrict human computers to the means that they can use, or the resources they are *allowed* to use.[41] But what does it mean to say that the human computer is allowed to use these resources? Who allows it? And why is it that the human is allowed to use only computational resources that are limited to these restrictive conditions? Why these restrictive conditions and not others? In examining how others have addressed these issues (more implicitly than explicitly), we find two very different answers to the questions, which we have labeled as the *cognitive* and *non-cognitive* approaches (Copeland and Shagrir 2013).

According to the *cognitive* approach, the (permitted) restrictive conditions 1–5 *essentially* reflect the upper limitations on humans' cognitive capacities— "essentially" in the sense that the correctness of the restrictive conditions is grounded in, or is justified by, those capacities. In this cognitive approach, the truth of the first premise in Turing's analysis (that the human computer operates under the restrictive conditions 1–5) depends on *empirical facts* about human cognition. The claim, however, is not that the restrictive conditions reflect the limitations of *general human mental processes*, or not even of those mental processes involved in mathematical thinking in general.[42] Rather, the claim is that

---

[40] The two quests for idealization can be equated with Kripke's Wittgenstein infinity and normativity arguments, which, Kripke argues, cannot be satisfactorily answered in dispositional terms (Kripke 1982). See discussion in Boghossian (1989).

[41] This section extracts from Copeland and Shagrir (2013).

[42] Gödel famously argued that Turing (1936) "gives an argument which is supposed to show that mental procedures cannot go beyond mechanical procedures" (1972: 306). But Turing clearly did not hold this strong cognitivist position (Piccinini 2003a; Copeland and Shagrir 2013; Sieg 2013b). Kreisel (1972) distinguished between human computers and machine computers, and apparently identifies the former with the wider class of constructive methods (p. 319). But this is an odd use of human effective computation: the constructive proposals Kreisel mentions (and which he attributes to Gödel) are considered non-effective, even by Gödel himself.

they reflect the limitations of the cognitive capacities involved in the process of calculation, or the *faculty of calculation* (which does not necessarily refer to a designed cognitive module, but to the capacities involved in calculating the values of functions or numbers).[43]

According to the *non-cognitive* approach, the restrictive conditions 1–5 need not reflect the limitations of the human condition. Whether or not they reflect the upper limits of the faculty of calculation is irrelevant to the analysis of computability. Conditions 1–5 are simply an explication of the concept of effective computation as it is used and functions in a certain context. The relevant context is that of logic and mathematics, in which the concept is tied to decision procedures, formal systems, and generating epistemically reliable, trustworthy results.

The cognitivist would agree, of course, that Turing explicated the "axioms of computability" by analyzing the concept of human computability as it is properly used and as it functions in the discourse of logic and mathematics. However, the cognitivist would argue that in addition to explicating the conditions, Turing grounds or justifies their correctness on the limitations of human cognitive-calculative capacities. In contrast, the non-cognitivist would maintain that the analysis offers no such justification for the conditions. In fact, according to the non-cognitivist, a call for further justification has no place whatsoever in the analysis of computability.

The difference between the two approaches can be made clear by considering the consequences for the extension of the concept of computability should the human faculty of calculation be found to violate one or more of conditions 1–5.[44] Imagine that scientists were to discover that human memory could involve an unbounded number of states, and, moreover, that this would result in hypercomputational mental powers—that is, in humans being able to calculate the values of functions that are not Turing machine computable. Would these discoveries threaten Turing's analysis of computability? The cognitivist and the non-cognitivist give different answers.

The cognitivist answers *yes*. If it is found that humans can, as a matter of cognitive fact, encode an infinite procedure, perform infinitely many steps in a finite span of time (a supertask), or even observe an unbounded number of symbols at any given step when calculating a value of a function, cognitivists would regard

---

[43] Also note that the approach is not "psychologistic" in taking those facts to be exactly those cognitive properties involved in the *performance* of calculation. The approach is *cognitive* in the sense that it ties effective calculability to the *competence* of the faculty of calculation (as discussed previously).

[44] Gödel explicitly challenges the (third) restriction about the boundedness of the number of states of mind (Gödel 1972). He also raises the possibility of accelerated processing in a somewhat enigmatic sentence (Wang 1974: 325).

this as undermining the analysis. If some of the constraints among 1–5 do not reflect actual upper limits on the faculty of calculation, then in the cognitive approach these constraints have no place in the analysis.

According to the non-cognitivist, however, the answer is *no*: discoveries about the human mind have no bearing on the analysis of computability. The non-cognitivist does not exclude the empirical possibility of the discovery that human memory is unbounded; the non-cognitivist simply denies that effective calculability is synonymous with this kind of cognitive calculability. Effective computability is calculability by finite means. Hence, the analysis of computability invokes a finite number of states of mind, because the analyzed concept is that of computation *by means of a finite procedure*. The focus is on what can be achieved by "finite means," not on whether human beings are actually limited to finite means.

There is not necessarily a clear delineation between the cognitive and non-cognitive approaches. One might contend, for example, that some of the restrictive conditions reflect the limitations on cognitive capacities, while others arise from the nature of anything properly describable as "finite means." Emil Post has one foot—or possibly even both feet—in the cognitive camp, saying that the purpose of his analysis "is not only to present a system of a certain logical potency but also, in its restricted field, of psychological fidelity" (1936: 105). Post refers to Church's identification of effective calculability with recursiveness as amounting "not so much to a definition or to an axiom but to a *natural law*" (p. 105), adding that "to mask this identification under a definition hides the fact that a fundamental discovery in the limitations of the mathematicizing power of *Homo Sapiens* has been made" (p. 105 n. 8).[45] Gandy appears to understand Turing's analysis in cognitivist terms, saying that Turing arrives at the restrictions "by considering the limitations of our sensory and mental apparatus" (2001: 11). Turing's position, however, is more nuanced. On the one hand, he says that "for the present I shall only say that the justification lies in the fact that the human memory is necessarily limited" (1936: 59), which looks very much like a cognitive position. On the other hand, he later comments that we can sidestep states of mind altogether "by considering a more physical and definite counterpart" (p. 79), such as written symbolic configurations, which might suggest that he is taking the non-cognitive route after all. Church, Gödel, and Kripke appear to be far closer to the non-cognitive camp.[46]

---

[45] See De Mol (2012) for a detailed discussion of Post's views about this.
[46] See Copeland and Shagrir (2013).

2.4 Effective Computability and Machine Computation

It has been claimed that Turing characterized effective computability in terms of Turing machine computability. The question is whether, and to what extent, this characterization captures the notion of machine computation in general and physical computation in particular. When addressing this issue, we are now in a position to prevent two potential mistakes.

One mistake is to identify *computability* with *computing* (this point pertains not only to effective computation, but to any kind of computation). The term *computability* refers to the *functions* (numbers, predicates) that can be computed.[47] The term *computing* refers to the *processes* by means of which these functions are carried out. As such, we can hardly expect computability to coincide with computing, namely, with the process of computation. Take a computable function, such as the zero function, whose output is zero for every input. It is *computable* in the sense that there is a computing process by which one can arrive at its values. But presumably there will also be many other, non-computing processes by which one arrives at the values of this (computable) function. Thus, computability cannot distinguish between the computing and non-computing processes that lead to the same (computable) function. Computability distinguishes between computable and non-computable functions. It does not distinguish between the computing and non-computing processes by which we can arrive at the values of the computable functions. As for *effective* computability, I do not think that it even provides a *necessary* condition for computing. As we have previously implied (Section 2.3.3), there are cases of *hypercomputation*—namely, of *computing* functions that are not Turing machine computable at all. In Chapter 3, we will discuss more cases of physical hypercomputation in detail.

The second mistake to avoid is the application of Turing's characterization of effective computation to machine computation in general.[48] As previously noted, Turing analyzed *effective calculation*—namely, calculation by a human being who is following an effective (that is, finite and mechanical) procedure. In particular, constraints that apply to the human computer do not apply to mechanical devices in general. One might challenge the human interpretation underpinning Turing's analysis. However, even those who do not accept this interpretation seem to agree that constraints 1–5 are too restrictive to be applied to computation in general. Gandy (1980), for example, stressed that Turing's analysis does not apply to machines that perform parallel computation, in which the

---

[47] Shapiro (1984) also notes that computability is a modal notion: it refers to what *can be* computed, rather than to what *is actually* computed.
[48] See Copeland (2015: sec. "Misunderstandings of the Thesis") for a long list of philosophers and computer scientists who have committed this error (some of whom are also mentioned in Chapter 3).

number of processing units is unbounded. A simple example of parallel computation is the well-known Game of Life (hereafter "Life"). Life is a potentially infinite two-dimensional grid of cells, each of which is in one of two possible states, *alive* or *dead*. Every cell interacts with its eight neighbors—that is, the cells that are horizontally, vertically, or diagonally adjacent. At each moment in time, the following transitions occur:

- A live cell with fewer than two live neighbors dies (loneliness).
- A live cell with two or three live neighbors stays alive (survival).
- A live cell with more than three live neighbors dies (overcrowding).
- A dead cell with exactly three live neighbors becomes a live cell (birth).

The initial pattern is any arbitrary arrangement of live/dead cells. The first generation is created by applying the preceding rules *simultaneously* to every cell in the grid; births and deaths occur simultaneously. The rules continue to be applied repeatedly to create further generations. It would seem that this evolution is a clear case of a computational process, although it does not conform to all of Turing's restrictions.

Specifically, Life does not satisfy Turing's fourth condition. As Gandy pointed out, Turing assumed that "a human being can only write one symbol at a time," and this assumption cannot be carried over to a parallel machine that "prints an arbitrary number of symbols simultaneously" (1980: 124–125). In Life, there is no upper bound on the number of cells that make up the grid, yet the symbols in all the cells are updated simultaneously. Thus, there is no upper limit on the number of changes ("change of cell") that can take place at each step. This is precisely the difference between the human computer and Life: even if the human computer can run computation in parallel, there is an upper limit on the number of parts that operate in parallel. Life is a parallel machine with no such limit.

Life therefore indicates that Turing's conditions 1–5 are too restrictive with respect to machine computation in general. Life performs computations but does not satisfy all of Turing's restrictive conditions. This does not mean that we cannot extend Turing's condition in a way that will include Life and other parallel machines. It also does not mean that Life cannot be seen as an effective or algorithmic computation in some extended sense. We will discuss precisely these possibilities in Chapter 3. The conclusion, rather, is that Turing's conditions on human computation cannot be taken as a characterization of machine computation. Assuming that Life, and other parallel machines, are instances of physical computation, we can also conclude that Turing's conditions fall short of capturing physical computation (I will elaborate on this as well in Chapter 3).

2.5 Summary

I have reviewed Turing's analysis of effective calculability, in which Turing formulates a set of constraints ("axioms") on effective calculability and shows that its scope does not exceed Turing machine computability. The conclusion is that Turing's analysis pertains specifically to human calculation, and that his analysis is too restrictive to apply to computing in general. One might be encouraged at this point to relax Turing's conditions so that they might apply to more computing systems. But while this expansion is appealing with respect to some kinds of computation, I suspect it is fruitless with respect to characterizing physical computation. In Chapter 3, I will develop this line of argument in detail.

3
Preamble to Machine Computation

There have been a number of attempts to characterize machine computation, both in philosophy and in computer science. As it turns out, however, some of these characterizations target different *kinds* of machine computation. In this chapter, I examine the inclusion relations between three kinds of machine computation (leaving the notion of *computation* very much unanalyzed at this point): generic computation (Section 3.2), algorithmic machine computation (Section 3.3), and physical computation (Section 3.4). My methodology is to compare different versions of the theses pertaining to the computational limits of these machines. The differences among the theses correlate with the differences among the kinds of machine computation.

My starting point is the account advanced by Robin Gandy in his paper "Church's Thesis and Principles for Mechanisms" (1980). Gandy provides a comprehensive and precise mathematical characterization of machine computation.[1] The gist of the account is summarized in Section 3.1. I use Gandy's account to distinguish between the different kinds of machine computation: generic computation, algorithmic machine computation, and physical computation. In particular, I argue that, conceptually speaking, the account falls ambiguously between the different kinds of machine computation and fails to fully capture any of them. As before, I advise readers who are only interested in the main argument as to which sections they can safely skip.

3.1 Gandy's Account of Machine Computation

Gandy, who was Turing's student, explicitly sought to expand his advisor's ideas from human calculation to machine computation by weakening certain constraints in Turing's analysis that did not apply to machines in general.[2] Specifically, he modified Turing's constraints so that they could be applied to parallel computations, such as the Game of Life.

Gandy starts with a very general thesis (to which we will return in Section 3.2):

---

[1] The account was subsequently simplified mathematically and explicated conceptually by Sieg (2002; Sieg and Byrnes 1999).
[2] This section is based on Copeland and Shagrir (2007).

Thesis M. *What can be calculated by a machine is [Turing machine] computable.* (1980: 124)

He immediately narrows this statement, stating that he will consider only *deterministic discrete mechanical devices*, which are, "in a loose sense, digital computers" (1980: 126). In particular, he says: "I exclude from consideration devices which are *essentially* analogue" (Gandy 1980: 125). Thus, Gandy is actually arguing for the following:

Gandy's thesis: Any function that can be computed by a discrete deterministic mechanical device is Turing machine computable.

The first step in Gandy's argument is to formulate the notion of a discrete deterministic mechanical device in terms of precise axioms, which he calls *Principles I–IV*. The first principle—"form of description"—describes a deterministic discrete mechanical device as a pair <S,F>, where S is a potentially infinite set of states, and F is a state-transition operation from S_i to S_{i+1}. Gandy chooses to define the states of the machines in terms of subclasses of hereditarily finite sets (HF) over a potentially infinite set of atoms that is closed under isomorphic structures (such subclasses are termed "structural classes"); he defines the transformations as structural operations over these classes. Leaving aside the technicalities of Gandy's presentation, the first principle can be approximated as follows:

I. *Form of description*: Any discrete deterministic mechanical device M can be described by <S,F>, where S is a structural class, and F is a transformation from S_i to S_j. Thus, if S_0 is M's initial state, then $F(S_0)$, $F(F(S_0))$, ... are its subsequent states.

In Life, for example, the state (configuration) at each moment is completely determined by the previous state and by the four simple rules of transformation that constitute F (see section 2.4).

Principles II and III place boundedness restrictions on S. They can be informally expressed as follows:

II. *Limitation of hierarchy*: Each state S_i of S can be assembled from parts, which can be assemblages of other parts, etc., but there is a finite bound on the complexity of this structure. In Gandy's terminology, this amounts to the requirement that the states of a machine be members of a fixed initial segment of HF.

In Life, the grid can be arbitrarily large, but the complexity of the structure of each state is very simple, and can be described as a list of pairs of cells—or, more generally, as a list of lists of cells, since each listed pair of cells is itself a list of cells. In general, we can picture a Gandy machine as storing information in a hierarchical way, such as lists of lists (Gandy 1980: 131), but Principle II states that for each machine there is always a finite bound on the structure of this hierarchy.

III. *Unique reassembly*: Each state S_i of S is assembled from basic parts of bounded size. There is, however, a bound on the number of types of basic parts (atoms) from which the states of the machine are uniquely assembled.

The grid of Life can be assembled from pairs of consecutive cells and their symbols (e.g., ["on," "off"], ["on," "on"], etc.). We need only a limited number of pairs like these to construct any configuration of the grid.

Principle IV, "local causation," puts restrictions on the types of transition operations available. It says that each changed part of a state is affected by a bounded neighborhood:

IV. *Local causation*: The parts from which $F(S_i)$ is assembled are causally affected only by their bounded "causal neighborhoods": the state of each part is determined solely by its local neighborhood.

In Life, the grid is assembled from parts—cells—each of which is either "on" or "off" at any given moment. A cell's state—"on" or "off"—is determined only by the bounded causal neighborhood consisting of its eight adjacent cells.[3]

The first three principles might be motivated by what is meant by a discrete deterministic device. Principle IV, according to Gandy, is an abstraction of two "physical presuppositions": "that there is a lower bound on the linear dimensions of every atomic part of the device and that there is an upper bound (the velocity of light) on the speed of propagation of changes" (1980: 126). If the propagation of information is bounded, an atom can transmit and receive information in its bounded neighborhood in bounded time. If there is a lower bound on the size of atoms, the number of atoms in this neighborhood is bounded. Taking these together, each changed state, $F(x)$, is assembled from bounded, though perhaps overlapping, parts of x. In Life, each cell affects the state of several cells, that is, the neighboring ones.

The second step in Gandy's argument is to prove a theorem asserting that any function computable by a device that satisfies Principles I–IV is Turing machine

---

[3] The *local environment* need not be understood geometrically or topologically. The important point is that the number of cells, or components, that affect the computation of each cell is bounded.

computable. The proof goes much further than the (relatively trivial) reduction of some given number of machines working in parallel to a single Turing machine that performs all of those machines' actions. The class of "Gandy machines" (i.e., machines conforming to Gandy's principles I–IV) includes machines with an arbitrary number of processing parts that work on the same regions—such as printing on the same cells of a tape.

We can summarize the argument as follows:

Premise 1: "Thesis P. *A discrete deterministic mechanical device satisfies Principles I–IV.*" (1980: 126)

Premise 2: "Theorem. *What can be calculated by a device satisfying Principles I–IV is [Turing machine] computable.*" (1980: 126)

Gandy's thesis: What can be calculated by a discrete deterministic mechanical device is Turing machine computable.

Overall, Gandy's argument has the same structure as Turing's. The first premise states axioms of computability for discrete deterministic mechanical devices (Turing formulated axioms for human computers). The second premise is a reduction theorem that shows that the computational power of devices constrained by these restrictive conditions ("Gandy machines") are bounded by the Turing limit (Turing put forward a reduction theorem with respect to machines that satisfy his restrictions 1–5). The conclusion ("Gandy's thesis") is that the computational power of discrete deterministic mechanical devices is bounded by Turing machine computability (Turing's thesis is the claim about the scope of human computability).

The main difference between Gandy and Turing pertains to the restrictive conditions. Gandy's restrictions are weaker than Turing's 1936 conditions on human computability, as they allow for state-transitions that result from changes in an arbitrary number of bounded parts (in contrast, Turing allows changes in only one bounded part). This way, Gandy's characterization encompasses *parallel computation*: "If we abstract from practical limitations, we can conceive of a machine which prints an arbitrary number of symbols simultaneously," and "proofs of Thesis M must take parallel working into account" (Gandy 1980: 124-125). Gandy's formulation takes into account computing systems whose state-transition involves simultaneous changes in an unbounded number of parts. As noted in Chapter 2, the grid of Life might consist of nine cells, or a hundred cells, or billions of cells, whereas in each of them all the cells are simultaneously updated. Thus, there is no upper bound on the number of parts (i.e., cells) that are updated at any given time.

Gandy, however, does set a boundedness restriction on the state-transition of each part. This is Principle IV of local causation. Each such state-transition is bounded by the local environment of this part. In Life, the configuration of each cell is completely determined by the previous configuration of itself and that of eight adjacent neighboring cells; thus the state-transition of each part—that is, each cell—is bounded to the configuration of nine cells (and four rules of operation). Turing, as we have seen, has a similar boundedness constraint. But while Turing's condition is related to human calculation, Gandy bases Principle IV on presuppositions about the *physical world*. The presupposition about the lower bound on the size of atomic parts derives from the premise that the system is discrete. The presupposition about the speed of propagation is a basic principle of relativity. Indeed, Gandy remarks that his Thesis P is inconsistent with Newtonian devices: "I am sorry that Principle IV does not apply to machines obeying Newtonian mechanics" (1980: 145). He points out that such machines may contain "rigid rods of arbitrary lengths and messengers travelling with arbitrary large velocities, so that the distance they can travel in a single step is unbounded" (1980: 145).

3.1.1 Gandy Machines, Turing Machines, and HUMAN Computers

Gandy machines are those that satisfy the set of Principles I–IV. We can think of these principles as mathematical axioms, and of Gandy machines as the devices that satisfy these axioms. This is in line with the approach to principles 1–5 as mathematical axioms abstracted from the limitations imposed on human computation (see the discussion in Section 2.3.1). I will use the label *HUMAN computers* for the class of machines satisfying the restrictive conditions 1–5, which includes human computers (since the restrictive conditions arguably model them) as well as other machines that satisfy these principles (see Chapter 2). Now, HUMAN computers constitute a proper subclass of Gandy machines (Figure 3.1), since, as we have seen, the class of Gandy machines clearly includes the class of HUMAN computers, while other machines, such as Life, are Gandy machines but not HUMAN computers (Chapter 2). The class of Turing machines is a proper subclass of HUMAN computers.[4] As noted in Chapter 2 (and indicated in Figure 2.1), Turing machines operate on one cell at a time, whereas HUMAN computers can operate on broader (but bounded) parts at a time.

---

[4] We can think here of *Turing machines* as the class of machines—abstract or not—that satisfy the mathematical definition.

54 THE NATURE OF PHYSICAL COMPUTATION

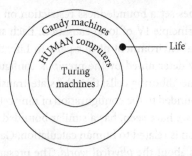

Figure 3.1 The inclusion relations between Turing machines, HUMAN computers, and Gandy machines. The class of Turing machines (machines satisfying the mathematical definition) is a proper subclass of the class of HUMAN computers (machines satisfying Turing's conditions 1–5), which in turn is a proper subclass of the class of Gandy machines (machines satisfying Gandy's Principles I–IV).

Notably, these inclusion relations—between Gandy machines, Turing machines, and HUMAN computers—pertain to the computing machines, rather than to the functions computed by the machines. Turing's and Gandy's theses actually assert that there are no such gaps when it comes to the computed functions. Turing's thesis asserts that the functions computed by HUMAN computers are also Turing machine computable. Gandy's thesis extends that result, asserting that the functions computed by Gandy machine are also Turing machine computable.

3.1.2 Summary

Gandy defines a class of computing devices ("Gandy machines") by formulating a set of Principles I–IV, and he proves that their computational power is bounded by Turing machine computability. He further argues that these principles explicate the notion of a *discrete deterministic mechanical device*, and hence that the class of Gandy machines coincides with the class of discrete deterministic mechanical devices. This argument is the basis of his thesis ("Gandy's thesis") that every function that can be computed by a discrete deterministic mechanical device is Turing machine computable. The real question is about the significance of the class of Gandy machines to machine computation. I will argue that Gandy's principles do not fully capture the notions of generic (Section 3.2), algorithmic (Section 3.3), and physical computation (Section 3.4). This discussion will enable us to clarify some of the relationships between these three notions (as summarized in Section 3.5).

3.2 Generic Computation

In his *Stanford Encyclopedia of Philosophy* entry on the Church-Turing thesis, Copeland (2015) remarks that the Church-Turing thesis has been roundly misinterpreted—including by theoreticians, practitioners, and philosophers. He cites some scholars who have even taken it as a thesis about computation in general. Dennett, for example, writes that "Turing had proven—and this is probably his greatest contribution—that his Universal Turing machine can compute any function that any computer, with any architecture, can compute" (1991: 215). This description is also found in the writings of computer scientists: "The *Church-Turing thesis* says that, from a theoretical standpoint, all computers have the same power. This is commonly accepted; the most powerful computers in the world compute the same things as Turing's abstract machine could compute" (Astrachan 2000: 397). Another bold statement of this sort is put forward by Allen Newell: "That there exists a most general formulation of machine and that it leads to a unique set of input-output functions has come to be called *Church's thesis*" (1980: 150).

Let us call this kind of description the *bold* Church-Turing thesis:

CTT-Bold (CTT-B): Any function that can be computed by any machine is Turing machine computable.

This bold thesis can be equated with Gandy's Thesis M (which states that whatever can be calculated by a machine is Turing machine computable). Gandy, however, does not equate Thesis M with the Church-Turing thesis. Moreover, he eventually argues for a more limited thesis—Gandy's thesis—that pertains to deterministic discrete mechanical devices. The *bold* thesis refers to any computing machine whatsoever. While neither Church nor Turing formulated this thesis, these and other misinterpretations reflect the dramatic shift in the sort of agents and problems associated with computation. The main difference is that the term *computer* is no longer associated, at least essentially, with humans, but rather with machines and systems in general.

Following Piccinini (2015), I will use the term *generic computation* to refer to this very general concept of machine computation. Piccinini himself may have confined the concept to physical computation, but I extend it to notional machines as well. A notional machine "abstracts from the issue of whether or not [it] ... could exist in the actual world" (Copeland 2000: 15). I will expand on the notional/physical contrast in Section 3.4.1.

It is generally agreed that CTT-B is false, at least under most accounts of computation. There are many examples of notional machines that compute non-recursive functions; these are *hypercomputers*. As Copeland puts it, "there are notional machines that generate functions that no Turing machine can generate" (Copeland 2000: 15).[5] Some have attributed to Turing (1939) the first example of hypercomputers, the so-called *oracle machines* (or *o-machines*).[6] I will describe another kind of hypermachine known as the *infinite-time Turing machine* (Hamkins and Lewis 2000; Hamkins 2002). These machines are of interest for two additional reasons: they are taken as a model of some physical supertask computation (Section 3.4.4), and they are, in some sense, deterministic discrete mechanical devices (Section 3.2.2).

3.2.1 Infinite-Time Turing Machines

Infinite-time Turing machines extend the operations of ordinary Turing machines to transfinite ordinal time.[7] The idea is to allow a Turing machine to carry out and complete computations that involve infinitely many computation steps. These computation steps proceed in time much like the ordinal numbers: if the computation does not halt at any of the finite stages 0, 1, 2, 3, . . . , then it arrives at the first infinite stage ω, continuing with stages ω +1, ω +2, ω +3, . . . and so on, eventually arriving at the second limit stage ω + ω, and continues through the ordinal numbers. The configuration of the machine at each stage is determined by the earlier configurations and the operation of a fixed finite program.[8]

Like a classical Turing machine, an infinite-time Turing machine features a head moving back and forth, reading and writing 0s and 1s on a tape according to the instructions of a fixed finitely-many-states program. In between the ordinal limit stages, the machine operates in the classical way, in the sense that "the classical procedure determines the configuration of the machine . . . at any stage α + 1, given the configuration at any stage α" (Hamkins 2002: 526). What is new is the behavior of the machine at the limit ordinal stages. At each such stage, the machine "is placed in the special *limit* state, just another of the finitely many states; and the values in the cells of the tapes are updated by computing a kind of limit of the previous values that cell has displayed" (p. 526). These new limit states enable the machine to compute beyond the Turing limit.[9]

Consider, for example, the famous halting problem—that is, the question whether there is a (classical) Turing machine that computes the halting function. The halting function accepts as an argument the pair (m,n) and returns 1 if the

---

[5] For a discussion of hypercomputation, see, e.g., Copeland and Sylvan (1999); Copeland (2002c); and Syropoulos (2008).
[6] See Copeland (2002c).
[7] The next two subsections draw on Copeland and Shagrir (2011).
[8] See also Cohen and Gold (1978).
[9] See Hamkins (2002) for a detailed presentation of these machines.

mth Turing machine halts on input n, and otherwise returns 0 (one may think of m as the code of the machine, or as the index of the machine in some enumeration). The halting function is an example of something that cannot be computed by a universal Turing machine. If the Church-Turing thesis is true, then the halting function is not computable by means of any algorithm whatsoever. A universal Turing machine can simulate the operations of the mth Turing machine operating on input n. It can also check, after each simulation step, whether the mth machine halted or not. Thus, our universal Turing machine can tell, within a finite number of steps, whether an arbitrary Turing machine m halts when operating on input n. Nonetheless, it does not compute the halting function: if the mth machine never halts, then the simulation keeps going forever, meaning that the simulating machine returns no output.

However, a universal infinite-time Turing machine that travels in transfinite time could "see" that the simulated machine, m, never halted. It could return the values of the halting function. By way of illustration, imagine a universal infinite-time Turing machine, which we supplement with a special limit state at the ordinal stage ω. This machine—which I will dub ITTM (infinite-time Turing machine)—*computes* the halting function as follows. Take a "designated square" on the tape to indicate whether or not m has halted (when operated on input n). This designated square is (for the sake of the argument) the first square to the left of the block of digits comprising the description of m, followed by the digits of the input number n. Once ITTM is set, its first action is to position the scanner over the designated square and print 0 (meaning "m does not halt"). The next step of ITTM, which is a universal machine, is to simulate m, performing every operation that m does and in the same order (albeit interspersed with sequences of operations not performed by m). If ITTM discovers that m halts, then ITTM returns to the designated square and changes the 0 written there to 1 (meaning "m halts"). Otherwise, the value in the designated square is calculated, in the limit ω stage, as the limit of the previous values that the square has displayed. If the square displayed the value 0 in all the (infinitely many) stages that preceded the limit stage, then the value at the limit-state is set to 0; ITTM halts at this point, returning the value 0 (meaning "m does not halt").[10]

Does ITTM disprove CTT-B? That depends on whether or not we consider the machine's activity computing. Those who devise such machines think that they compute. Hamkins, for example, writes that infinite-time Turing machines "provide a natural model of infinitary computability" and are "computing machines" (2002: 521); Cohen and Gold (1978) titled their paper "ω-Computations on Turing Machines," Löwe (2001) entitled his "Revision Sequences and Computers

---

[10] See Hamkins and Lewis (2000, sec. 2) for a discussion of the power of infinite-time Turing machines.

with an Infinite Amount of Time," and Koepke (2005) called his "Turing Computations on Ordinals." Moreover, the claim that ITTM computes is consistent with most accounts of computation—including the semantic account (Shagrir 2006b; Sprevak 2010), the mechanistic account (Miłkowski 2013; Fresco 2014; Piccinini 2015), the causal account (Chalmers 2011), and the BCC (broad conception of computation) account (Copeland 1997).

3.2.2 Why Infinite-Time Turing Machines Are Not Gandy Machines

Infinite-time Turing machines raise interesting questions in the context of Gandy machines (those machines that satisfy Principles I–IV). On the one hand, infinite-time Turing machines are obviously not Gandy machines, since they can exceed the Turing limit. On the other hand, they appear, at least in some sense, to be discrete deterministic mechanical devices. They are *discrete* in the sense that they operate on separable elements and according to a finite program with finitely many separable states. They are *deterministic* in that the configuration of the machine at each stage is completely determined by the earlier configurations and the operation of a fixed finite program. They appear to be *mechanical* in the sense that they operate according to a fixed rule or program. Gandy says that he will "distinguish between 'mechanical devices' and 'physical devices' and consider only the former" (1980: 126). Mechanical devices are those that satisfy Principle IV, the principle of local causation. ITTM does satisfy this condition—at least if, instead of "accumulating" the values in the designated square, there is some indicator of the number of changes that occurred in this square since the first step ("print 0") of the computation process; ITTM would then return 0 if the indicator shows no changes.

So why would the infinite-time Turing machine, ITTM, not be considered a Gandy machine? Merely appealing to the fact that infinitely many steps are involved does not provide an answer. Gandy's postulates allow for processes comprising infinitely many steps; so do classical Turing machines. Rather, the difference is that ITTM allows *terminating* processes that consist of infinitely many steps, whereas Gandy's proof assumes that processes consisting of infinitely many steps do not terminate. The fact that ITTM exhibits such processes is apparent: while the simulated Turing machine does not halt, ITTM halts after the infinitely-many-steps simulation, producing 0 as its output. But this raises the question of which of Gandy's principles is violated by ITTM.

The answer lies in an ambiguity in the term *deterministic* (Copeland and Shagrir 2007, 2011). Gandy says that by *deterministic* he means that "the subsequent behavior of the device is uniquely determined once a complete description of its initial state is given" (1980: 126). In that sense, ITTM is certainly deterministic: its halting state, whether it halts on 0 or 1, is uniquely determined once a complete description ("configuration") of its initial state ("stage") is given. But Gandy assumes more than that: he requires that the behavior at each stage (except the first one) is to be uniquely determined by the configuration of the *previous* stage. This stronger requirement is present in the formulation of Principle I, which requires that the process can be described as a sequence $S_0, F(S_0), F(F(S_0)), \ldots$ (where S_0 is the initial state, and F is the state-transition function).

If we define *deterministic* thusly, ITTM is *not* deterministic. Consider the halting state of ITTM. If ITTM halts before reaching the limit stage, then its subsequent behavior is deterministic in Gandy's sense; but if it reaches the limit stage, its behavior is no longer *Gandy-deterministic*. To count as Gandy-deterministic, its behavior at the limit stage must be completely determined by the configuration of the *previous* stage. However, there is no such last previous stage—in other words, there is no stage that is the one that comes just before the limit stage. After each (classical) stage of ITTM, there are infinitely many others that precede the limit stage.

For precisely this reason, ITTM is also *not* a Turing machine. A Turing machine is Gandy-deterministic: what it does at each stage is completely determined by the configuration at this stage (and thus the configuration of the machine at each stage α+1 is completely determined by the configuration of the machine at the preceding stage α). It is therefore misleading to call ITTM an infinite-time Turing machine. If anything, it is an example of a *non*-Turing machine.

As we have noted, there is a thoroughly reasonable account of determinism according to which ITTM *is* deterministic. It is deterministic in that its configuration is uniquely determined by the machine's initial configuration at every moment. In particular, its configuration at the limit stage is a *limit* of previous configurations. On this account, ITTM is deterministic in that its end stage is the limit of the previous configurations (in the other stages, it is Gandy-deterministic). This sense of determinism is in good accord with the physical usage, whereby a system or machine is said to be deterministic if it obeys laws that invoke no random or stochastic elements.

The upshot is that Principles I–IV do not cover all instances of generic (machine) computation. They might not even cover all instances of deterministic discrete mechanical machines. Infinite-time Turing machines may be seen as deterministic discrete mechanical devices, but they are not deterministic in the sense of Principle I. Thus, the class of Gandy machines—that is, machines that satisfy Principles I–IV—comprises a proper subclass of generic (computing) machines, namely, the class of computing machines in general (Figure 3.2).

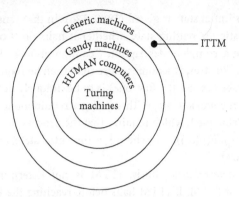

Figure 3.2 Generic machines. The class of generic machines includes the class of Gandy machines (and its proper classes).

3.3 Algorithmic Computation

Within computer science, in the disciplines of computability theory, theories of automata and formal languages, computational complexity, and so forth, computing is explicitly associated with *algorithms*.[11] In computer science, however, algorithms are primarily associated with *machines*, not with humans (more precisely, human computers are considered one type of machine). Even the Church-Turing thesis is formulated in terms of machines in most computer science textbooks. Consider the following:

> Today the Turing machine has become the accepted formalization of an effective procedure. Clearly one cannot prove that the Turing machine model is equivalent to our intuitive notion of a computer, but there are compelling arguments for the equivalence, which has become known as Church's hypothesis. (Hopcroft and Ullman 1979: 147)

> Because the Turing machines can carry out any computation that can be carried out by any similar type of automata, and because these automata seem to capture the essential features of real computing machines, we take the Turing machine to be a precise formal equivalent of the intuitive notion of "algorithm." (Lewis and Papadimitriou 1981: 223)

> The claim, called *Church's thesis* or the *Church-Turing thesis*, is a basis for the equivalence of algorithmic procedures and computing machines. (Nagin and Impagliazzo 1995: 611)

---

[11] Parts of this section are excerpted from Copeland and Shagrir (2019).

In light of these and many other statements, we can reformulate the thesis in terms of algorithmic computation by machines:

CTT-Algorithm (CTT-A): Any function that can be algorithmically computed by a machine is Turing machine computable.

This formulation can be contrasted with the original thesis, which ties algorithmic computation to a human computer:

CTT-Original (CTT-O): Any function that can be algorithmically computed by an idealized human computer is Turing machine computable.

As implied by the textbook statements just cited, CTT-A is widely believed as well. The main arguments for the thesis, however, no longer tie it to human calculation, nor to formal derivation (as noted in Chapter 2, Turing's analysis is rarely mentioned in computer science textbooks). The two arguments for the thesis that appear in most textbooks are the argument from confluence and the argument from non-refutation. The argument from non-refutation, we recall, states that the thesis has not been refuted, even though it is refutable.[12] The argument from confluence asserts that many characterizations of computation that differ in their goals, approaches, and details nonetheless encompass the same class of computable functions.[13]

This shift in argumentation also implies that the notion of algorithmic machine computation encompasses a wide range of algorithms, including some that do not satisfy Turing's restrictive condition on human computation. The Game of Life is one example of algorithmic computation that does not satisfy the (humanly) restrictive conditions 1–5, but there are many others: "In addition to classical sequential algorithms, in use from antiquity, we have now parallel, interactive, distributed, real-time, analog, hybrid, quantum, etc. algorithms" (Gurevich 2012: 32).[14] This expansion of the concept of algorithm is not an unusual phenomenon; in fact, concept expansion is pervasive, including in

---

[12] A common formulation of the argument in computer science textbooks runs as follows:

Church's thesis could be overthrown at some future date, if someone were to propose an alternative model of computation that was publicly acceptable as fulfilling the requirement of "finite labor at each step" and yet was provably capable of carrying out computations that cannot be carried out by any Turing machine. No one considers this likely (Lewis and Papadimitriou 1981: 223).

[13] Here is one formulation of the argument:

Turing machines can be imitated by grammars, which can be imitated by μ-functions, which can be imitated by Turing machines. The only possible conclusion is that *all these approaches to the idea of computation are equivalent* (Lewis and Papadimitriou 1981: 224).

[14] See also Brabazon, O'Neill, and McGarraghy (2015), who discuss algorithms inspired by systems and phenomena (e.g., genetic algorithms) in the natural world.

mathematics itself.[15] The concept of *number*, for example, was once limited to positive integers—including zero—and then the negative integers. Gradually, it has expanded to include rational, real, and now complex numbers. Even the concept of *human computation* has undergone an expansion: initially, the founders of computability defined it with respect to total functions alone, such as the general recursive functions (Gödel 1934; Church 1936a; Kleene 1936), and only later expanded it to partial functions (Kleene 1938). In the latter case, mathematicians realized that some algorithmic processes do not terminate.[16] Yet another example is the extension of *effective calculability* to cover functions over the real numbers (see Section 3.4.2).

How broad is the notion of algorithmic machine computation? Here we find at least two dramatically different approaches. Within computer science—at least in the textbook characterizations cited earlier—it is assumed that the functions computed by algorithms (in the expanded sense), at least to date, are all Turing machine computable. This approach limits algorithms to machines that stay within the ballpark of Turing's conditions—namely, Turing's conditions are relaxed, but only up to a degree (as in, e.g., Gandy 1980 and Dershowitz and Gurevich 2008).[17] We still think of algorithmic process in terms of state-transition: each state consists of atomic parts, the number of atomic part *types* is bounded, the transition function is defined across these atomic types, and the number of transitions in each terminating process is finite.[18] When talking about *algorithmic machine computation* in this chapter, I refer to this approach. A second approach identifies *algorithms* with very general rules of operation. According to this approach, even hypercomputers—machines that compute non-Turing computable functions—execute algorithms. This approach maintains that algorithms define computation in general, namely, generic computation. It is adopted by Copeland, who writes that "to compute is to execute an algorithm" (1996: 335).[19]

The labels matter little here. My substantive claim is that the notion of *algorithm* is not constitutive of physical computation under either approach. With

---

[15] See Buzaglo (2002) and Shapiro (2013) for a pertinent discussion about the evolution of mathematical concepts; Shapiro refers to human computation in particular.

[16] In Gödel's opinion:

The precise notion of mechanical procedures is brought out clearly by Turing machines producing partial rather than general [i.e., total] recursive functions. In other words, the intuitive notion does not require that a mechanical procedure should always terminate or succeed. A sometimes-unsuccessful procedure, if sharply defined, still is a procedure, i.e., a well determined manner of proceeding. (Quoted in Wang 1974: 84)

[17] Thus Gurevich (2012: 33) writes that "none of the other known kinds of algorithms seem to threaten the thesis."

[18] For earlier characterizations, see Kolmogorov (1958); Kolmogorov and Uspensky (1963); Knuth (1973: 4–6); and Gurevich (2000).

[19] Gurevich (2019), too, seems to adopt this approach.

respect to the former approach, I argue (in Section 3.4) that although highly important in some contexts (e.g., computer science), algorithmic computation does not entirely coincide with physical computation: some algorithmic computations are not physical, and some physical computations are not algorithmic. With respect to the latter approach (Copeland), I argue (in Chapter 4) that the notion of *algorithm*, even in its broader sense, does not help in distinguishing between computing and non-computing physical systems. First, however, I want to further clarify the notion of *algorithmic machine computation*.

3.3.1 What Is an Algorithm?

The nature of algorithms is a matter of debate within computer science. One question concerns the ontology of algorithms.[20] The dominant view is that algorithms are abstract mathematical entities—but the question as to which abstract entities are algorithms is moot.[21] The difficulty is that the very same algorithm can be "executed" by different machines using, say, different marks or memory systems. Similarly, the same algorithm can be "expressed" by different programs—say, one written in Lisp and another in C++. Therefore, most theoreticians view algorithms as abstract entities that are invariant under isomorphism.[22] Moschovakis (1984, 1998, 2001) defines algorithm in terms of abstract recursion,[23] Gurevich (2000) in terms of abstract-state machines, and others in terms of equivalent classes of abstract machines (Milner 1971; Moschovakis 1998) or programs (Yanofsky 2011).[24] Vardi suggests that an algorithm is *both* abstract-state machine and recursor (Vardi 2012). We can say that the notion of *algorithmic computation* refers to machines that "implement" an algorithm, or that "execute" a program that "expresses" the algorithm.[25]

It is debatable whether an algorithm should be *physically* implementable, at least in principle. The dominant view is that it does not need to be. Thus, Moschovakis and Paschalis say that their approach adopts "a very abstract

---

[20] The nature of algorithms is addressed informally by, e.g., Knuth (1973: 1–9), Rogers (1987: 1–5), and Odifreddi (1989: 3), alongside the others mentioned in what follows.

[21] See also Dean (2016) for a criticism of the view that algorithms are essentially abstract.

[22] Gandy's (1980) characterization also invokes this concept of invariance under isomorphism.

[23] However, see Gurevich (2012) for a criticism of this view.

[24] For a criticism of this view—namely, that algorithms are equivalence classes of programs—see Blass, Dershowitz, and Gurevich (2009).

[25] Within computer science, this distinction is sometimes highlighted in terms of the *abstract/concrete* distinction—*abstract* referring to a high-level description of a machine (such as a specification that is invariant under isomorphism), and *concrete* usually pertaining to a lower-level description of a machine, e.g., a specification that is not invariant under isomorphism. Importantly, a *concrete* machine can be quite abstract (in the non-physical sense); see, e.g., Pnueli, Siegel, and Singerman (1998) and Tucker and Zucker (2004).

notion of algorithm that takes recursion as a primitive operation and is so wide as to admit 'non-implementable' algorithms" (2008: 87).[26] But others do mention physical implementation—even if only as a theoretical or feasible possibility. David Harel, for example, maintains that

> any algorithmic problem for which we can find an algorithm that can be programmed in some programming language, *any* language, running on some computer, *any* computer, even one that has not been built yet but *can* be built . . . is also solvable by a Turing machine. This statement is one version of the so-called *Church/Turing thesis*. (1992: 233)[27]

Another question about algorithms concerns definability. Gurevich (2012) argues that the notion of algorithm cannot be rigorously defined in full generality, mainly because it keeps evolving: "New kinds of . . . algorithms may be introduced" (2012: 32). Nonetheless, Gurevich claims, the open-ended nature of algorithms does not imply that a rigorous definition of algorithms is hopeless: "Some strata of algorithms have matured enough to support rigorous definitions" (p. 33). Turing (1936) provided a rigorous, if highly restricted, definition of one stratum of sequential (non-parallel) algorithms.[28] Gurevich (2000) himself provides a rigorous definition of sequential algorithms, which was later extended to parallel algorithms, and which is arguably more general than Gandy's.[29] But, according to Gurevich, even this characterization is not the final one. The concept of algorithm keeps evolving, creating ever more strata that cannot be captured by a single characterization—at least for the time being.

The last comment is about the relations between *effective computation* and algorithmic computation. Here we find three approaches. One is to keep effective procedures and algorithms together, even in the broader context of machine computation. In their reference to effective procedures (mentioned earlier), Hopcroft and Ullman seem to take this route. A second approach is to understand effectiveness in the sense of a practical or physical procedure.

---

[26] See also Gurevich (2019), who distinguishes between algorithms that are "real-world" ("effective") and those that are not ("non-effective").

[27] The requirement of physical implementability is prevalent among theorists of quantum computing. See discussion in Section 3.3.2.

[28] Gurevich associates Turing's notion with "symbolic" (symbol-pushing, digital) sequential algorithms.

[29] The axiomatic definition was extended to synchronous parallel algorithms in Blass and Gurevich (2003), and to interactive sequential algorithms (Blass and Gurevich 2006, 2007a, 2007b). Dershowitz and Gurevich (2008) derive the Church-Turing thesis from axioms of sequential algorithms; see Boker and Dershowitz (forthcoming) for further discussion of the axiomatic definitions.

This approach is taken by Gurevich (2019), who equates effectiveness with something that works in practice or that can be found in the real world, and to some extent by Cleland (1993), who analyzes effective procedures in terms of mundane procedures.[30] According to this approach, algorithms and effective procedures are different entities. Gurevich talks specifically about non-effective algorithms; he offers oracle algorithms as an example. Cleland says that Turing machines are abstract entities, and therefore cannot carry out mundane ("effective") procedures. A third approach is to reserve *effective procedures* to the notion analyzed by Turing, Church, and the founders of computability—namely, that of human computation. This approach is advocated by Copeland (2015; Copeland and Shagrir 2019). Like Gurevich (2019), Copeland adopts a broader understanding of the notion of algorithm. But unlike Gurevich, who equates effectiveness with real-world procedures, Copeland maintains the original meaning of effectiveness in its logical-mathematical context. I side here with Copeland and prefer to reserve the term *effective computation* for human computation.[31]

Before comparing algorithmic computation with other kinds of machine computation, we will take a short detour through computational complexity, with its extended Church-Turing thesis. The uninterested reader can skip this and go directly to Section 3.3.3.

3.3.2 Computational Complexity

Much of theoretical computer science today is about complexity, not computability. The birth of *computational complexity* dates back to the early 1960s (Fortnow and Homer 2003).[32] Its main concern was the functions (or problems) that have an *efficient* or *feasible* algorithm. This is in contrast to functions (or problems) that are algorithmically computable, but none of whose algorithms is efficient. The assumption here is that *efficient* means

---

[30] Cleland defines a mundane procedure as a fixed and finite set of instructions that *reliably* produces a certain kind of outcome when followed (see Cleland 1993: 186ff.; 2002: 167). Cleland argues that mundane procedures, such as recipes, are effective in this sense.

[31] Maintaining the original sense of effective procedures, we can state the original version of the Church-Turing thesis as follows (Copeland and Shagrir 2019):

CTT-Original (CTT-O): Any function that can be computed by the idealized human computer—i.e., can be effectively computed—is Turing machine computable.

[32] The issue of complexity was raised by Gödel in a 1956 letter to von Neumann (by the time von Neumann received it, Gödel had already passed away); for discussion, see Urquhart (2010). In the letter, Gödel notes that if the complexity of a derivation is not polynomially bounded, then a human (computer) would not be able to complete the derivation, even if it exists. For this and other philosophical implications of complexity, see also Aaronson (2013) and Dean (2019).

polynomial time (namely, that its time complexity—say, the number of computation steps it takes to produce the output) is upper-bounded by a polynomial of the size of the input. The main issues and open questions in computational complexity pertain to (non-)reducibility relationships between complexity classes of computable functions.

I will not say much about complexity in this book. However, I want to distinguish CTT-A—which is about the bounds of algorithmic computability—from another thesis, which is about complexity. In complexity theory, the time complexities of any two general and reasonable models of computation are assumed to be polynomially related: if a problem's time complexity is t in some (general and reasonable) model, then its time complexity is assumed to be poly(t) in the single-tape Turing machine model (Goldreich 2008: 33). This premise has various names in the literature: Goldreich (2008) called it the *Cobham-Edmonds thesis* (following Cobham 1964 and Edmonds 1965), while Yao (2003) introduced the term *extended Church-Turing thesis*. The thesis is of interest only if P ≠ NP, as otherwise it is trivial.[33]

Quantum computation researchers also use a variant of this thesis, expressed in terms of probabilistic Turing machines. Bernstein and Vazirani say:

> Just as the theory of computability has its foundations in the Church-Turing thesis, computational complexity theory rests upon a modern strengthening of this thesis, which asserts that any "reasonable" model of computation can be *efficiently* simulated on a probabilistic Turing machine (an efficient simulation is one whose running time is bounded by some polynomial in the running time of the simulated machine). (Bernstein and Vazirani 1997: 1411)

Aharonov and Vazirani (2013) offer the following formulation of this premise, which they dub the *extended Church-Turing thesis*—although it is not quite the same as Yao's earlier thesis of the same name, which did not refer to probabilistic Turing machines:

> **CTT-Extended (CTT-E):** "Any reasonable computational model can be simulated efficiently by the standard model of classical computation, namely, a probabilistic Turing machine." (Aharonov and Vazirani 2013: 329)

---

[33] P is the class of problems (functions) for which there is a deterministic Turing machine polynomial solution. NP is the class of problems for which there is a nondeterministic Turing machine polynomial solution. The term *non-deterministic* can be interpreted as referring to a type of deterministic (branching) machine; see Goldreich (2008) and Arora and Barak (2009) for discussion.

We can think of the extended thesis in terms of invariance. Just as the Turing machine is a general model of algorithmic computability (the subject of CTT-A), the Turing machine is also a general model of time complexity (the subject of CTT-E). The computational complexity of a "reasonable" model of computation can be very different from that of a Turing machine, but only up to a point: the time complexities must be polynomially related.

What is a *reasonable* model of computation? Bernstein and Vazirani say that they take "reasonable to mean in principle physically realizable" (1997: 1411). Aharonov and Vazirani (2013) interpret "reasonable model of computation" as "*physically realizable* model of computation" (p. 331). This requirement of physical realizability is prevalent among theoreticians of quantum computation, and for a reason: quantum computing has attracted much attention in theoretical computer science due to its potentially dramatic implications for computational complexity. In their paper, Bernstein and Vazirani proceed to give formal evidence that a *quantum Turing machine* violates CTT-E. Another renowned example from quantum computation that might violate the extended thesis is Shor's factoring algorithm (Shor 1994, 1997).

3.3.3 Algorithmic Machine Computation and Generic Computation

The notion of algorithmic machine computation is narrower in scope than that of generic computation. As implied in Section 3.3.1, it does not encompass infinite-time Turing machines such as ITTM; otherwise, ITTM would have been understood to refute CTT-A. As we have seen, however, CTT-A is considered by almost everyone in theoretical computer science to be true. I have also previously indicated that ITTM is not algorithmic, as one of its terminating processes consists of infinitely many steps. If the simulated machine does not halt, then the (terminating) process that indicates this fact by returning 0 consists of infinitely many steps.

This is not to say that infinite-time Turing machines do not act in accordance with any rule. They surely do: ITTM is governed by a fixed "program" with finitely many states. My suggestion is that these rules are *not* considered "algorithms" in the sense of that word in theoretical computer science today ("the first approach"). In this respect, algorithmic computation is more restrictive than generic computation. There are systems (such as infinite-time Turing machines) that compute, but not algorithmically. It is an open question whether this or other rules will be further relaxed, changed, or finalized at some point. To a large extent, the answer appears to depend on developments in the theory of algorithms, and perhaps on other factors as well.

3.3.4 Algorithmic Computation and Gandy Machines

The notion of an algorithmic computation is broader in scope than that of HUMAN computation. There are algorithmic machines, such as Life, that do not satisfy all of the restrictive conditions 1–5. The more interesting question concerns the relationship between algorithmic computation and Gandy machines (those that satisfy Principles I–IV). Gandy machines are algorithmic: Gandy (1980) provides a characterization of (some) parallel synchronous algorithms. The question is whether Gandy's characterization captures the notion of algorithm in its full generality—in other words, whether every algorithmic process satisfies Principles I–IV.

I agree that Gandy provides a very general characterization of algorithmic computation.[34] Moreover, as I have previously implied, I also believe that the expansion strategy that Gandy adopts (namely, weakening constraints) reflects the evolution in the concept of algorithm. But it is precisely for these reasons that Gandy's characterization is apparently not the last word in this expansion process. Indeed, there are asynchronous algorithms that do not satisfy Principle I.[35] We can think of a machine that we might call Life*—which is almost identical to Life, with the exception that at each step, only a fixed number of cells are updated, and the identity of the updated cells is chosen according to different clocks or even at random. This machine, Life*, is just a simple example of an asynchronous algorithm. Another example is that of interactive or online (algorithmic) machines that constantly interact with the environment while they compute.

There might also be parallel synchronous algorithms that violate Principle IV. Gandy himself mentions *Markov normal algorithms*: "The process of deciding whether a particular substitution is applicable to a given word is essentially global" (1980: 145). The term *global* here means that there is no bound on the length of a substituted word; hence the demand for *locality* (Principle IV) is not satisfied. The theory of neural networks provides further examples. In Chapter 4, we discuss in detail a network for solving the *n-queens problem* (Shagrir 1992)— the problem of locating n queens on an $n \times n$ chessboard, such that there is no more than one queen on each row, column, and diagonal. In this network, each "cell" receives "global" information at each update from *all* other cells, of which there are unboundedly many. This is in contrast to Gandy machines, whose parts are affected by their local environment.

---

[34] See also Sieg (2002: 390; 2009: 528).
[35] See Copeland and Shagrir (2007) and Gurevich (2012) for examples.

In short, it seems that Gandy appears to have characterized only a proper subclass of algorithmic machine computation. This should not be too surprising. As previously noted, Gandy was concerned with discrete deterministic (synchronous) mechanical devices, and links *mechanical* to the "two physical suppositions" that underpin Principle IV, the principle of local causation. It will be of interest, then, to see whether Gandy captures the notion of *physical computation*. But first, a short summary is in order.

3.3.5 Summary

We will conclude this section by stating that the notion of algorithmic machine computation is more restrictive than that of generic computation, but less restrictive than that of computation by Gandy machines. On the one hand, some infinite-time Turing machines lie outside the scope of algorithmic computation; on the other, there are apparently non-Gandy machines that compute algorithmically (Figure 3.3).

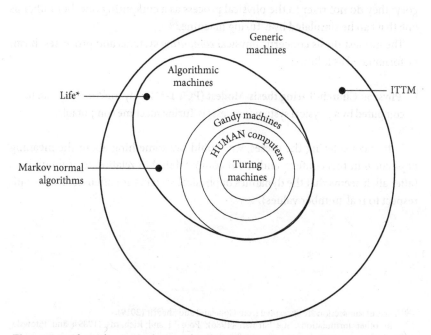

Figure 3.3 Algorithmic machines. The class of algorithmic machines is more restrictive than the class of generic machines, but less restrictive than the class of Gandy machines.

3.4 Physical Computation

Questions about the computational limits of the physical world were raised more explicitly in the mid-1980s.[36] Stephen Wolfram described "a physical form of the Church-Turing hypothesis" (1985: 735), and stated it as follows: "universal computers are as powerful in their computational capacities as any physically realizable system can be, so that they can simulate any physical system" (p. 738). David Deutsch, who laid the foundations of quantum computation, provides a somewhat similar formulation, calling it "the physical version of the Church-Turing principle" (1985: 99).[37]

Piccinini (2011) distinguishes between *bold* and *modest* versions of the *physical* Church-Turing thesis. The bold thesis is about physical systems and processes in general—not necessarily restricted to the computing ones:

Physical Church-Turing thesis-Bold (PCTT-B): "Any physical process is Turing [machine] computable" (p. 746).

The formulations by Wolfram, Deutsch, and others fall within this category: they do not refer to the physical process as a computing one, but rather as one that can be simulated by a Turing machine.[38]

The modest thesis concerns physical *computing* systems and processes. It can be formulated as follows:

Physical Church-Turing thesis-Modest (PCTT-M): Any function that can be computed by a physical system (process) is Turing machine computable.[39]

Before considering the theses, we should say something about the meaning of *physical* in this context, and about computability in relation to real numbers (after all, it seems that the dynamics of physical systems are often described with respect to real-number values).

---

[36] Parts of this section are excerpted from Copeland and Shagrir (2019).
[37] For other formulations, see Earman (1986); Pour-El and Richards (1989); and Pitowsky (1990, 2002).
[38] Piccinini emphasizes, however, that the "bold" versions proposed by different writers are often "logically independent of one another," and exhibit "lack of confluence" (2011: 747–748).
[39] Piccinini (2011) formulates the thesis as follows: "Any function that is physically computable is Turing-computable" (p. 734). I have adapted the formulation to the format shared by the other versions of the Church-Turing theses mentioned previously.

3.4.1 What Is Physical?

Our overall concern is with the meaning of *physical computation*, and in particular with the distinction between physical computing and non-computing systems. In this section, I will say something about the physicality of computation, and about the distinction between physical and non-physical computation. In one sense, there is no particular problem with the physicality of computation: whether or not a computation is physical depends on whether or not the computing system in question is physical. One condition of being physical is that the system has physical properties (or descriptions). Another condition, which perhaps follows from the first, is that the dynamics of the system conform to physical laws. We also want to include in this class idealized physical systems—namely, systems that operate in certain idealized conditions. The idealizations invoked in the context of physical computing systems usually pertain to unbounded resources and computing time. We usually allow the physical system to use, for instance, unbounded memory and unbounded time when necessary—even though the actual machine under consideration is bound to exhaust memory resources and break down before the computation is completed.

We also want to include in this class not just actual machines, but also *physically possible* machines—namely, physical machines (whether idealized or not) that do not exist in the actual world, but that did exist, will exist, and even *could* have existed (or that do exist in some physically possible world). Of course, there is no consensus as to what is considered an actual, idealized, or possible physical system. But this is less of a concern for us here, as it is no more problematic for the physicality of computation than it is for physicality in general.

Things get a bit murkier when we try to contrast physical computation with non-physical computation. The most common contrast is with abstract computation. Smith, for example, invokes the physical/abstract contrast when he asks whether computation is "a concrete (physical) or abstract notion" (2010: 1). In the section entitled "Abstract Computation and Concrete Computation," Piccinini (2017) also identifies *physical* with *concrete*, saying that concrete computation is "computation in concrete physical systems such as computers and brains." Concrete—that is, physical—computation is then contrasted with *abstract computation*, which is associated with algorithms, Turing machines, and other (abstract) automata. The two are related, according to Piccinini, through the notion of implementation: "We speak of physical systems as running an algorithm or as implementing a Turing machine" (Piccinini 2017). In other words, a computing physical system is one that implements an abstract computation.

Leaving aside this particular characterization of physical computation for now, I would like to make a few comments about the *physical/abstract* contrast. First, I take it that physical systems or machines are those that have physical properties. However, there is no question that we apply abstract descriptions to these systems; some would further say that physical systems have abstract properties. The *physical/abstract* contrast hinges on the assumption that *abstract* entities (presumably, algorithms and Turing machines) have only abstract—that is, mathematical and logical—properties, and no physical ones.

Second, the identification of the term *concrete* with *physical* is somewhat misleading. The *concrete/abstract* distinction is sometimes invoked in different contexts, where *concrete* refers to a lower-level and fairly detailed description of a system, whereas *abstract* refers to a higher-level and less detailed description that omits or ignores some or even many of the properties. In other words, *abstract* means taking into account only certain properties or features of a system, whereas *concrete* means taking in account more, most, or all of these features (the distinction, of course, being a matter of degree). However, in principle, an abstract description might take into account some, or even only, physical properties (while omitting others)—in contrast to the stronger sense of *abstract* adopted here, which is restricted to mathematical and logical, non-physical properties (see also the discussion in Section 2.3.1). Similarly, the *concrete* description might refer to some, or even only, mathematical and logical properties. Indeed, the term *concrete* is often invoked in computer science to denote a lower-level specification of a computing system, which in itself can be abstract in the stronger sense—namely, it may have only mathematical properties (see note 25). I shall therefore refrain from using the term *concrete* in the sense of strictly physical, and use the term, if at all, in the detailed-description sense. By way of distinction, when referring to a specific system as defined by its physical description alone, I will use the term *concrete physical*.

Third, Piccinini's statement that "we speak of physical systems as running an algorithm or as implementing a Turing machine" calls for a distinction between three classes of computing systems, not two (Figure 3.4). The first class is that of *physical* (computing) *machines*—including actual, idealized, and possible physical systems. A second class is that of *physically realizable* or *implementable machines* (with the notions of *implementation* and *realization* unanalyzed at this point). This class is more inclusive than the class of physical machines: it includes physical systems that are realized in the actual world or in some possible world, as well as *non-physical* computing machines, such as abstract machines that, though they have no physical properties and are therefore considered non-physical, are physically realizable, at least in principle. Piccinini mentions in this context the Turing machine (when viewed as a purely mathematical entity). These are the machines that "*can* be built" (Harel 1992: 233) or that are "in principle physically realizable" (Bernstein and Vazirani 1997: 1411). The Gandy machines (those satisfying Principles I–IV) also belong here. Some of the Gandy machines may be

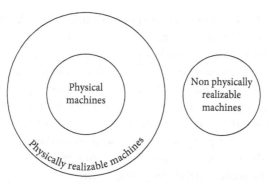

Figure 3.4 Physical and non-physical machines. The class of *physical* machines—including actual, idealized, and possible physical systems—is more restrictive than the class of *physically realizable* machines, which might also include abstract machines. The latter class is to be separated from machines that are *not physically realizable* at all.

abstract (e.g., Turing machines), yet they are physically realizable ("mechanical," in Gandy's terms) in the sense that they conform to *local causation*.

The third class is the systems that are not physically realizable. One example of a machine that belongs to the third class is the "infinitely accelerating Turing machines," of which there is no "evidence that they can be constructed."[40] Piccinini calls them "purely notional" machines. Copeland (2000), too, invokes these notional machines in order to distinguish between realizable and non-realizable computing machines (although he does not use the latter terms). He distinguishes between two senses of a machine. The narrow, this-worldly sense refers to "a machine that conforms to the physical laws (if not to the resource constraints) of the actual world" (p. 15). The broader sense "abstracts from the issue of whether or not the notional machine in question could exist in the actual world" (p. 15). If a machine conforms to the physical laws, then it is physical—or at least physically realizable. If a notional machine could not exist in our world, then presumably it does not conform to physical laws and hence is not physically realizable. It is less clear, however, whether purely notional machines coincide with abstract ones. It appears that notional machines are, as their name suggests, merely notions or concepts of machines. In addition, they appear to be concepts not of physically realizable machines, but of something that is *almost* physically realizable—the "almost" meaning that many of their properties are physically realizable (as in the case of the accelerating Turing machines discussed later).

---

[40] By *constructability* Piccinini means "physical constructability" (2011: 744; 2017).

Lastly, the *abstract/physical* computation distinction is far from universally accepted. Smith, who asks whether computation is "a concrete (physical) or abstract notion" (2010: 1), replies that "computing is fundamentally and ineliminably *concrete*: a direct consequence of the physical nature of patches of the world" (2010: 23). We can have abstract mathematical *theories*—such as computability theory—of these concrete physical systems. But these theories are no more mathematical than theories in physics. They use mathematical structures, Smith says, to "model that physical reality"; they are mathematical models of a concrete, physical computation. There is no reason, however, to think of these modeling structures as (abstract) computing devices.

While I am sympathetic to much of Smith's claim, I think that he may have taken his conclusions too far. First, I agree that many of the mathematical structures, including the abstract machines, that constitute computability theory are models of other domains. This is a fairly widespread view within theoretical computer science, where we find a lot of talk about models of computation. It is also common to think that the modeled target domains are real-world processes. Thus, Aharonov and Vazirani (2013) say that some conceive of a Turing machine as "an idealized model for a mathematical calculation (think of the infinite tape as an infinite supply of paper, and the Turing machine control as the mathematician, or for our purposes a mathematical physicist calculating the outcome of an experiment)" (p. 330), whereas others understand them to "represent the evolution of physical systems in the classical world" (p. 331). In both cases, the Turing model is viewed as describing, representing, or modeling some other domains: one is the calculating mathematician, and the other is the evolution of physical systems. However, this view is consistent with the claim that the Turing machine model itself is an abstract computing machine. This dual role—of a model and a computer—is in no way unique to abstract systems. There are many models of real-world processes where the modeling system itself is a physical computing device.[41]

Second, the claim that "computing is fundamentally and ineliminably *concrete*" does not entail that every computing device must be physically realizable. As stated previously, there might be notional machines such that most—but not all—of their operations are constrained by the physical laws of our world. Accelerating computing machines are a nice example: they violate the special relativity principle regarding the constancy of the speed of light, since at some point the speed of signal propagation exceeds this limit. If special relativity theory is true, then the accelerating machines are not physically realizable. Yet in many other respects, the operations of these machines are constrained by (or are a

---

[41] See, e.g., Frigg and Hartmann (2020).

"direct consequence of") how real-world computing devices operate. They are constrained by "the physical nature of patches of the world."

The next two sections focus on the extensions of Turing computability to real-value functions (Section 3.4.2) and on whether the bold physical Church-Turing thesis is true (Section 3.4.3). The material in these sections is pertinent but not essential to the rest of the chapter. The uninterested reader can skip these sections and move directly to Section 3.4.4, which discusses the modest thesis.

3.4.2 Computability over the Reals

Some physical computing systems presumably operate on real-valued magnitudes. The best-known example is that of analog computers that operate on real-valued variables.[42] It is thus advisable to consider computability over the reals. The extension of computability to real-value domains (or non-denumerable domains more generally) requires some explanation.[43]

The starting point is the notion of a *computable number*. The basic definition is already found in Turing's paper "On Computable Numbers" (1936), in the statement that the "'computable' numbers may be described briefly as the real numbers whose expressions as a decimal are calculable by finite means" (p. 58). A real number is effectively computable only when there is an algorithm for calculating its nth expansion (in its expression as a decimal) for any n. Turing then suggests that "there is an algorithm" can be replaced with "there is a Turing machine"; hence, the effective computability of real numbers can be replaced with Turing machine computability.[44]

The next step is to define real-valued computable functions.[45] Here, there are basically two approaches. One is in terms of approximation computability, which we will consider here;[46] another is in terms of the real-RAM machine

---

[42] Piccinini and Bahar (2013) clarify that analog computers operate on real-valued *variables*. The difference is that the values of variables are always to some degree approximate.

[43] As Papayannopoulos (2018) notes, however, the extensions of computability to real-value domains were not developed as theories of analog computation. Conversely, there are theories of analog computation that look very different from theories of real computability. Only more recently have scholars proved (e.g., Bournez et al. 2006) that certain characterizations of real computability and of analog (GPAC) computability are equivalent. See Papayannopoulos (2018) for further discussion.

[44] Gherardi (2011) highlights the problems with this definition (e.g., it represents the numbers in reverse order used for standard algorithms, e.g., for addition). He also points out that Turing refrained from this definition and presented the *admissible representation* in the correction note for the 1936 paper (Turing 1937).

[45] Gherardi (2011: 418ff.) notes that Turing (1936) provided a (problematic) definition of computable real functions alongside the Turing machine (approximation) model.

[46] Definitions that fall within this model are provided by Grzegorczyk (1955, 1957); Lacombe (1955); Mazur (1963); and Pour-El (1974; Pour-El and Richards 1989). More recently, definitions have been offered by Weihrauch (2000), whose characterization is known as *Type 2 theory of*

model.[47] We will examine the first approach. One condition of a *computable function* is that it maps sequences of computable numbers to sequences of computable numbers. There is no computable number(s) input i whose output $f(i)$ is not computable number(s). However, this condition cannot suffice. The set of computable real numbers is (obviously) enumerable, since the set of Turing machines is enumerable. Thus, real-valued (computable) functions also map many non-computable numbers to non-computable numbers. For example, the real-valued (computable) function *plus* maps computable number pairs (x,y) to a computable $x + y$—but it also maps even more non-computable number pairs (x,y) to a non-computable $x + y$. We therefore need another constraint, one that can distinguish between computable and non-computable mappings of non-computable numbers (assuming that the first constraint is satisfied).

For example, we want to distinguish the computable *plus* function from the non-computable *plus** function: *plus** behaves like *plus* with respect to the computable numbers but is absolutely erratic and arbitrary with respect to the non-computable numbers. We cannot invoke the idea of Turing machine approximation once again, as this, as we have seen, can support only enumerably many mappings. What may do the trick, however, is some continuity constraint. There are several (non-equivalent) characterizations of this constraint. Let us look at the one offered by Grzegorczyk (1955, 1957).[48]

We start with numbers:

Definition 1: A sequence of rational numbers $\{x_n\}$ is said to be effectively computable if there exist three Turing computable functions (over N) a,b,c such that $x_n = (-1)^{c(n)} a(n) \div b(n)$.

Definition 2: A real number r is said to be effectively computable if there is an effectively computable sequence of rational numbers that converges effectively to r. (*Converges effectively* means that there is an effectively computable function d over N such that $|r - x_n| < 1 \div 2^m$ whenever $n \geq d(m)$.)

Now to functions:

---

effectivity (TTE), and Braverman and Cook (2006), who extend it to what they call *bit computability*. These definitions are related but not equivalent. See Avigad and Brattka (2014) for a detailed review of these and other definitions and the relations between them.

[47] An early definition that falls within this approach is provided by Blum et al. (1998). Gherardi (2011: 423ff.) notes that this approach is also founded on Turing's work (Turing 1948). See also Feferman (2013) and Papayannopoulos (2020a), who discuss the relationship between the two approaches.
[48] The exposition here is adopted from Earman (1986).

Definition 3: A function f is an effectively computable function of the reals if: (i) f is *sequentially computable*: for each effectively computable sequence $\{r_n\}$ of reals $\{f(r_n)\}$ is also effectively computable

(ii) f is effectively *uniformly continuous* on rational intervals: if $\{x_n\}$ is an effective enumeration of the rationals without repetitions, then there is a three-place Turing computable function g such that $|f(r) - f(r')| < 1 \div 2^k$ whenever $x_m < r, r' < x_n$ and $|r - r'| < 1 \div g(m,n\ k)$ for all $r,r' \in R$ and all $m,n,k \in N$.

If we confine f to a closed and bounded interval with computable endpoints, then Definition 3 simplifies, no enumeration is necessary, and g is only a function of k. Earman suggests extending the definition to allow some discontinuities (1986: 119–120). The basic idea is that we converge on the discontinuous function through a sequence of Grzegorczyk computable functions.

3.4.3 Is the Bold Physical Church-Turing Thesis True?

The bold thesis (PCTT-B) states that the behavior of every physical system can be *simulated* (to any required degree of precision) by a Turing machine. Deutsch, Wolfram, and others who formulate the bold thesis clearly mean that the physical processes (might) involve arbitrary real-number values. They therefore ask whether the mathematical equations (functions) describing these processes are real-valued computable functions. Speculation that there may be physical processes whose behavior cannot be calculated by the universal Turing machine stretches back over several decades (Scarpellini 1963; Komar 1964; Kreisel 1965, 1967). Roger Penrose (1989, 1994) conjectures that some mathematical insights are non-recursive. Assuming that this mathematical thinking is carried out by certain physical processes in the brain, PCTT-B must then be false. But Penrose's conjecture is highly controversial.[49]

Interestingly, it appears that by and large, the (known) physical laws give rise to Turing machine computable (real) functions. A well-known exception was discovered by Pour-El and Richards (1981), who showed that the wave equation produces non-computable sequences for certain computable initial conditions (input computable sequences). In this respect, the wave equation violates the first condition in the definition of a real computable function (Definition 3 in the previous section). Pour-El and Richards also show an example where the solution

---

[49] See also Bringsjord and Zenzen (2003) and Bringsjord et al. (2006), who advance an argument for the claim that the human mind is hyper-computational.

to the equation maps computable sequences of reals to computable sequences of reals—yet the mapping violates the second constraint (continuity) in the definition of a real computable function. But their results are at the mathematical level. It is an open question as to whether the initial conditions, in both examples, are physically realizable (see, e.g., Pitowsky 1990). Whether the initial conditions properly encode all the relevant information is also a matter of debate—it has been argued that when the inputs include all the information, then the Pour-El and Richards example does not refute PCTT-B.[50]

Piccinini (2011) raises two challenges to PCTT-B that are also relevant to PCTT-M. One challenge derives from the postulate of genuine physical randomness (as opposed to quasi-randomness). A random element is one that generates random sequences of bits. It is argued that if physical systems include systems capable of producing unboundedly many digits of an infinite random binary sequence, then the bold thesis is false (see also Copeland 2000, 2004c; Calude and Pavlov 2002; Calude and Svozil 2008; Calude et al. (2010). Copeland further argues that a digital computer under unboundedness conditions and using a random element would constitute a counterexample to PCTT-M (Copeland 2002c). However, it is an open question as to whether or not genuine random elements—which are able to generate unboundedly many digits of random binary sequences—exist, or even could exist, in the physical universe.

A second challenge to PCTT-B stems from the continuous nature of physical magnitudes. Piccinini (2011) notes that "if our physical theories are correct, most transformations of the relevant physical properties are transformations of Turing-uncomputable quantities into one another" (p. 748). He then states that "a transformation of one Turing-uncomputable value into another Turing-uncomputable value is certainly a Turing-uncomputable operation" (pp. 748–749). It follows from these two premises that PCTT-B is false (assuming that some computers operate on Turing uncomputable values, this argument also challenges PCTT-M). Piccinini is right in some sense. A Turing machine cannot transform a specific, exact Turing uncomputable input into a specific, exact Turing uncomputable output. In fact, it can receive at most a denumerable number of different inputs. But the definitions of real-valued computable functions aim to address the cardinality gap between the physical functions (defined over non-denumerable domains) and the Turing computable functions (defined over denumerable domains). According to Grzegorczyk's definition of computability stated earlier, the transformation of one Turing uncomputable value into another Turing uncomputable value can be a Turing computable operation. The real-valued function of *plus* is Turing machine computable, even though it maps some Turing uncomputable arguments to Turing uncomputable

---

[50] See Weihrauch and Zhong (2002) and Gherardi (2008).

values. When x and y are computable, so is $x + y$ (satisfying condition (i) in Definition 3). In addition, *plus* is *effectively uniformly continuous* on rational intervals (hence satisfying condition (ii) in Definition 3). Thus, contrary to the second premise, the fact that there are transformations of Turing uncomputable values into other Turing uncomputable values does not mean that these transformations are Turing uncomputable operations.

Lastly, Copeland, Shagrir, and Sprevak (2018; Copeland and Shagrir 2019) have recently introduced a stronger form of physicality thesis that is related to undecidability in physics. Unlike the bold thesis, this *super-bold physical Church-Turing thesis* concerns not only the ability of the universal Turing machine to simulate the behavior of physical systems to any required degree of precision, but also further decidability questions about this behavior. Examples of such decidability questions are "Is the solar system stable?" and "Is the motion of a given system, in a known initial state, periodic?" (Pitowsky 1996: 163). The physical processes involved in these scenarios—the motion and stability of physical systems—may, so far as we know at present, be Turing computable: it is possible that the motions of planets in the solar system can be simulated by a Turing machine to any required degree of accuracy. However, the answers to certain physical questions about the processes—namely, if the motion is periodic (under ideal conditions) or will terminate at some point—are, in general, uncomputable. The situation is similar in the case of the universal Turing machine itself: the machine's behavior (consisting of the physical actions of the read/write head) is always Turing machine computable, since it is produced by the Turing machine's program, but the answers to some questions about the behavior—such as whether or not the machine halts given certain inputs—are not computable. The interested reader is referred to our papers (mentioned previously) for a formulation of the super-bold physical thesis and the arguments for and against it.

3.4.4 Relativistic Computation

Let us turn now to the *modest* thesis (PCTT-M), which states that every physical *computing* processes yields only Turing computable functions. Is the thesis true? In this section, I discuss the relativistic machines that are considered the main threat to PCTT-M. These machines are of interest for several reasons. First, broadly speaking, they belong to the class of the infinite-time Turing machines we mentioned earlier. Second, they are (arguably) non-algorithmic—and therefore drive a wedge between the concepts of *algorithmic* computation and *physical* computation. Third, they are "mechanical" in Gandy's sense of satisfying Principle IV. Whether or not these devices are truly physically realizable is debatable (see Section 3.4.5).

Relativistic machines are a type of *supertask* machines. A machine performs a supertask if it completes infinitely many operations in a finite span of time (Manchak and Roberts 2016). Russell (1915) and Blake (1926) discuss the potential realization of an arbitrary process in which each step takes half the time of the previous step. Weyl considered a machine that is capable of completing

> an infinite sequence of distinct acts of decision within a finite amount of time; say, by supplying the first result after 1/2 minute, the second after another 1/4 minute, the third 1/8 minute later than the second, etc. In this way it would be possible ... to achieve a traversal of all natural numbers and thereby a sure yes-or-no decision regarding any existential question about natural numbers. (1949: 42)

In this instance, Weyl describes an *accelerating* machine that completes the first operation in 1/2 of a given time period, the second in 1/4 of that period, the third in 1/8 of a period, and so on. Since $1/2 + 1/4 + 1/8 + \ldots + 1/2^n + 1/2^{n+1} + \ldots$ is less than 1, an accelerating machine can perform infinitely many operations within a moment of operating time.[51] More recently, Ian Stewart (1991) mentioned accelerating Turing machines, asking us to "imagine a Turing machine with a tape that accelerates so rapidly that it can complete an infinite number of operations in one second" (1991: 664).[52]

An accelerating machine is perhaps the most obvious example of a machine that performs a supertask. But there are other kinds of supertask machines. For instance, shrinking machines produce another machine that is (say) half the size of the original one as part of their operations; the speed of information propagation is constant, but the distances between the units are shorter in the offspring machine.[53] I focus here on *relativistic machines*. These machines are compatible with the two "physical presuppositions" that underpin Gandy's principle of local causation (which, as we recall, are an upper bound on the speed of signal propagation, and a lower bound on the size of atomic constituents). The accelerating machines are incompatible with the first presupposition, since they presume no upper bound on the speed of signal propagation. The shrinking machines are incompatible with the second presupposition, since there is no lower bound on the size of atomic constituents. The accelerating and shrinking machines are

---

[51] The concept is also implicit in Boolos and Jeffrey (1989), who envisaged Zeus attacking problems in mathematical logic by enumerating infinite sets "in one second by writing out an infinite list faster and faster" (p. 14).

[52] Copeland was perhaps the first to coin the term *accelerated Turing machine* (Copeland 1998: 151); he subsequently replaced the term *accelerated* with *accelerating* (Copeland 2002b). However, the term *accelerated* is still common—e.g., Calude and Staiger (2010) and Potgieter and Rosinger (2010).

[53] Intriguing setups are suggested in Davies (2001) and Beggs and Tucker (2006). There are also examples of quantum mechanical supertasks (e.g., Norton 1999).

arguably compatible with (idealized) Newtonian mechanics[54]—however, Gandy excludes them, on the grounds that "Principle IV does not apply to machines obeying Newtonian mechanics" (1980: 145). Rather, he considers relativistic machines—namely, those that obey the principles of relativity theory. In particular, the signal propagation in these machines is bounded by the speed of light.

The construction of a relativistic machine was proposed by Pitowsky (1990), who described a machine with extreme acceleration that functions in accordance with special relativity. He suggested that similar setups could be replicated by spacetime structures in general relativity.[55] Hogarth (1992, 1994) and Malament provided examples of such spacetime structures (e.g., anti–de Sitter spacetimes). Hogarth also pointed out the non-recursive computational powers of such devices, and suggested that the class of computable functions (in the broad sense) depends on the properties of the spacetime.[56] More recently, Etesi and Németi (2002), Hogarth (2004), Welch (2008), Button (2009), and Barrett and Aitken (2010) further explore the computational powers of these devices, within and beyond the arithmetical hierarchy.

In essence, these setups are based on the observation that there are solutions to Einstein's equations whereby there are spacetimes that possess a future endless curve λ with a past endpoint p and a point q, such that the entire stretch of λ is included in the chronological past of q (Figure 3.5). Such spacetimes facilitate what is known as *bifurcated supertasks*. The bifurcation is at the point p, where two agents start their travel—one along the endless curve λ, and the other along the trajectory from p to q. The latter agent can detect a signal sent from the first agent. Assume that the first agent performs some non-terminating computation that consists of infinitely many steps: this entire computation process is included in the chronological past of the second agent, who is traveling on the edge from p to q. Assuming that the second agent is somehow informed about the infinite computation, we have a supertask. The supertask is defined with respect to the infinite computation performed by the first agent, which takes a finite span of time in relation to the second agent.

I discuss a machine for computing the halting function (Figure 3.6). The machine, RM (relativistic machine), consists of a pair of Turing machines, or of their physical implementations. The two Turing machines—T_A and T_B—are in communication with each other. T_B moves along λ, and T_A moves along a future-directed curve that connects the beginning point p of λ with q. The time it takes T_A to travel from p to q is finite, while during that period T_B completes the infinite-time trip along λ. This physical setup permits the computation of the halting function. One

---

[54] See Copeland (2002b: 289) and Earman (1986: 34) for a discussion of accelerating machines, and Davies (2001: 672) for a discussion of shrinking machines.
[55] Pitowsky (1990) is based on a lecture he gave in 1986—about the same time that Istvan Németi proposed his construction of relativistic computation (Andréka et al. 2018).
[56] See Hogarth (1994: 127–133).

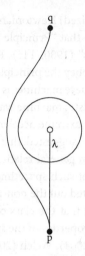

Figure 3.5 A setup for relativistic computation (adapted from Hogarth 1994: 127). In this relativistic Malament-Hogarth spacetime, the entire stretch of λ is included in the chronological past of q. (From Hogarth, Mark L. 1994. "Non-Turing Computers and Non-Turing Computability." *Proceedings of the Biennial Meeting of the Philosophy of Science Association 1*: pp. 126–138. Published by The University of Chicago Press on behalf of the Philosophy of Science Association, Copyright 1994, The Philosophy of Science Association).

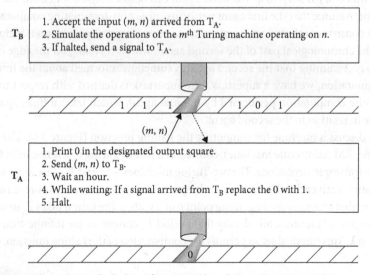

Figure 3.6 Computing the halting function. The relativistic machine RM, consisting of two communicating standard Turing machines T_A and T_B, computes the halting function.

feeds T_A with input (m,n). T_A prints 0 in its designated output cell, then sends a signal with the pertinent input to T_B. T_B is a universal machine that mimics the computation of the mth Turing machine operating on input n. In other words, T_B calculates the Turing-computable function $f(m,n)$ that returns the output of the mth Turing machine (operating on input n) if this Turing machine halts—and no value if this Turing machine does not halt. If T_B halts, it immediately sends a signal back to T_A; if T_B never halts, it never sends a signal. Meanwhile, T_A "waits" during the time it takes T_A to travel from p to q (say, one hour). If T_A has received a signal from T_B, it prints 1, replacing the 0, in the designated output cell. One way or the other, the output cell shows the value of the halting function after one hour (of T_A). It is 1 if the mth machine halts on input n, and otherwise it is 0.

3.4.5 Does Relativistic Computation Refute the Modest Thesis?

I have described a physical system that computes beyond the Turing limit, that is, the halting function. It arguably violates PCTT-M. Or does it? The answer depends on whether RM *computes*, and on whether its entire operations are *physically possible*. As for the computation issue, what has been said about the infinite-time Turing machines applies here as well: RM computes in the sense of the term *compute* as set out in the semantic, the mechanistic, the causal, and the BCC (broad conception of computation) accounts. According to all these accounts, *if* RM is physical, it represents a counterexample of PCTT-M. This shows that the *concept* of physical computation accommodates relativistic computation (and, indeed, hypercomputation). This conceptual possibility drives a wedge between the *concepts* of *physical* and *algorithmic* computation—for even if it transpires that RM is not physically realizable, RM indicates that executing an algorithm (in the sense just described) is not necessary for the concept of physical computation. This is because, as I will go on to argue, RM does not perform algorithmic computation.

What about the physicality of RM? Németi and his colleagues provide the most physically realistic construction, placing machines like RM in setups that include huge slow-rotating Kerr black holes (Andréka et al. 2018), and emphasizing that the computation is physical in the sense that "the principles of quantum mechanics are not violated" and RM is "not in conflict with presently accepted scientific principles" (Andréka, Németi, and Németi 2009: 501). They suggest that humans might "even build" their relativistic computer "sometime in the future" (Andréka, Németi, and Németi 2009: 511).

Naturally, all this is controversial. Earman and Norton (1993) point out that the physical plausibility of relativistic computation depends on "a resolution of some of the deepest foundations problems in classical general relativity, including the nature of singularities and the fate of cosmic censorship" (1993: 40–41). They note that communication between the agents is no trivial matter and is subject to blue-shift effects. One problem is that the halted signal would destroy the receiving agent.[57] Another problem is that infinitary computation requires infinite—and not merely unbounded—memory.[58] Yet another related problem is that the infinitary computation would require an unbounded amount of matter-energy, which appears to violate the basic principles of quantum gravity.[59] These issues cast heavy doubt on the physical possibility of relativistic computation. At this point, we can say that it is unclear whether relativistic computation is physically possible—and, accordingly, whether PCTT-M is true or false.

3.4.6 Supertasks and Algorithmic Computation

I have suggested that RM presents a challenge to PCTT-M. Does it also challenge the algorithmic version of the Church-Turing thesis (CTT-A)? I will suggest that it does not—simply because RM does not carry out *algorithmic* computation in the sense assumed in this chapter.

At first blush, RM appears to be merely a pair of communicating Turing machines—and, as such, carries out an algorithmic computation. A pair of communicating Turing machines, *under standard time scales*, computes no more than what is computable by a single Turing machine. Our RM computes beyond the Turing limit, it seems, because it performs a supertask. In other words, performing a supertask appears to be the only difference between RM and a standard communication between Turing machines; this difference is the feature responsible for the computational leap from the computable to the uncomputable.

This first impression, however, is misleading. Performing a supertask is *not* the only difference. Let us look more carefully at the computational structure of RM. One feature of RM is that it always reaches an end stage. After one hour, the halting state of the simulated machine is displayed in the designated output cell of T_A; at this moment, as we recall, T_B no longer exists. A second feature of RM is that it consists of two machines, T_A and T_B, and some communication devices. Thus, when talking about the configurations of RM, we must take into account the configurations of T_A and T_B. We cannot ignore the configurations of T_B, for example; if we do, we cannot say that RM has performed a supertask and computed the halting function. Likewise, we must take into account communication between the two machines—in particular, how and when the configuration of one machine determines the configuration of the other.

---

[57] See Etesi and Németi (2002), Németi and Dávid (2006), and Andréka, Németi, and Németi (2009: 508–509) for proposed solutions to the signaling problem.

[58] The memory problem is discussed in Shagrir and Pitowsky (2003: 88–90).

[59] See Németi and Dávid (2006) for a proposed solution to the energy problem.

A third feature of RM is implied by the first two. It is that the end stage of RM cannot be described as a stage α + 1 whose configuration is completely determined by the preceding stage, α. This is simply because there is no such preceding stage, α—at least not when the simulated machine never halts. What would be such a preceding stage α? It cannot be the initial "print 0" stage, as it was followed by infinitely many stages of T_B. And it cannot be some stage of T_B, since each such stage was followed by infinitely many others. One could stipulate a "decision moment" of T_A, in which it is decided whether or not to replace the 0 with 1 (as in Shagrir and Pitowsky 2003). However, this merely shifts the problem to the "decision moment"; now it is this "decision moment" that cannot be described as an α + 1 stage.

We may conclude, therefore, that RM cannot be described in terms of standard communication between Turing machines. When the communication is standard, each stage of the communicating machines is an α + 1 stage, whose configuration is completely determined by that of the previous stage α (only the initial stage need not satisfy the requirement). For this reason, RM is neither a Gandy machine nor an algorithmic one. These concepts, as we have seen, assume a very specific notion of determinism, in which the configuration of each α + 1 stage is determined by that of the previous α stage. RM is certainly deterministic in *another* sense—namely, that the state of T_A-halting-on-0 is uniquely determined by the initial state of the machine. This is because the state of T_A-halting-on-0 is a *limit* of previous states of T_B (and T_A), of which the relevant feature is their not-sending a signal to T_A. In this account, RM is deterministic in that the T_A-halting-on-0 state is in part the limit of the previous no-signal-being-sent states of T_B. This sense of determinism accords well with the physical usage, whereby a system or machine is said to be deterministic if it obeys laws that invoke no random or stochastic elements. RM is deterministic in the same sense that an infinite-time Turing machine is deterministic. In fact, RM may be seen as a physical realization of an infinite-time Turing machine. We can therefore conclude that even though RM behaves according to a fixed and finite rule, it is not algorithmic in the sense of *algorithm* as described earlier—and is therefore not a counterexample to CTT-A.[60]

3.4.7 Physical Computation and Gandy Machines

Gandy admits that his account does not encompass physical computation in general, and he states that "Principle IV does not apply to machines obeying

---

[60] Note that I do *not* claim that the performance of supertasks is incompatible with being algorithmic; on the contrary, the point is that we must distinguish between supertask machines that are algorithmic and supertask machines that are not algorithmic (Copeland and Shagrir 2011).

Newtonian mechanics" (1980: 145). Relativistic machines are of interest because they apparently satisfy Principle IV (aside from the problems of physical realization, there is a lower bound on the size of atomic constituents and an upper bound on the speed of signal propagation). As just noted, these machines violate Principle I ("form of description"), and therefore they are not Gandy machines.

3.4.8 The Relationship Between Physical and Other Notions of Computation

Figure 3.7 sums up the inclusion relationship between the various notions of computation that we have considered. On the one hand, physical computation

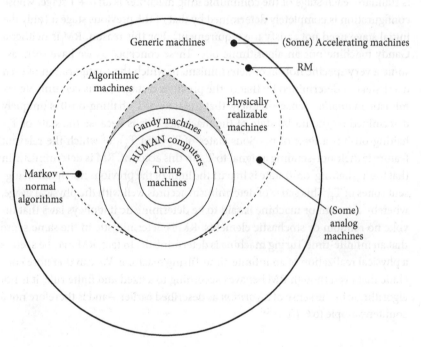

Figure 3.7 Relations between classes of computing machines. The class of physically realizable machines might include RM (if physically realizable) and some analog machines that are not algorithmic in the sense assumed here. The class of algorithmic machines might include systems ("global algorithms") that are not physically realizable. Both classes are more restrictive than the class of generic computation, which includes accelerating machines, some of which are arguably neither algorithmic nor physically realizable. Both classes are more inclusive than that of Gandy machines.

(the class of *physically realizable computation*) is more restrictive then generic computation, assuming that there are computations that are not realizable. On the other hand, physical computation includes physical machines that are not Gandy machines (such as analog computers). The more interesting conclusion, however, is that there is only a partial overlap between physical and algorithmic computation. On the one hand, the general *concept* of algorithm is so broad today "as to admit 'non-implementable' algorithms" (Moschovakis and Paschalis 2008: 87). By way of example, we examined a few algorithms whose basic steps are "global"—namely, that at each step each component takes into account information from every other component (note that these machines are not realizable in their "global" form—but of course every "global" step can be broken into several physically realizable steps). On the other hand, there are physically realizable computations that are not algorithmic—such as, arguably, some analog computers, as Gandy points out. My focus here was relativistic machines, some of which compute functions that are not Turing machine computable. Whether these machines are physically realizable is debatable—but even if they are not, their conceivability indicates that being algorithmic is not a requirement of a physical computation.

3.5 Summary

I have used Gandy's account of machine computation to discuss and differentiate between three notions of computation: generic computation, algorithmic computation, and physical computation. As it turned out, the class of machines that Gandy characterized—that is, the Gandy machines—is more restrictive than the classes of generic, algorithmic, and physical computation. He captured none of these notions (the extent to which Gandy fully captures the physically realizable algorithmic computation is left an open question). But his account helps us to see more clearly the differences between generic computation, algorithmic computation, and physical computation.

4
Computation as Step-Satisfaction

This chapter deals with Robert Cummins's account of computation as set out in *Meaning and Mental Representation* (1989; see also 1983, 1996). In this work, Cummins devotes only a few pages to the characterization of computation.[1] This is unsurprising because the book is about mental representation, not computation. Despite the brevity of this account, computation plays a central role in the book for two reasons. One is that Cummins aims to account for the notion of mental representation as it appears in *computational theories of cognition*. The second reason is that the specific account of computation he puts forward is pivotal in his positive account of mental representation (whereas most of the book is devoted to presenting and rejecting other accounts of mental representation). Cummins calls this account of representation *interpretational semantics*.

Cummins's account is typical of earlier accounts of physical computation. He associates computation with a certain notion from computation theory—namely, *program execution* (this is the *logical dogma* mentioned in the Introduction). He then associates program execution with a select (architectural) feature of the causal structure of the physical system—namely, *step-satisfaction* (this is the *architectural dogma* mentioned in the Introduction). In previous chapters, I challenged the first stage, attempting to dissociate physical computation from the canonical notions of computation theory. My aim in this chapter is to challenge the second stage: I will argue that step-satisfaction is not helpful in meeting the classification criteria. Depending on how it is understood, step-satisfaction either excludes important cases of computing physical systems, or is empty and applies to virtually every physical system. Either way, this shows that step-satisfaction is not an essential condition for computation, as it plays no role in distinguishing computing from non-computing physical systems. I believe that this strategy of argumentation applies to other accounts that ground computation in certain architectural properties.

---

[1] See mainly Cummins (1989: 91–92ff.). See also a detailed discussion of some of the relevant notions (mainly of instantiation and program execution) in Cummins (1983: 34–51 and 163–191).

4.1 Cummins's Account of Computation

Cummins is perhaps the first philosopher who explicitly aims to account for *physical computation*—namely, computation in physical systems. He develops his account in two steps. The first step is to say how physical systems *satisfy* and *instantiate* (but do not necessarily compute) functions. The second step is to say how physical systems also compute functions.

4.1.1 Satisfaction and Instantiation

According to Cummins, a physical system *satisfies* a function g if it produces o as its output on input i just in case $g(i) = o$—where the term *produce* refers to some causal process that starts from an input i and terminates with the output o. This scope of satisfaction is broad in one sense and narrow in another. It is broad in that satisfaction applies to *any* physical system whose dynamics or mechanism proceeds in accordance with the laws of nature. The digestive processes in the stomach, the trajectories of planets in the solar system, and the cycles of washing machines are all examples of *satisfying* functions. In short, I think it is safe to say that every physical system—no matter whether its processes are algorithmic, continuous, digital, or analog—satisfies a *mapping function* from its input-physical properties to its output-physical properties.

The scope of satisfaction is narrow in that the mapping function g relates only *physical* entities and properties. In other words, physical systems satisfy only *physical functions* that map input physical properties i to output physical properties o. They do not—and cannot—satisfy, say, such mathematical functions as *plus*, which relates a pair of *number* arguments $<m,n>$ and the *number* value $m + n$. The *plus* function, according to Cummins, relates abstract mathematical properties, which are not physical properties and as such cannot be satisfied by a physical system. A (satisfied) function, g, only relates *physical* properties.

At first blush, the fact that the notion of *satisfaction* is tied to physical functions appears to render it redundant if we want to account for physical systems that deal with, let alone compute, mathematical functions such as *plus*. If the notion of *satisfaction* is tied to physical functions—that is, functions that map physical properties—then how can we use it to account for physical systems that somehow deal with mathematical functions? At this point, Cummins (like many others) turns to the notion of *instantiation*. True, the satisfied function g links together physical, non-abstract properties—but these related physical entities can be *representations*. They can be *numerals*—namely, physical entities that represent numbers. Thus, a physical system is an adding machine not because it operates on numbers. A physical system only satisfies g, which maps physical properties to physical properties. A physical system is an adding machine when it satisfies a function *g* that maps representations of numbers. It is an adding machine, according to Cummins, when it *instantiates* the *plus* function:

> What an adding machine does is *instantiate* the plus function. It *instantiates* addition by *satisfying* the function *g* whose arguments and values represent the arguments and values of the addition function, or in other words, have those arguments and values as interpretations. (1989: 89)

The picture we get is what Cummins dubs the *London-Tower Bridge* (Figure 4.1). The bottom span is the satisfied function, *g*, which maps physical states of the system (e.g., button-pressing sequences) to other physical states of the system (e.g., displays of certain states). The top span is the mathematical function *plus*, which maps pairs of numbers <*m,n*> to the values *m + n*. The vertical arrows stand for the so-called *interpretation function*, which maps the physical states of the system to numbers. Specifically, it maps the pair of physical buttons <*M,N*>, which are numerals, to the pair of numbers <*m,n*>, and the physical display *D*, another numeral, to the value *m + n*. According to this interpretation, the system is an adding machine in the sense that it instantiates the top-span function *plus* by satisfying the bottom-span function *g*, whose arguments and values represent the arguments and values of the *plus* function.

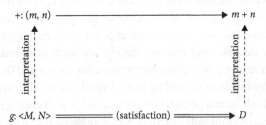

Figure 4.1 The London-Tower Bridge. The bottom span is the satisfied function *g*, which maps physical inputs (e.g., button-pressing sequences) to physical outputs (e.g., displays of certain states). The top span is the mathematical function *plus*, which maps pairs of numbers <*m,n*> to the values *m + n*. The vertical arrows stand for the *interpretation function*, which maps the physical states of the system to numbers.

4.1.2 Step-Satisfaction

Up to this point, Cummins talks about satisfaction and instantiation, and even about representation and interpretation, but not about *computation*.

Computation, he argues, is a special way of satisfying functions. In his words, "to compute a function g is to execute a program that gives o as its output on input i just in case $g(i) = o$. Computing reduces to program execution" (p. 91). Program execution, in turn, "involves *steps*," and so "program execution reduces to step-satisfaction" (p. 92). The bottom line, then, is that computation is a step-satisfaction process: it proceeds by means of steps. Each single step is satisfied, but the entire process—which consists of series of (satisfied) steps—is not only satisfaction but also computation.

As with satisfaction, computation operates on a function g that relates physical inputs and outputs. It does not operate on abstract entities such as numbers. However, a physical machine can still compute an abstract function such as *plus*. A physical computing machine is an adding machine when the (physical) arguments and values of g are representations of numbers. It is an adding machine when it *instantiates* the plus function—in which case the system *computes* a function g, whose arguments and values represent the arguments and values of the *plus* function, the way it does earlier (Figure 4.2). The only difference between the two figures is that the bottom span in Figure 4.2 stands for step-satisfaction, hence computation, whereas the bottom span in Figure 4.1 stands for satisfaction, but not necessarily for *step*-satisfaction.

A major motivation of Cummins's account is to link this notion of computation to cognition. Here, his general idea is that "what works for addition will work for cognition" (p. 111). A cognitive system is merely one that "computes representations whose contents are the values of the cognitive function, and computes these from representations of the function's arguments" (p. 109). According to Cummins, we can have the London-Tower Bridge scheme for describing cognitive systems, whereby the bottom span consists of a sequence of physical states that involve representations. We might think of these representations, for example, as physical symbols whose syntactic structure enables us to infer certain conclusions (such as q) from certain premises (e.g., p

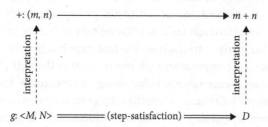

Figure 4.2 Computing addition. The bottom span stands for step-satisfaction, hence computation. The rest is as in Figure 4.1. The arguments and values of g are *interpreted as* representing the arguments and values (numbers) of the *plus* function.

and $p \to q$). The upper span is a sequence of interpretations of these bottom-span events, such as the corresponding propositions. This cognitive system computes in the sense that the process at the bottom span is state-transition ("step-satisfaction")—that is, it satisfies certain epistemic (e.g., rational) constraints. Cognition, in other words, is computation of a physical function that instantiates a cognitive function.

4.1.3 The Essentials of the Account

Let me highlight the essentials of Cummins's account. First, Cummins accounts for computation in the context of *physical* systems. Moreover, he stresses that such an account should meet the *classification criteria* (see Section 1.2.1). Specifically, in his terms, the account should distinguish between two kinds of physical systems that satisfy functions: those that satisfy functions by computing them, and those that satisfy functions without computing them. As previously noted, virtually every physical system satisfies some sort of function, but "functions need not be computed to be satisfied. Set mousetraps satisfy a function from tripping to snapping without computing it, and physical objects of all kinds satisfy mechanical functions without computing them" (p. 91). The aim of an account of computation is to state the conditions under which a physical system not merely satisfies a function, but also computes it.

Second, Cummins draws the distinction between computing and non-computing by invoking the theoretical notion of a *program*. Specifically, he claims that the difference between a physical system that satisfies a function by computing it and one that satisfies a function without computing it is that the former "executes a program" and the latter does not. But what exactly does it mean to say that one physical system executes a program while another does not? At this point, Cummins appeals to the notion of *step-satisfaction*: a physical computational process consists of a disciplined stepwise process, whereas a stand-alone step on its own is only satisfied, not computed. A non-computational physical process apparently proceeds in a non-stepwise, perhaps continuous fashion.

Cummins's characterization of computation falls within the first two dogmas I identified earlier, in the Introduction. The first stage involves the *logical dogma*, which is to associate computation with one or another theoretical notion from logic or computer science (an algorithm, program, procedure, formal proof or rule, automaton, etc.). Cummins's favorite is *program execution*. The second stage involves the *architectural dogma*: Cummins associates the theoretical notion—that is, the execution of a program—with a distinct architectural property that is characteristic of computing physical systems. This architectural property is *disciplined step-satisfaction*, which apparently refers to processes that proceed in a stepwise fashion. This architectural property is a necessary feature of computing systems: processes that do not possess it are not of the computing type.

This pattern of characterization is typical of earlier accounts of computation. As I noted in the Introduction, it is found in the characterizations put forward by Newell and Simon (1976), Haugeland (1978), Fodor (1980, 1994; Fodor and Pylyshyn 1988), Stich (1983), Copeland (1996), Crane (2016), and others. Thus, Copeland (1996) associates computation with the theoretical notion of an *algorithm* ("To compute is to execute an algorithm" [p. 335]), and he understands an algorithm to typically consist of "step-by-step applications of a certain propagation rule" (p. 337). However, Copeland (and others) also think that unless we add to this step-by-step feature a certain type of mapping relationship (which he dubs *honest modeling*) that is stronger than Cummins's instantiation, we will end up with unlimited pancomputationalism.

Third, elsewhere Cummins (1975, 1983, 2000) presents a certain style of explanation—which he calls *functional analysis*—and argues that it is typical of psychological, specifically computational explanations. In functional analysis, a capacity or disposition of a system is explained by means of an analysis that breaks down the capacity into a set of simpler capacities that are arranged in a way that allows the capacity to be explained. In some cases, these subcapacities may be assigned to the components of the system—but in other instances, they are assigned to the system as a whole. One example of the latter is a Turing machine: "Turing machine capacities analyze into other Turing machine capacities" (Cummins 2000: 125). Another example can be found in some of our arithmetical abilities: "My capacity to multiply 27 times 32 analyzes into the capacity to multiply 2 times 7, to add 5 and 1, and so on, but these capacities are not (so far as is known) capacities of my components" (Cummins 2000: 126).

Cummins does not mention functional analysis in his account of computation, but there are notable affinities between the two accounts. As we have just seen, he uses examples of computation to explain the idea of functional analysis.[2] More importantly, the notion of a *program* plays a key role in his account of functional analysis. Cummins says:

> Functional analysis consists in analyzing a disposition into a number of less problematic dispositions such that programmed manifestation of these analyzing dispositions amounts to a manifestation of the analyzed disposition. By "programmed" here, I simply mean organized in a way that could be specified in a program or flowchart. (2000: 125)

---

[2] Computations, however, are by no means the only examples of functional analysis. Another example is a cook's ability to bake a cake (Cummins 2000: 125).

If computation is an *execution of a program*, as Cummins suggests, then computational processes are natural candidates for functional analysis (as the preceding examples indicate). Moreover, by describing a physical system as a computing system, we naturally explain it by means of functional analysis. For example, when describing a cognitive system as a Turing machine, we tend—or perhaps are even bound—to explain its capacity in terms of a programmed manifestation of subcapacities. Thus, computational descriptions have an important explanatory role in cognitive science.

Fourth, Cummins characterizes computation as *program execution*, and further asserts that "program execution is surely disciplined step satisfaction" (1989: 92). But what is meant by "disciplined"? On the one hand, *discipline* looks like just another constraint on computation, in addition to the steps; on the other, when seeking to explain the discipline, Cummins says that the "discipline takes care of itself" (p. 92)—implying that a computing system need not have a special control or program unit, as in a Turing machine. It is enough, he argues, for the causal relations to be such that the satisfaction of one step follows the satisfaction of another (as in a flowchart). If that is the case, the discipline does not appear to add a substantial constraint on computation beyond step-satisfaction (this observation is discussed in more detail later).

Lastly, Cummins distinguishes between computation (and satisfaction) of different *kinds* of functions. A system can satisfy/compute a *physical* function g that links together physical inputs and outputs—but it can also satisfy/compute an *abstract* or *cognitive* function. The difference is that the computation of mathematical, cognitive, and perhaps other functions invokes the notion of instantiation. A system computes an *abstract* or a *cognitive* function f by computing a physical function g that instantiates f, whereas instantiation (etc.) stands for a mapping relation between g and f. The important point is that the distinction between *computation* and *satisfaction* does not hinge on the distinction between the kinds of functions that are being computed. Physical systems can either *compute* or *satisfy* abstract/cognitive functions: a physical system may satisfy an abstract/cognitive function by *computing* it, but it may also merely *satisfy* that function without computing it. This suggests that the notion of instantiation does not play a role in the difference between computation and mere satisfaction. The difference between computation and satisfaction is wholly dependent on the feature of disciplined step-satisfaction.

It is worth noting that Cummins's notion of *instantiation* is broader than the notion of *implementation*, which is the focus of Chapter 5 of this book. Instantiation is much like implementation, in that it assumes an isomorphism relationship between g and f. Instantiation is broader than implementation, in that the instantiated function f need not be mathematical (abstract): f can be a non-abstract (e.g., cognitive) function. This is perhaps why Cummins also uses the terms *interpretation* and, subsequently, *simulation representation* to describe this instantiation relationship; following Ramsey (2007), I will use the term *input-output representation*. This does not, however, render the account of computation a semantic one. As we have seen, the notion of *instantiation* (etc.) is not necessarily invoked in the contexts of physical computation. According to Cummins, a physical system computes just in case the process that is mediating inputs and outputs is of the step-satisfaction type. The notion of instantiation is invoked in contexts where the computed function is mathematical or cognitive, and perhaps for other functions as well.[3]

My view is that Cummins got things exactly backward. Input-output representation plays a central role in characterizing computation, especially in cognitive and neural systems, whereas step-satisfaction plays no role in distinguishing between computing and non-computing. I return to the former claim, about input-output representation, in Chapter 9. The rest of this chapter is devoted to my negative claim about step-satisfaction.

4.2 Is Step-Satisfaction Necessary for Computation?

Cummins argues that computation is step-satisfaction. In assessing this claim, we need to ask whether step-satisfaction constitutes a *necessary* and *sufficient* condition for computation. I think that it is clear that step-satisfaction is not sufficient for computation: there are many physical processes that are of the step-satisfaction type but do not constitute computation, even when described this way. Cooking a pie, the evolutionary development of traits, the workings of the human body, and manufacturing processes in factories are all described in terms of sequential step-satisfaction operations—but they are not computations. Cummins himself mentions some of these processes (such as cooking) as ones that are explained through functional analysis, and, hence, as step-satisfaction. But even when they are explained this way, we do not see them as computation.

One might suggest that adding instantiation to step-satisfaction would result in a sufficiency criterion. But this suggestion has two difficulties. One—just mentioned—is that, according to Cummins, some physical systems compute without instantiating at all. Another difficulty is that some physical systems will not compute even when instantiating in Cummins's sense (Copeland 1996). In Chapter 5, I argue that even stronger notions of instantiation would not yield a

---

[3] Moreover, the role of *simulation representation*, according to Cummins, is to naturalize the semantic content that occurs in the context of a computing system. This is made possible, he argues, because simulation representation is defined not in semantic terms, but in terms of isomorphism. For further discussion on this subject, see chap. 8 of Cummins's book (1989).

sufficiency criterion. In the rest of this chapter, I focus on the necessity element—namely, the claim that step-satisfaction is necessary for computation—and argue that it is in fact not necessary. Step-satisfaction plays no role in delimiting physical computation.

According to Cummins, step-satisfaction appears to draw the line between computing and non-computing based on the number of steps satisfied. An input-output process that consists of two steps appears to be computation—or at least the beginning of computation, because it involves the satisfaction of two steps—and the more steps you have, the merrier. However, drawing the *computation/non-computation* boundary along the *one-step/two-steps* distinction—namely, that a one-step process is not computation, but a process of two or more steps is—has its problems. This one-step/two-steps distinction runs counter to how computation is perceived in computer science, as there is no pertinent difference between (say) a Turing machine that operates in one step and those that work in two or more steps. They all compute, even if in the former case the computation is trivial. More importantly, we can easily think of quite a few cases of physical computation—such as analog computation—that satisfy a function in one go, as it were.

One might suggest, at this point, including one-step processes in the definition of computing. But this route has its problems as well. If we consider one-step processes computation (albeit trivial), then virtually every physical process, it seems, is computation, since every physical process satisfies a physical function by means of a process comprising at least one step. If so, the criterion of step-satisfaction is pointless, as it does not help to distinguish computing from non-computing. Indeed, the whole point of the distinction between *satisfaction* and *computation* is that satisfying a function in one fell swoop is not computation—rather, computation might start when the process consists of two or more steps that are being satisfied.

To understand the nature and scope of this dilemma, let us examine a simple yet compelling example put forward by Itamar Pitowsky (1990).

4.2.1 Pitowsky's Average Machine

Consider the following machine for averaging three numbers (Figure 4.3). Imagine we have an insulated container divided into three equal sections by insulated removable barriers. We put a thermometer in each section, and set the temperature in each section to a temperature equal to the corresponding three numbers k, m, and n. We now remove the barriers simultaneously and wait until the temperatures equalize. This process is a thermodynamic one, and

COMPUTATION AS STEP-SATISFACTION 97

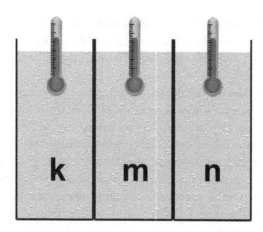

Figure 4.3 Pitowsky's machine for averaging three numbers. The device is divided into three equal sections by insulated removable barriers with a thermometer in each section. The temperature in the sections is set to correspond to k, m, and n. Removing the barriers will result (output) in the average $(k + m + n)/3$.

usually described by a set of differential equations. However, its output is always $(k + m + n) / 3$.

Consider another scenario. We repeat the process with k, m, and n—but this time we (or some other mechanism) remove only one barrier, and wait for the temperature to equalize $((k + m) / 2)$. Only then do we remove the second barrier. The final output is clearly the same as in the first scenario, but this time it involves a two-step process:

Basic step 1. Average k and m:
$[(k, m) \rightarrow (k+m) / 2]$;
Basic step 2. Operate on the output of step 1, $\text{output}_1(k,m)$, and n:
$[(\text{output}_1(k,m), n) \rightarrow 2/3 \cdot \text{output}_1(k,m) + n / 3]$.

In other words, in the first scenario the device averages k, m, and n in a single-step process, whereas in the second scenario it does so through a two-step process.

Now, let us ask whether the one-step process, as described in the first scenario, is computation. Consider, first, that the answer is *no*—meaning that the averaging, one-step process is not computation. Assuming that the two-step averaging process, as described in the second scenario, is deemed to be computation, we get the arguably absurd conclusion that computing depends on moving the barriers of the thermal machine twice instead of once. It seems very odd to deem the two-step process to be computation and the one-step averaging process not to be. It is much more reasonable to think that if the two-step process is computation, so too is the one-step process. Indeed, the device serves to compute the average of three numbers, regardless of whether the barriers are moved once or twice.

Now let us consider the *yes* answer—namely, that the averaging, one-step process is worthy of being considered computation.[4] The trouble here is that if this thermodynamic one-step process is computation, then we surely must accept that other thermal processes—such as tornado storms and boiling water, which also are described by thermodynamic equations—are forms of computation as well. Indeed, by that definition, it is hard to see what physical dynamics is *not* computation. But if all these processes are also computation, then the concept of step-satisfaction plays no effective role in distinguishing computation from other physical dynamics.

I maintain that this kind of dilemma challenges the accounts that associate computation with a particular architecture. One horn of the dilemma is that the select architectural profile (e.g., step-satisfaction) excludes important classes of computing systems; the other horn is that attempting to include such computing systems by weakening the architectural constraint causes too many physical systems to meet that constraint, and therefore the chosen architectural profile plays no role in distinguishing computing from non-computing.

4.2.2 A Way Out of the Dilemma?

Let us review several potential attempts to escape the dilemma—beginning with those that take the second-horn escape route, in which the one-step process is counted as computation. One could say that the averaging process is computation in that it is (one)-step satisfaction. Nevertheless, the reply goes, there are many other processes that do not involve steps in any essential way—such as processes that have no inputs or outputs, or processes that are not deterministic and involve some sort of randomization. The argument should be that all these processes are instances of satisfaction, but not of computation, as they involve not even a single step.

The difficulty with this proposal is that it is not clear why such no-step processes cannot be computation too. Many instances of computation do not involve inputs and outputs in any essential way—for example, finite-state automata without inputs and outputs, as well as various kinds of cellular automata, such

---

[4] Copeland (1996) appears to suggest this when he characterizes *computation* as the execution of an algorithm that comes down to "the performance of some sequence of the primitive (or 'atomic') operations made available in the architecture" (p. 337). He adds that "the sequence may consist of a single operation" (p. 337).

as some instances of the Game of Life, neural networks, and so forth. There are non-halting computing machines that receive inputs but emit no defined output. The same goes for processes that are not deterministic. There are many instances of probabilistic computation,[5] and ones involving physical randomness.[6] With the rise of Bayesian approaches in cognitive science, randomness is also considered an integral part of neural computation.[7] These counterexamples indicate that we cannot contrast step-satisfaction (as computation) with processes that lack inputs/outputs and/or that are not deterministic (as non-computation). Some computations also lack inputs/outputs and/or are not deterministic.

Another attempt to differentiate the one-step computing process of averaging from non-computing processes is grounded in the instantiation relationship. Thus, one could point out that from Cummins's point of view, the averaging process satisfies a particular mapping physical function—mapping input physical magnitudes to output physical magnitudes—and this physical function instantiates the mathematical function of averaging—namely $(k,m,n) \rightarrow (k + m + n) / 3$. However, the physical processes that take place in stomachs, tornadoes, and mousetraps, though they can be described by mathematical functions, do not instantiate these functions, so they do not compute the mathematical functions. There are constraints on instantiation that the averaging process meets, but the other processes do not.[8]

This proposal, however, is also not very helpful. First, Cummins talks about the computation of *physical* functions—which does not invoke the notion of instantiation. We can therefore describe Pitowsky's device as operating on physical properties without introducing the mathematical function of averaging. We will thus face the same dilemma with respect to the physical function. Second, let us assume, for the sake of argument, that there are constraints on instantiation that differentiate the (computing) averaging machine from other physical processes, at least with respect to mathematical functions. The problem with this proposal is that it nullifies the notion of step-satisfaction. Hitherto, we thought that step-satisfaction was capable of distinguishing computing from non-computing—but it now turns out that it does no such thing. Stomachs, tornadoes, and mousetraps also involve one or more step-satisfaction processes—but the reason that they do not compute, whereas the averaging machine does, is not step-satisfaction, but

---

[5] For a review see, e.g., Gurari (1989: chap. 6). Probabilistic computation involves certain random choices. Thus, there is a certain probability (chance) that the program gives wrong answers; the trade-off is a reduction in the speed of computation.
[6] See, e.g., Calude and Pavlov (2002).
[7] See, e.g., Chater, Tenenbaum, and Yuille (2006); Griffiths et al. (2010); and Clark (2013).
[8] This might be Copeland's approach when he suggests that the required instantiation relationship should be that of adding to this step-by-step feature a certain kind of mapping relationship that he calls *honest modeling* (see also the discussion in Chapter 5).

rather (arguably) instantiation. Thus, step-satisfaction is an empty notion with respect to the distinction between computing and non-computing.

Let us turn now to the attempts to escape the dilemma by the first horn—namely, by insisting that the averaging one-step process is not computation. One tempting move in this direction would be to insist that the two-step process is also not computation—as then there would be no need to solve the puzzle of why the two-step process is computation but the one-step process is not. The difficulty with this proposal is that it undermines the role of step-satisfaction in delimiting computation. The number of steps satisfied—be it one, two, or more—does not help to capture the computing processes. In other words, it does not really matter whether a given process consists of one, two, or more steps: what makes it computation or not is determined by yet other features.

One might contend that the one-step process is non-computing, and the two-step process is computing, on the grounds of the *analog/digital distinction* (the term *analog* here used in the sense of "continuous").[9] Thus, the one-step, analog process would be deemed non-computing, as it consists of a single continuous dynamic, while the second process is less analog (and perhaps more digital) in that it consists of two somewhat discrete, continuous dynamics. In the two-step process, then, we have the beginning of computation—and when many steps are involved, we get a full-fledged computation.

Crucially, I do not deny that step-satisfaction might capture an important difference between different *kinds* of computation—the one-step process being an instance of analog computation and the two-step process an instance of digital computation. If this were the case, then step-satisfaction would play a role in differentiating analog from digital computation. However, my point is that this difference between one-step and two-step processes is immaterial to the difference between computation and non-computation. It is counterintuitive—not to say absurd—to say that the two-step averaging process is computation but the one-step averaging process is not, since moving the barriers one at a time does not make the process more of a computation than moving them simultaneously. Either way, we can use the device (or not) to compute the average of three numbers.

This point can be generalized to other attempts at characterizing computation based on the *analog/digital* distinction.[10] The problem of distinguishing between digital and analog is notoriously challenging, and there are several

---

[9] The other sense of *analog*, in terms of mirroring, is discussed in Chapter 9.
[10] Gandy (1980) provides an account of digital computation, which Sieg (2008, 2009) sees as a general account of machine computation. Piccinini (2007) and Fresco (2014) offer accounts of computation that mainly apply to digital computation; both of them have subsequently attempted to extend their accounts so that they apply to at least some cases of analog computation (Fresco and Wolf 2014; Piccinini 2015).

intriguing proposals as to how it may be done.[11] I have no qualms here about such proposals. Moreover, I think that the fact that the digital/analog distinction is difficult to make should not discourage us from making it. While the distinction is perhaps not as clear as we would like it to be, this does not necessarily indicate that it makes no sense or that it is insignificant. My claim against step-satisfaction is that even if it plays a role in distinguishing analog from digital (and it may certainly do so here), it plays no role in distinguishing between computation and non-computation. The same claim, I maintain, applies to other accounts that seek to characterize computation based on the digital/analog distinction. Architectural features—be they step-satisfaction or something else—might play a role in distinguishing digital from analog, but they do not play the same role in distinguishing computing from non-computing. The two distinctions—analog versus digital and computing versus non-computing—do not overlap. The averaging machine is only one of many analog devices that we view as computing.[12]

One might point out that an essential difference between the one-step and two-step processes lies in how they are described. We describe the averaging results of the one-step process in terms of thermodynamic equations. We describe, and explain, the averaging results of the two-step process not only through equations, but also by appealing to the recursive relations between the inputs and outputs of the first step and the inputs and outputs of the second step. We emphasize that the outputs of the first step are the inputs of the second step, and we describe the recursive relationship between the arguments and values of each step. Moreover, one could relate the different descriptions to what Cummins (1975, 2000) sees as different styles of explanation: when we describe the one-step process in terms of thermodynamic equations, we are explaining the average result in terms of the initial conditions and thermodynamic equations, which might fit with the deductive-nomological model. The description of the two-step process, however, might be the beginning of a functional (task) analysis. We explain the capacity of the device—of averaging three numbers—by appealing to the subcapacities of averaging two numbers, and how they are arranged to yield the average of three numbers (the "program"). In this way, the difference between computing and non-computing is grounded, at least to some extent, in the way we describe and even explain the process—and it is this difference that is captured by the notion of *step-satisfaction*.

---

[11] Goodman (1968) and Haugeland (1981a) draw the *analog/digital* distinction along the distinction between *continuous* and *discrete*. In Haugeland's terms, a digital system is a set of (input, output, and perhaps other) types, whereby the reading and writing procedures of the tokens of those types are positive and reliable—see also Pylyshyn (1984) and Goel (1991) for further discussion. Under this proposal, the one-step average process is analog, whereas a one-step flip-flop operation is digital. Lewis (1971), Fodor and Block (1973), and Demopoulos (1987) associate *digital* with high-level physical properties and *analog* with the basic physical properties expressed by physical laws.

[12] See Maley (2018) for a useful survey of analog computers; see also Chapter 9.

In replying to this claim, one could dismiss the attempt to link computation to description and explanation by pointing out that every process could be described and explained as step-satisfaction—namely, as a process consisting of a series of steps—and hence as computation. At some level, we could also describe the first process (of moving the two barriers at once) as stepwise—and hence, too, as computation. We could also describe it in terms of moving and bouncing particles, whereby each change in the direction of a particle would be considered a new step. Notably, however, this reply—regardless of its merit—is *not* my reply. I agree that it may well be that, at some level, everything could be described as step-satisfaction, and therefore as computation. But this is no reason to automatically dismiss such an account of computation. Rather, I would respond to the claim by asserting that the differences among descriptions and explanations at best reflect a difference among various *types* of explanation, but not a difference between computing and non-computing. We might explain the two-step process by means of a functional analysis, whereas we might not do this with respect to the one-step process. This difference may also be related to the claim that the two-step process is more digital than the one-step process. But this difference in explanatory styles, important as it is, does not mean that the one-step process is not computation—only that computation can be associated with different styles of explanation. Some computations are described and explained through functional analysis, others in terms of dynamical equations ranging over real-valued variables, and yet others might combine these different explanatory styles. Put more succinctly, moving the barriers one at a time might call for a different style of explanation, but that does not change the computational status of the process. I return to discuss this issue in Chapter 6, in the context of mechanistic explanations.

There may be other reasons for counting the one-step process as non-computation. It could be claimed that one-step processes are much like lookup tables—perhaps the most known example being the slide rule, which is arguably not computation (Churchland and Sejnowski 1992: 71).[13] The problem with this proposal is that the one-step averaging process is *not* a lookup table—because the average is not read off, but generated by the thermal process. It could also be argued that one-step analog processes are not really computations. We call the averaging device a *computer* simply because we use it to obtain the average of three numbers, but we do not really view the (analog) process itself as computing. So there is a difference between the term *computer*—which refers to devices we use to systematically produce results—and the term *computation*, which refers to how those results are achieved. In other words, analog devices are

---

[13] See Ulmann (2013) and Papayannopoulos (2020b) for further examples; of course, we can say that *we* compute addition, multiplication, etc. through the use of slide rules and other instruments.

computers, but they do not compute. However, I see little reason to accept this proposal. Although we use slide rules to get certain results, we do not see them as computers (they *are* lookup tables). Finally, one can bite the bullet and simply declare that one-step analog processes are not really computing. In Section 4.3, I will attempt to show that this stance is implausible, at least in the context of neural computation.

4.3 Neural Computation

In the final section of this chapter, I discuss the difficulties faced by Cummins's account in the context of neural computation. I will focus on a special kind of networks known as *attractor neural networks*. After reviewing several theoretical issues about neural networks in general (Section 4.3.1), I highlight the essentials of attractor neural networks (Section 4.3.2), then describe an attractor neural network that solves the n-queens problem (Section 4.3.3). I then turn to discuss the dilemma raised by attractor neural networks with regard to Cummins's account (Section 4.3.4). In some sense, the dilemma is a reiteration of the one discussed in Section 4.2. Finally, I will conclude with attempts to solve the dilemma, after discussing replies by Cummins and others (Section 4.3.5).

4.3.1 Neural Networks

Neural networks have played a key role in the computational theory of cognition at least since McCulloch and Pitts (1943).[14] Over the years, these studies have received various names: *neural computation, neural networks, neurocomputing,* and more.[15] The field saw a certain decline in the 1960s and 1970s but reemerged with renewed vigor in the 1980s under the titles of *PDP* (*parallel distributed processing*) and *connectionism*.[16] It was seen as a new paradigm, challenging the classical computational approaches in AI and cognitive science.[17] Today, neural networks dominate the fields of AI, computational neuroscience, and computational cognitive science. More importantly for our purposes, these networks have posed a challenge to accounts of computation that associate computation with a particular architectural feature.

---

[14] See Piccinini (2004a) for a discussion, including of the work on neural networks that preceded McCulloch and Pitts (1943).
[15] For a useful collection of the classical papers, see Anderson and Rosenfeld (1988) and Anderson, Pellionisz, and Rosenfeld (1990).
[16] Many attribute this decline to the criticism by Minsky and Papert (1969).
[17] See the two volumes issued in 1986 by the PDP research group (Rumelhart and McClelland 1986; McClelland and Rumelhart 1986).

Crucially, the term *neural network* does not necessarily refer to actual biological neurons or to biological networks. Rather, neural networks are "neural" in the sense that their basic units ("neurons") display information-processing behavior that is similar to the information-processing behavior of real neurons. Each unit typically receives different information from various input channels, but sends the same information values (output) to many other units (Figure 4.4). In addition, the signals received from other units are modulated through inhibitory or excitatory ("synaptic") weights. The total input to each unit is typically measured by the sum $\Sigma_i w_{ji} \cdot a_i$—where w_{ji} is the value of the "synaptic" weight from unit$_i$ to unit$_j$ (the value is positive if the connection is excitatory, and negative if inhibitory), and a_i is the activation value ("action potential") of unit$_i$. The output of each unit, a_j, is some function defined over this input; typical functions are *threshold*, *linear*, *sigmoid*, and others. In a threshold function, for example, the output of the unit a_j is 1 ("fires") if the total input, $\Sigma w_{ji} a_i$, exceeds a certain threshold (θ), and 0 otherwise. The important point is that these information-processing, many-to-one-signal units need not be biological at all. They can be made out of any material that satisfies these conditions.

Another important distinction is between *abstract* and *physical* neural networks. This is not very different from the distinction between abstract and physical Turing machines. The abstract network is often seen as a description, representation, or model of the physical system, and sometimes the physical system is seen as an implementation or realization of the abstract system (these relationships between the abstract and the physical will be further discussed in Chapter 5). Constructing physical neural networks is often a tedious, even unfeasible task, as they might consist of many units, and many more connections between them. Nor is it always easy to come up with the suitable hardware to implement an abstract neural network. Instead, we often use standard digital

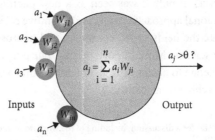

Figure 4.4 Computational properties of a "neural" cell. The cell receives different information from various input channels that are modulated by the "synaptic" weights, and emits a single output value that is propagated to other cells.

computers (e.g., desktops and laptops) to *simulate* the behavior of an abstract neural network. This is indeed the case with the network described in Section 4.3.2, which solves the n-queens problem. In other words, what we do is use digital computers to produce the output of the network for a given input, although the mediating (computing) processes may be very different from the process taking place within the (simulated) network.

This discrepancy between abstract and physical networks might pose a difficulty. Our concern here is the computational properties of *physical* systems—hence of physical neural networks. We do not ask whether the desktop computers and laptops that simulate the abstract networks compute. They surely do: the desktop computer that simulates the abstract n-queens network does compute a solution for the problem. Our question is whether a *physical neural network* computes a solution to the n-queens problem. In particular, we ask whether the physical network reaches the solution by means of a computing process or by means of another, non-computing physical process. But if all we have available is the abstract network and the simulating desktop, then the question about the physical neural network may not make sense.

Fortunately, this difficulty can be easily overcome. The queries about the physical networks do make sense, at least in most cases. The fact that we do not always construct physical networks does not mean that they are not physically constructible. Often, we do not construct a physical network even though we can—because it simply takes too many resources to construct one, at least one that can be operated efficiently. There may be physical networks whose constructions are beyond our technological capabilities—but even in those cases the networks are physically constructible, in the sense that their construction is in accordance with the laws of physics. Thus, when asked whether neural networks compute, we can always refer to physical neural networks (unless explicitly specified otherwise), or at least to physically constructible neural networks.

A final, and related, concern is the term *neural network model*, which is widespread in cognitive neuroscience. A neural network model is seen as a model that is itself a neural network. But there are some complications. First, neural network models can be abstract (e.g., mathematical) or physical objects (this is arguably true of other scientific models too).[18] Second, to achieve various data and results, we very often run a simulation of the abstract or physical model on a standard digital computer—and sometimes refer to this as a model as well. Third, when we use a neural network model to describe cognitive and neural processes, we often view the modeled system itself—namely, the cognitive/nervous

---

[18] For a review and discussion of these issues, see Weisberg (2013) and Frigg and Hartmann (2012).

system—as a neural network, too. In one sense, this is not very surprising, at least if we assume that the modeling and the modeled systems must be similar in some sense.[19] What is surprising is this: some neural network models assume that the nervous system itself is a neural network, even though it is also assumed that the modeling network is not "biologically plausible." Nonetheless, the assumption is that the modeled system itself is a biologically plausible neural network, in that it employs some kind of biological mechanism that is not represented by the model.[20]

These issues deserve further discussion that digresses from the topic of this chapter. For our purposes, it is important to maintain the distinction between the neural network model and the (modeled) neural network system. When, in the context of neural network models, we ask whether the system computes, we are referring (unless otherwise specified) to the *modeled neural network*, rather than to the modeling one.

The neural networks that captured the most attention in the 1980s were the *feed-forward nets* that used various learning methods (such as *back-propagation*) to model various cognitive tasks.[21] These learning methods have significantly improved over the years. Today's networks, which use deep learning and other techniques, are at the forefront of computational neuroscience, machine learning, and AI.[22] Our focus here, however, is on another type of network that was reinitiated in the 1980s, known as *attractor neural networks* (ANNs). The seminal work in this area was performed by the physicist John Hopfield (1982), who introduced the notion of *energy function*.[23] But as Hertz, Krogh, and Palmer (1991: 7–8) put it, the real power of statistical mechanics was brought to the forefront in the work of Amit, Gutfreund, and Sompolinsky (1985; Amit 1989). These networks bear close links to dynamical systems such as the spin-glass magnetic systems studied in statistical mechanics. I will use this linkage to undermine the attempts to draw a distinction between computing and dynamical systems by appealing to program execution, step-satisfaction, and their logical and architectural siblings.

---

[19] See Frigg and Hartmann (2020) for a discussion of these issues.
[20] E.g., Zipser and Andersen (1988) invoke the back-propagation method to train a neural network model on the behavior of cells in area 7a of the posterior parietal cortex (PPC) of monkeys (the model is discussed in Chapter 9). But they caution: "That the back-propagation method appears to discover the same algorithm that is used by the brain in no way implies that back propagation is actually used by the brain" (p. 684).
[21] See Rumelhart, Hinton, and Williams (1986).
[22] See Krizhevsky, Sutskever, and Hinton (2012); LeCun, Bengio, and Hinton (2015); and Schmidhuber (2015).
[23] See Hertz, Krogh, and Palmer (1991) for a brief review of the earlier history.

4.3.2 Attractor Neural Networks

ANNs are typically fully recurrent networks. All the units ("cells") have the same role. There are typically no designated input, hidden, or output units. Every unit is typically interconnected to all other units, and is updated according to the information it receives from all other units. Much like in the feed-forward networks, the total information is modulated through the synaptic weights and is often presented by the term $\Sigma w_{ji} a_i$, and the activity of the cell is then determined by a certain function (e.g., threshold), with $\Sigma w_{ji} a_i$ as its argument. The total activity of *all* cells, at any given moment, is known as the *state* of the network, and the state-space is described by the so-called *energy function*. This energy function is the evolution of the network with respect to time, and captures *possible* trajectories, depending on the initial state of the network.

An initial state of ANN can be seen as an "input state," after which the states often change with each iteration (each state consisting of the cells and their respective activity at a given moment). An iteration involves the updating of one or more cells at a time ("asynchronous") or all cells at once ("synchronous"); in the asynchronous case, the updated cells are typically chosen at random. The network does not have a designated output state and, in principle, can change states ad infinitum. Hopfield (1982), however, showed that if an ANN satisfies certain constraints, it will *always* reach an equilibrium point ("attractor")—namely, its state will never change unless we stimulate the cells with new, external stimuli.[24] Moreover, Hopfield showed that these attractors are precisely the minima points in the energy function/space of the network. As previously noted, Hopfield imported the idea from statistical mechanics. In the spin-glass magnetic system, particles (which are analogous to cells) spin in certain directions (analogous to activities), depending on the magnetic force of other cells (analogous to the weighted input to the cell). If certain conditions are met, the system eventually relaxes into an equilibrium point that is a minimum point in its energy space (Figure 4.5). The energy is often the temperature of the system, so the equilibrium points are the states in which the system has cooled off.

4.3.3 A Neural Network for the n-Queens Problem

ANNs have been used to model various cognitive and AI tasks having to do with memory, problem-solving, and so forth.[25] Here I shall briefly describe a

---

[24] More specifically, it is known (following Hopfield 1982) that when the weights of the net are fixed, symmetric ($w_{ij} = w_{ji}$), and a-reflexive ($w_{ii} = 0$), and the values of the units are updated one at a time (asynchronously), it is guaranteed that the net will always relax.

[25] For memory networks, see Hopfield (1982); Amit and Fusi (1994); and Seung (1998). For problem-solving, see Hopfield and Tank (1985) and Rumelhart, Smolensky, McClelland, and Hinton

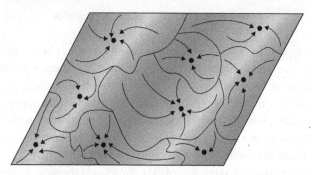

Figure 4.5 The dynamics of an attractor neural network. Each arrow is a potential trajectory of the network towards an attractor. Each point on an arrow is a "total state" of the network, and the emphasized black points are the attractors. Starting from an initial state (a point on one of the arrows), the dynamics of the network proceeds along this arrow toward the attractor.

problem-solving network designed to solve an AI task known as the *n-queens problem* (Shagrir 1992). I chose this example because it is relatively easy to comprehend, but any other ANN could be used to illustrate my claims. The *n*-queens problem is the task of placing *n* queens on an $n \times n$ chessboard so that no two queens are placed on the same row, column, or diagonal (Figure 4.6).

It is known that the problem has at least two solutions for every $n \geq 4$; the task is to find one solution at a time. The problem has a simple backtracking algorithm that requires an exponential time. More efficient solutions have also been proposed.[26] A natural way to approach the problem is with an ANN. The $n \times n$ board is represented by a fully recurrent net of $n \times n$ units, with each unit representing one square on the board. An activated unit represents a queen on the cell, and an inactivated unit represents an empty square. The designer's task is to devise the weights such that when the network starts from an initial arbitrary state, it will advance toward a stable state (attractor), which is a solution to the *n*-queens problem. In other words, the dynamical evolution of the network should gradually lead to an attractor state in which exactly *n* units are activated, and those units represent squares in different rows, columns, and diagonals.

The space state of the network can be represented by an energy function such as the following:

(1986). For these and other tasks—such as learning, control, classification, and so forth—see also Amit (1989); Eliasmith and Anderson (2003); and Eliasmith (2013).

[26] See Sosic and Gu (1990, 1991) and Fernau (2010).

Figure 4.6 A solution to the 8-queen problem. Eight queens are placed on the board such that no pair of them is on the same row, column, or diagonal.

(1) $\quad A \cdot \sum_x \sum_k \sum_{j \neq k} a_{xk} \cdot a_{xj} + B \cdot \sum_k \sum_x \sum_{y \neq x} a_{xk} \cdot a_{yk} + C \cdot (\sum_x \sum_k a_{xk} - n)^2 + D \cdot \sum_x \sum_k \sum_{m \neq 0} a_{xk} \cdot a_{x+m, k+m}$

$\quad + E \cdot \sum_x \sum_k \sum_{m \neq 0} a_{xk} \cdot a_{x+m, k-m}$

Here, a_{xk} stands for the activation value of the unit in row x and column k. When the constants A, B, C, D, and E are positive numbers, the global minimum of (1) is 0. It occurs when, and only when, all the terms are 0. The first triple sum is 0 when there are no two queens in the same row; the second is 0 when there are no two queens in the same column; the third is 0 when there are exactly n queens on the board; and the fourth and fifth are 0 when there are no two queens in any diagonal line.

I showed that no simple Hopfield net provides a general solution to the problem—the reason being that there are many local, non-zero, minima points of the energy function, (1), and these points are not solutions to the problem (Shagrir 1992). The level of energy in Hopfield nets, however, constantly decreases. Thus, the net stabilizes when it reaches a local minimum. To overcome this difficulty, I suggested allowing every unit to inhibit itself—that is, $w_{ii} < 0$. With this additional feature of self-inhibition, the energy level can also go up, thus enabling the net to escape the local minima. I proved that, under appropriate parameter values, the level of energy remains low enough so that the net always rapidly finds a solution to the n-queens problem, irrespective of the initial state or the dimension n. I also showed that the density of global minima (solutions) among the minima points in the energy space of the net is $1/n^2$. Thus, after arriving at a low-energy space in about $\log(n)$ iterations, the net randomly travels through that space until it stabilizes at a global minimum point (in about n^2 iterations).

4.3.4 Do Attractor Neural Networks Compute?

Focusing on ANNs, we can now turn to ask: Do neural networks compute? Does the n-queens network compute a solution to the problem? Let us assume that the abstract ANN is implemented in some appropriate physical, fully interconnected network with n^2 physical cells. Using Cummins's scheme, we can say that the "input" state of the physical network can be interpreted as representing the initial state of the n-queens network, and the "output" state can be interpreted as a solution to the problem. We can also assume that the mediating physical process implements the evolution of the abstract network. The question is whether this physical process computes. More generally, the question is whether attractor neural networks compute.

I suggest that the question poses a real dilemma to accounts of computing that at least partly identify computing with a particular architectural profile. Let us assume that neural networks compute.[27] This means that neural networks possess a select architectural profile, such as step-satisfaction, that differentiates them from at least some non-computing systems. However, the architecture (e.g., functional organization) of the networks looks similar to the architectures of dynamical systems that we do not consider computing. As we saw, ANNs are designed with an eye to exploiting and reflecting the kind of dynamics found in random magnetic systems. These systems also consist of a lattice of particles ("cells") and the magnetic forces between them ("weights"). Moreover, the dynamics of the network is described and explained in terms of dynamical equations (e.g., energy function) found in statistical mechanics. For example, the behavior of the n-queens network is described and explained in terms of finding the global minima points that constitute a solution to the problem. Even the proofs that the n-queens network solves the problem are based on an analysis of the energy function, (1), described previously. This means that *if* neural networks possess the required architecture, then virtually every dynamical system possesses it too. But then the select architectural profile does not play a role in differentiating computing from non-computing. It is vacuous.

Let us assume, then, that neural networks do not compute.[28] This assumption is quite costly. First, almost all the computational work today in AI and

---

[27] An assumption supported by many, including Churchland (1989, 2007); Churchland and Sejnowski (1992); Bechtel and Abrahamsen (2002); O'Brien and Opie (2006); and Hinton and Anderson (2014).

[28] Among those claiming that neural networks do not compute are Pylyshyn (1984); Fodor (1994, 2000); and Gallistel and Gibbon (2002).

computational and cognitive neuroscience incorporates neural networks of one kind or another.[29] Excluding these networks from the computational domain runs against the current practices in these computational sciences. An account of computing that disregards neural networks as computing seems irrelevant to the modern developments in AI and computational neuroscience. Moreover, the architecture of neural networks is similar in many interesting ways to the architecture of finite-state automata and other computing systems.[30] Excluding neural networks from the computational domain also endangers the status of these other systems as computing. There are some who are willing to consider all of these systems non-computing.[31] But this is an extreme view that is detached from current computational approaches.

In summary, neural networks challenge accounts that identify computation with architectural profile. If neural networks satisfy the architectural condition, then the architectural profile seems to have no role in distinguishing computing from non-computing. If neural networks do not satisfy the architectural condition (and thus do not compute), the architectural profile seems to exclude important classes of computing systems. In the rest of the chapter I discuss in more detail some attempts to address this challenge, focusing on accounts that associate computation with program execution.

4.3.5 A Way Out of the Dilemma?

Cummins characterizes computation in terms of program execution, which is in turn reduced to step-satisfaction. The dilemma mentioned earlier is summarized by Piccinini as follows: "Either connectionist systems execute programs, or they do not compute" (Piccinini 2008c: 312). Cummins (1989: chap. 11; Cummins and Schwarz 1991), who foresees the difficulty, offers two optional answers to the question of whether neural networks compute. One is that neural networks do execute programs and, hence, compute. Cummins calls this option "conservative connectionism," and appears to opt for this route.[32] The second option is that

---

[29] See also Koch (1999); O'Reilly and Munakata (2000); and Shadmehr and Wise (2005).

[30] In his seminal book *Finite and Infinite Automata*, Minsky (1967) even defines finite-state automata in terms of McCulloch and Pitts networks (which he considers computing).

[31] Pylyshyn (1984: 70–71), e.g., draws the line between computers and non-computers along the architectural property of possessing a functional distinction between *program* and *memory*. A Turing machine, he says, is a computer because it possesses this property; finite-state automata and the "new connectionist" machines (his term), however, do not, and so they are not computers.

[32] Cummins writes: "What I really hope is that the conservative connectionist will turn out to be on the right track" (1989: 155).

neural networks do not compute.[33] Another route, discussed later on, is to deem some neural networks as computing and others as non-computing.

We will examine the first two options in turn. Following Cummins, I will focus on connectionist systems (which are one style of neural networks). My aim in this examination is not to undermine Cummins's conclusions about connectionist computation; my aim is to show that Cummins's step-satisfaction condition leads to further undesirable, even fallacious results, so this condition cannot really support his conclusions, even about connectionist computation.

The first option proposed by Cummins is that connectionist systems compute because they execute programs. However, the claim that connectionist systems execute programs is somewhat odd, because we usually attribute program execution to machines that maintain a functional distinction between program and memory, such as Turing machines.[34] Friends and foes of connectionism, however, repeatedly emphasize that connectionist systems typically do not maintain a program/memory distinction; the program and memory are blended in the same functional components.[35] In the n-queens network, for example, both "program" and "memory" are encoded in the very same ("synaptic") weights that link the units together. So, in what sense do connectionist systems execute programs?

It follows, then, that Cummins clearly holds a notion of program execution that is not committed to a functional distinction between *program* and *memory*.[36] As previously noted, program execution ultimately boils down to "disciplined" step-satisfaction (1989: 92)—the "discipline" in this context being the requirement that the satisfied functions ("steps") are carried out in the right order. However, this discipline need not come from a different program-stored unit. As Cummins puts it, "the discipline takes care of itself" (p. 92): all that is required is for there to be some mechanism ensuring that the correct order of operations is maintained. If we think of the program as a flowchart with boxes and arrows (p. 92), and there is an arrow from box1 to box2, then the mechanism makes sure that the output of box1 is the input of box2 (presumably, each box satisfies each function in a single step).

According to this notion of program execution, the n-queens network might be viewed as executing a program. Each iteration of the network can be viewed as

---

[33] This option is proposed in Cummins and Schwarz (1991). Cummins (1989) proposes a third option: that connectionist systems execute programs (compute) but do not compute *cognitive* functions. I will not discuss this option, as it does not differ from the first option with respect to computation. I will only comment that the proposal is fairly radical, and that Cummins and Schwarz themselves remark that they "know of no connectionist research that consciously seeks to exploit this possibility" (1991: 68).
[34] See, e.g., Pylyshyn (1984).
[35] See, e.g., Fodor and Pylyshyn (1988: 34–35) and Schwarz (1992).
[36] See Cummins (1989: 165–166 n. 5), where he discusses various notions of program execution.

a single step and the process as a whole as step-satisfaction, and therefore as computation. True, the discipline in the network is more relaxed than that described earlier, since the unit being updated is chosen at random. In addition, the flowchart diagram with boxes and arrows is not a perfect model for the network. In a boxes-and-arrows flowchart, each box usually stands for a component in the target system. In the n-queens network, a single-step, state-transition satisfaction is from one total state ("energy level") of the network to another. Nevertheless, the overall dynamics can be seen as disciplined step-satisfaction, in the sense that what we have here is a series of iterations in which one iteration's "output" (which is a total state of the network) is the next iteration's input.

Putting aside for the moment the question of whether this characterization is faithful to the notion of program execution (I shall return to this point when discussing Roth's proposal later), I argue that this characterization of program execution is a non-starter as a criterion for distinguishing computing from non-computing. One problem, as we mentioned earlier, is that if we count the n-queens network as performing step-satisfaction ("program execution"), then, it seems, we must consider virtually every physical dynamics phenomenon to be executing a program. The dynamics of particles on a lattice is also step-satisfaction, as it proceeds from one energy state to another. The same goes for the stages in the digestive process or of planetary movements, or the successive cycles of a washing machine: they all appear to execute programs, and thereby to compute functions. Indeed, almost every physical process would appear to fulfill the criterion of disciplined step-satisfaction in this sense. If that is the case, the notion of executing a program plays no real role in distinguishing computing from non-computing.

Another problem is that sometimes we do want to count a single-step process as computation. This problem surfaced earlier, when we noted that drawing the distinction between computing and non-computing along the boundary between one and two-or-more steps does not make much sense. In the context of neural networks this problem is even more acute, as there are networks that complete their operations in a single step. Indeed, these are the kind of networks that Cummins has in mind when discussing connectionist computation—feed-forward connectionist networks with only two layers, input and output. In these networks, the dynamics consists of a single step, whose input is the activation values of the network's input units, and whose output is the activation values of the network's output unit. Rumelhart and McClelland's network for past-tense acquisition (1986) is a well-known example: in their (trained) network, the inputs encode the verb's present-tense form, and the outputs encode the verb's past-tense form. Yet this single-step input-output mapping is often viewed as a paradigm example of connectionist computation (Buckner and Garson 2019).

Martin Roth (2005) proposes to deal with this problem by offering a different construal of *program execution*. According to Roth, a system might execute a program even if the satisfaction process does not consist of successive steps, and it might execute a program even if it generates outputs from inputs in a single step. Roth relaxes the requirement that going through the temporal sequence of states that corresponds to the order of functions specified by the program is necessary for the execution of a program. Rather, he proposes the requirement that the weights of the connectionist system mirror the functional dependencies specified by the program; in the flowchart diagram, this means that functional dependencies refer to the dependencies of the inputs of one subroutine ("box") on the outputs of another subroutine. But the weights need not mirror the causal-temporal order of the program. All that is required is that the weights be derived from the functional dependencies of the program. If the weights are derived from a partial product program, the network computes multiplication by virtue of executing a partial product program; if they are derived from successive additions, the network computes multiplication by executing the successive-additions program. As long as the weights are derived from a program, the network executes the program, regardless of whether it does so in a single step or in two or more steps. The technique of deriving weights from a program is based on the work of Smolensky, Legendre, and Miyata (1992; Smolensky and Legendre 2006).

Piccinini (2008c) criticizes Roth's construal on several grounds. His main complaint is that under Roth's construal, the notion of *program execution* is completely detached from its original meaning, as conceived in computer science.[37] He says that "the connectionist system computes the function defined by the program without executing the program" (p. 314).[38] In this respect, I would add that Roth provides an account of how a network implements a program, not of how a network executes a program. However, I am not sure that proponents of connectionist computation would be happy with the view that connectionist theory is a theory of implementation.[39]

Piccinini (2008c) also contends that Roth's proposal applies only to a special case of networks, not to all networks—and therefore his notion of *program execution* cannot be taken as a comprehensive account of connectionist computation. On this point, I would add that Roth classifies computing systems according to the types of programs they execute. However, he does not delineate clearly between systems that execute programs and those that do not, nor does

---

[37] Piccinini (2008c) notes that "acting in accordance with a program is hardly sufficient for program execution" (p. 314).
[38] Piccinini also notes (2008c: 314 n. 5) that this is exactly how Smolensky and Legendre (2006: 72) themselves describe the situation.
[39] See, e.g., McClelland, Rumelhart, and Hinton (1986: 10–11), who insist that the PDP models belong to the cognitive level.

he tell us under what conditions a physical system does or does not execute a program. Therefore, he does not really account for the difference between computing and non-computing systems. Returning to the n-queens network, we can derive (in one way or another) the weights of the network from a given program. The problem is that we can also derive the "weights" of a magnetic (presumably non-computing) system from a program—and the same goes for virtually every physical system. In a nutshell, Roth does not account for physical systems that do *not* execute programs. Thus, his construal of program execution cannot be accepted as a satisfactory account of physical computation.

The point of this discussion is not that neural networks do not compute—I think that they do. The point is that the notion of *program execution*—at least when reduced to a designated architectural feature—does not provide the required account of computation. It does not help us in differentiating computing networks from other, non-computing systems. So far, I have discussed one architectural account (Cummins's) at some length, and provided a brief discussion of another (Roth's). But as I have already mentioned, the difficulties in accounting for connectionist computation are, I believe, symptomatic of the attempts to associate *computation* with some single architectural feature or another.

Let us turn, then, to the second option put forward by Cummins—namely, that neural networks do not compute. Cummins and Schwarz supply the following reasoning for this option:

> Representational states, while causally significant, are states in a dynamical system whose characteristic function—the function defined by its dynamic equations—is not itself computable. This, of course, is more than a mere possibility. A network whose representational states are real-valued activation vectors and weight matrices, and whose dynamics is given by a set of differential equations, is in general, going to be just such a system. (1991: 69)

The reasoning here seems to be this: (1) a pertinent connectionist system that operates with states that are "real-valued activation vectors and weight matrices, and whose dynamics is given by a set of differential equations" is likely to be characterized by a function that is not Turing machine computable; (2) if the characteristic function of the system is not Turing machine computable, the system does not compute. Hence, the connectionist system in question does not compute.

The premises in the argument are flawed, however. With regard to the first premise, the fact that a characteristic (input-output) function is defined over real-valued parameters does not mean that the characteristic function is not Turing-computable. There are extensions of Turing computability to real-valued magnitudes; when these extensions are in place, it is in fact very likely that the characteristic (real-valued) function of a physical system is Turing computable (see Section 3.4.2). As for the second premise, there may be physical (hypercomputing) systems that compute functions that are not Turing computable (see Chapter 3). In the present context, I would cite the work of Hava Siegelmann (1995, 1999), who introduces real-valued (analog) neural networks whose characteristic function is not computable. Many do count these networks as computing non-Turing computable functions.[40] Thus, as they fail to establish the premises, Cummins and Schwarz's conclusion is not supported.

A third possibility is to argue that we are presented with a false dilemma. It is a mistake to assume that either all neural networks compute or they do not. In truth, some neural networks compute, and some do not. Networks that operate with discrete values, for example, compute; networks that operate with analog, continuous values do not.[41] I have already discussed the proposal to exclude analog systems from the domain of computing. But I would like to highlight two points that were made in the previous section about analog and digital in the context of neural networks. One is that it is hard to draw a clear line between analog and digital networks. A neural network system might be analog in some parameters and digital in others. One parameter is the activation function of cells: threshold and bi-stable functions might be conceived of as digital, while other (e.g., sigmoid) functions might be seen as analog. Things get more complicated if we add spiking rates (which might be seen as digital) and stochastic elements to the equation.[42] Another parameter is the weight values of the system. In some networks we find only two values—*inhibitory* (−1) and *excitatory* (+1)—which makes the scale digital; in other networks the values can assume any rational or even real value between −1 and +1, which makes the scale more analog. Yet another parameter is the process itself. As we noted earlier, some processes are described as consisting of "steps" (e.g., iterations), while others comprise a single step; the former process, then, might be seen as digital, and the latter as analog. Some of these processes are described in terms of differential equations, but others might be described in terms of automata theory (as in Minsky's 1967 book). Here, too, there are complications pertaining to stochastic elements, such as synchronous and asynchronous updating. Another parameter has to do with the energy landscape. Some energy landscapes consist of separable point attractors; these are typical of Hopfield networks, and can be seen as digital. But other landscapes can consist of line, ring, or plane attractors.[43]

---

[40] For a pertinent review see, e.g., Copeland (2002c).
[41] Piccinini (2008c) makes this claim more explicitly, but he has apparently since changed his mind (2015). I discuss his views in Chapter 6.
[42] Further examples are provided by Maley (2018), who shows that the (analog) time and rate between spikes serve as representations.
[43] See Eliasmith (2007) for a useful review.

In Chapter 9 I discuss a neural network with a line attractor, in which all the (infinitely many) fixed points lie on a continuous line. This landscape can be seen as analog—and here, too, there are many more varieties of attractor (and non-attractor) networks.

For example, the n-queens network can be seen as digital: the activation values are determined by a threshold function; the weights can be either +1 or -1; the process consists of a series of iterations; and the landscape function consists of separable fixed-point attractors. Nonetheless, the overall dynamics is described in terms of differential equations. Hopfield and Tank (1985) describe a network for the "computation of decisions in optimization problems" (p. 141), such as the traveling salesman problem. Their network is digital, in that it consists of separable units and separable attractors—but it is analog in its activation response and weight values. The oculomotor memory system (to be discussed in Chapter 9) can be seen as mostly analog: it consists of real-valued activation and weight values; the process is governed by differential equations; and the energy landscape incorporates a line attractor. Nonetheless, the process ranges over spike trains and consists of iterations. One could also construct a network that has bi-stable activation values and a continuous scale of weight values. In this network, the process consists of a series of steps, yet it is described by a set of differential equations. Such a network is neither analog nor digital, but rather (so it seems) somewhere in between. There is no clear line between analog and digital neural networks.[44]

The second point concerns the claim that the digital/analog distinction does not correspond to the computing/non-computing distinction. The reason for this disparity is *not* the vagueness of the digital/analog distinction: had the digital/analog distinction aligned with the computing/non-computing distinction, then the latter distinction would be vague too. We certainly cannot rule out the possibility that there is no fine line between computing and non-computing in the physical world. Rather, the reason for the disparity is something else. It does not seem right that when you turn one parameter from digital to analog, you make the system less computing. Take, for example, the n-queens network: turning the activation-value function to non-linear and analog (as in the Hopfield-Tank network) makes the network less digital and more analog—but it does not make the network the slightest bit less computing, as the Hopfield-Tank network computes a solution for optimization problems just the same. Changing parameters from digital to analog might make the system less robust, more sensitive to noise, and sometimes

---

[44] Piccinini and Bahar (2013) argue that computation in the brain is neither analog nor digital: "Current neuroscientific evidence indicates that typical neural signals, such as spike trains, are graded like continuous signals but are constituted by discrete functional elements (spikes); thus, typical neural signals are neither continuous signals nor strings of digits" (p. 453).

intractable. But nowhere in the scientific literature on neural networks has the network been considered less computing. The same goes for the neural oculomotor memory network (Chapter 9): it is more analog than the Hopfield-Tank network, in that it is a line attractor network—but it is not considered less computing than any other neural network.[45] One could, *by fiat*, decide to exclude all networks from the computational domain—but, as we noted earlier, such an approach is not very interesting with regard to understanding the role of computation in the computational sciences.

4.4 Summary

This chapter has focused on Cummins's account of computation. As in many earlier accounts, Cummins defines computation in two stages: first, he associates computation with a theoretical notion in computer science (*program execution*), then he reduces that notion to a select architectural property in the physical world (*step-satisfaction*). In Chapter 3, I disassociated physical computation from theoretical notions of logic and computer science. In this chapter, I addressed the second stage, arguing that the architectural feature of step-satisfaction plays no role in distinguishing computing from non-computing. Under some interpretations of step-satisfaction, it excludes important instances of computing systems, while under other interpretations it is meaningless, in that it can be applied to virtually any physical system. To support this claim, I described two cases of computing systems: Pitowsky's averaging machine, which is representative of many analog computers, and attractor neural networks, which are used extensively in AI and in the cognitive and neural sciences. While I have dealt mainly with Cummins in this chapter, my argument—that architectural features play no essential role in characterizing physical computation—can be extrapolated to other architectural accounts of computation.

---

[45] A similar point is made by Piccinini and Bahar (2013) about computation in the brain. As mentioned in note 44, processes in the brain are computations, regardless of whether they are analog, digital, or something else (*sui generis*).

5
Computation as Implementation

Many accounts of computation associate it with the implementation of some abstract structure such as an automaton, algorithm, or program. David Chalmers (1996, 2011) offers the most detailed account of this approach. In developing it, he sought to undermine the *triviality results* put forward by Hilary Putnam (1988) and John Searle (1992). These results are often thought to amount to *unlimited pancomputationalism*—that is, the claim that every physical object performs every computation. Many have responded to Putnam and Searle, arguing that their notion of implementation is far too liberal. Whether these responses are successful remains a matter of dispute.

This chapter addresses Chalmers's account of implementation and computation (Chalmers 2011).[1] My own view about this account is nuanced. On the positive side, I suggest that his notion of implementation—perhaps with some modifications—successfully circumvents the dire consequences of Searle's and Putnam's triviality results. However, while I believe that implementing some formalism (in Chalmers's sense) is necessary for computing, I will argue that it is not sufficient for computing.

I start the chapter by reviewing Searle's and Putnam's triviality results and their implications (Section 5.1). I then discuss Chalmers's response to these results—suggesting that, by and large, it manages to avoid overly strong triviality (Section 5.2). I next ask whether Chalmers's notion of implementation can serve to define physical computation (Section 5.3). I suggest that implementation is a necessary condition of computation—with two qualifications, which Chalmers appears to accept (Section 5.3.1). One qualification (which runs against the logical dogma) is that the implemented structure can be any kind of dynamical formalism, even if that formalism is not part of computability theory. The other qualification (which runs against the architectural dogma) is that the implemented formalism is not to be located in certain architectural properties of the implementing physical system. I finally argue that implementation is not a sufficient requisite for computing (Section 5.3.2); many physical systems—such as rocks, stomachs,

---

[1] Chalmers developed his response to triviality results (1994, 1996) into an account of computation that provides foundations for the study of cognition. The unpublished account was aired in 1994 and received significant attention. It was later published in a special issue of the *Journal of Cognitive Science* (Chalmers 2011), followed by twelve articles and a reply (Chalmers 2012).

The Nature of Physical Computation. Oron Shagrir, Oxford University Press. © Oxford University Press 2022.
DOI: 10.1093/oso/9780197552384.003.0006

and hurricanes—do not compute, even when they implement a formalism of some sort.

5.1 Triviality Results

An early version of triviality results (from the 1970s) is attributed to Ian Hinckfuss. Hinckfuss points out that under a suitable categorization of states, a bucket of water sitting in the sun ("*Hinckfuss's pail*") can be taken to implement the functional organization of a human agent.[2] However, triviality results are mainly associated with Hilary Putnam (1988) and John Searle (1992). Putnam (1988) put forward the claim that every physical system that satisfies some minimal conditions implements every finite-state automaton. Searle (1992) asserted that the wall behind him implements the Wordstar program, and that a large enough wall implements any program one would wish. Putnam and Searle took these results to undermine a number of theses about the relationship between computation and mind, in particular the view that "the brain is a computer and mental processes are computational" (Searle 1992: 198). We will return to the more general arguments in Section 5.1.3. Let us begin, however, with the triviality results themselves.

5.1.1 Searle's Triviality Results

Here is how Searle puts the results in *The Rediscovery of the Mind*:

> On the standard textbook definition of computation, it is hard to see how to avoid the following results:
> 1. For any object there is some description of that object such that under that description the object is a digital computer.
> 2. For any program and for any sufficiently complex object, there is some description of the object under which it is implementing the program. Thus for example the wall behind my back is right now implementing the Wordstar program, because there is some pattern of molecule movements that is isomorphic with the formal structure of Wordstar. But if the wall is implementing Wordstar then if it is a big enough wall it is implementing any program, including any program implemented in the brain. (1992: 208–209)

---

[2] See Lycan (1981: 39) and Cleland (2002). See also the discussion in Sprevak (2018), who provides a useful summary of various triviality arguments.

Searle's argumentation is somewhat loose. First, he does not clearly distinguish between two different results. One states that every object under some description is a digital computer. If Searle assumes that a digital computer is a universal machine, then the result amounts to the claim that every object can simulate the operations of every algorithm. But if Searle understands a digital computer to be an object that implements (or executes) an algorithm (or a program), then the result amounts to the claim that every object implements at least one algorithm (or program). A second result refers to specific objects: Searle refers to the wall behind him, which does not appear to involve many changes or complex dynamics. He claims that even this fairly simple object implements a complex Wordstar program. Searle next says that a big enough wall implements any program—here, the size apparently enables the implementation of more operations and/or memory.

Second, Searle does not say much about the notion of implementation. He mentions in brief that the implementing physical structure should be *isomorphic* to the implemented formal structure of the program. Presumably, he means that there is a structure-preserving mapping relationship between the pertinent physical object (such as the wall) and the pertinent program (e.g., Wordstar). Another way to put it is that the implementing physical object and the program have "the same structural or formal features" (Swoyer 1991: 457): they share a high-order formal or mathematical structure called a *shared structure*.

To be a little more precise: Let D and $\underline{D}$ be two domains, each of which comprises individuals and relations. We would say that the two domains are isomorphic just in case the following two conditions hold: (a) there is a one-to-one function f from the full domain of relations of D onto that of $\underline{D}$ which maps each R-relation of the first to some $\underline{R}$-relation of the same type in the second, and (b) the function f is *structure-preserving*—that is, for every n-ary relation R and n-tuple of individuals (in D) the following is true:

$$<x_1, \ldots, x_n> \in R \text{ if and only if } <f(x_1), \ldots, f(x_n)> \in f(R).$$

When talking about implementation, one should add two caveats. One is that isomorphism should be replaced with the weaker condition of *homomorphism*. Like isomorphism, homomorphism is structure-preserving, but it need not cover the full domain of relations of the implementing physical object. Some relations in the physical object might not play any role in implementing the program. It is mandatory, however, that the function f covers the full domain of relations of the program—and that f is structure-preserving. Another caveat is that the typical relations in the implemented program must be of the state-transition type. Thus, a state-transition relation of the program from

P to *Q* (sometimes represented as *P* → *Q*) is mirrored by a transition of states from *p* to *q*—whereby *p* and *q* are states of the implementing physical object, and $f(p) = P$ and $f(q) = Q$.

Searle does not justify these results. His argument goes as follows (Searle 1992: 205–210): Philosophers of mind invoke the multiple realization argument to support computationalism (the view that "the brain is a computer and mental processes are computational"). The multiple realization argument is the claim that mental states can be realized in a great many different physical structures, and thus mental states are not reducible to any one of their realizing physical structures.[3] One well-known version of this claim associates mental states with computational states ("computationalism"). But this notion of realization—according to which a physical system implements a program—also leads to triviality results that undermine computationalism.[4] Searle, however, provides no proof of this result.

Copeland (1996) explicates and directly proves Searle's (stronger) result under certain assumptions. Copeland's idea is roughly this: Let SPEC be a specification of an architecture of a machine (such as axioms describing an architecture) and of an algorithm, α, for that architecture. Let <*e, L*> be an ordered pair, where *e* is some entity, and *L* is a labeling scheme for that entity. We will say that *e* implements (executes) the algorithm α iff <*e, L*> is a model of SPEC, a *model* here being basically a structure-preserving function. Finally, we will say that *e* computes the function *f* if and only if it implements (executes) an algorithm α for *f*. In section 4 of his paper, Copeland proves Searle's theorem: For a given entity *e* (with a sufficiently large number of discriminable parts) and for any architecture-algorithm specification SPEC, there exists a labeling scheme *L* such that <*e, L*> is a model of SPEC. Copeland shows that if SPEC is the architecture of register machines, and if the registers can occupy different parts of the physical object at different times, then there will always be a one-to-one mapping ("model") of SPEC to the labeling *L* of *e*. We will not get into further details here, but rather focus on Putnam's argument for (his) triviality results. Suffice it to say that Copeland, who is not happy with the triviality results, goes on to criticize the notion of implementation that underlies Searle's result.

---

[3] The multiple realization argument is one of the most-discussed topics in philosophy of mind. See Putnam (1967) and Fodor (1975) for early versions of the argument in which mental states are considered as computational states; see Aizawa and Gillett (2009) for a more recent defense of multiple realization. See also Polger and Shapiro (2016) for an extensive and sober overview of the argument and its scope.

[4] I will assume (following Searle and Putnam) that implementation is the realization of formal/abstract properties by a physical system; I will use only the term *implementation* unless discussing the arguments of others.

5.1.2 Putnam's Triviality Results

Putnam notes early on that "everything is a Probabilistic Automaton under *some* description" (1967: 435). Like Searle, he maintains that the multiple realization argument is a double-edged sword. We used to think that multiple realization supplied more reasons to believe in computationalism, but it now appears that multiple realization of an automaton extends to the implementation of every automaton—which undermines computationalism. Putnam provides an extensive argument for the universal realization claim. In fact, in the appendix to *Representation and Reality* (1988: 121–125), he proves that every ordinary open system that satisfies two minimal principles is an implementation of every abstract finite automaton. The *principle of continuity* states that electromagnetic and gravitational fields are continuous; this principle is fairly natural, at least in classical physics. The *principle of non-cyclical behavior* states that the physical system is in different maximal states at different times; "a 'maximal' state is a total state of the system, specifying the system's physical makeup in perfect detail" (Chalmers 1996: 310). The truth of this principle is less obvious. But even in its absence, the result applies to a great number of systems whose behavior is noncyclical.[5]

Putnam's theorem addresses finite-state automata (FSA) without inputs/outputs (I/O). Here is an outline of the proof. Take the FSA that runs through the state-sequence PQPQPQP at a given time interval. Here, P and Q are the states of the FSA. Let us see how a rock realizes this run in a 6-minute interval—say, between 12:00 and 12:06. Assume that the rock is in a maximal physical state p_0 at 12:00, p_1 at 12:01, and so forth. Assume also that the states differ from each other (this is Putnam's principle of non-cyclical behavior). Now let us define a physical state **a** as $p_0 \vee p_2 \vee p_4 \vee p_6$, and define **b** as $p_1 \vee p_3 \vee p_5$. The rock implements the FSA in the sense that the causal structure of the rock "mirrors" the formal structure of the FSA. The physical state **a** corresponds to the logical state P, the physical state **b** corresponds to the logical state Q, and the transitions from **a** to **b** correspond to the computational transitions from P to Q. A complete proof would require further elaboration and assumptions.

I will add three quick comments about Putnam's result. One is that—much like Searle—Putnam assumes that implementation is exhausted by some homomorphism between the implementing physical system and the implemented automaton. The second is that Putnam's proof takes the (implementing) states of the physical objects to be total, "maximal" states of the objects. The state-transition in the physical object is between two total states. Searle and Copeland, in contrast, appeal to the inner structure of the physical object, such as registers that

---

[5] See Chrisley (1994, sec. 5), who discusses the scope of the principle of noncyclical behavior.

change places within the physical object. Lastly, Putnam observes (p. 124) that the proof cannot be immediately extended to FSA with inputs/outputs (I/O). If the I/O are functionally individuated, then the I/O can be treated much like abstract internal states, and the extension is natural. In that case, we get the result that (almost) every physical object implements every finite automaton. If the I/O are individuated by their physical biological properties (as some functionalists might hold), then we slide into behaviorism: in that case, what distinguishes a rock from a brain is only I/O (since both implement every automaton without I/O).[6]

To recap: both Searle and Putnam aim to show that (almost) every physical object that satisfies certain minimal conditions implements (almost) every program (Searle), algorithm (Copeland), or automaton (Putnam). Their arguments for these (triviality) results are similar. They equate implementation with a simple structure-preserving mapping function (homomorphism) and argue that (almost) every physical object is homomorphic to every automaton (etc.).

The *triviality argument*, then, bears the following structure:

1. Every physical object is homomorphic to every automaton (etc.).
2. A physical object s implements an automaton S_C if s is homomorphic to S_C.
3. **Triviality:** Every physical object implements every automaton (from 1 and 2).

Putnam justifies the first assumption by providing a proof that pertains to finite-state automata. Searle does not provide such a detailed proof for his construction, but Copeland does later on.

5.1.3 Implications of Triviality Results

What follows from these triviality results? If true, these results have far-reaching and devastating consequences for the notions of *implementation*, *computation*, and *mind*. Let us review them in turn.

Implementation. One implication pertains to the notion of implementation. If every physical object implements every program (Searle), algorithm (Copeland), or automaton (Putnam), then the notion of *implementation* is in danger of being rendered trivial (hence the name *triviality results*). If a rock, a chair, a laptop, or a mind are all the same at implementing every automaton (etc.), then the notion of implementation does not appear to have the important theoretical and practical role that we thought it had. We used to think that the notion of implementation

---

[6] See Putnam 1988: 124–125.

was what differentiated rocks and chairs from laptops and minds—namely, the former did not implement the kind of automata implemented by the latter. But it now transpires that there is no such difference—they all implement the same class of (i.e., all) automata.

This outcome not only is counterintuitive, but also undermines a major working hypothesis in the theory and practice of computer science. Computer scientists make a huge effort to find ever more efficient algorithms for solving theoretical and practical problems, with the aim of implementing them in objects (such as laptops) that will then solve those problems. The exhaustive-search algorithm for factorization—the problem of decomposing an integer into the (unique) product of primes—is exponential. There is apparently no efficient, polynomial-time algorithm for the problem, but there are sub-exponential algorithms that improve the search considerably.[7] The backtracking algorithm for the n-queens problem is exponential. But there are also efficient polynomial-time algorithms that can solve the problem fairly quickly (see the discussion in Chapter 4). However, if triviality is true, then this effort is redundant, since even rocks implement all these algorithms. A rock, therefore, solves the n-queens problem with the exponential-time algorithm, and simultaneously solves the same problem with a polynomial-time algorithm. The same goes for factorization and other problems. Indeed, a rock simultaneously provides all the possible solutions to all these problems! The notion of implementation therefore loses its theoretical and practical significance.

Computation. A second consequence of triviality results concerns computation. Let us start with the accounts of computation that assume that computation *is* the implementation of an automaton, a program, or an algorithm.[8] If triviality is true, and assuming that performing a computation C corresponds to implementing an automaton S_C, then every physical object performs every computation (unlimited pancomputationalism).

The reasoning, in a nutshell, goes as follows:

3. Every physical object implements every automaton (triviality).
4. A physical object s performs a computation C iff s implements an automaton S_C.
5. Every physical object performs every computation (unlimited pancomputationalism) (from 3 and 4).

---

[7] Computational complexity is discussed in Section 3.3.2.
[8] This is an assumption held by Putnam and Searle. Searle takes his characterization of implementation to be the "standard textbook definition of computation" (1992: 208).

Today, we distinguish between *limited* and *unlimited* pancomputationalism.[9] *Limited pancomputationalism* is the claim that every physical object (system) performs at least one computation. *Unlimited pancomputationalism* is the claim that every physical object (system) performs every computation. Limited pancomputationalism seems to undermine the distinction between computing and non-computing physical systems, as it implies that every physical system computes (but see the discussion in Section 5.3; see also Chapter 1). Unlimited pancomputationalism seems to undermine, in addition to the computing/non-computing distinction, the notion of computational equivalence—namely, the distinction between different *types* (kinds) of computing systems. It implies that rocks, chairs, desktops, and minds are all computationally equivalent, as they perform exactly the same—in fact, all—computations.

Triviality results also endanger accounts of computation that assume that implementing an automaton, a program, or an algorithm is *necessary* for computing. These accounts do not immediately face the challenge of pancomputationalism: there might be other features that exclude rocks and chairs from the computational domain and/or from computing everything. However, triviality results imply that implementation is an empty condition—that is, a condition that appears to provide no additional information as to whether or not something is a computer, or whether or not objects are computationally equivalent. After all, every physical object satisfies the condition.

Triviality results are also worrisome for those who think that implementing an automaton (etc.) is neither sufficient nor necessary for computation. Even the proponents of such accounts usually accept that there are many physical computers that do implement some automata (etc.), and that this fact plays at least some role in determining the computational identity (and hence, the computational equivalence) of physical objects. But, as we have just seen, triviality implies that implementing automata (etc.) contributes nothing to determining computational equivalence.

The upshot, then, is that triviality results, if true, have dire consequences for practically every account of physical computation.

Mind. Triviality results have another important set of implications pertaining to certain theories of mind and cognition that associate mentality with computation. One cluster of theories assumes what Chalmers (2011) calls the *computational sufficiency thesis* (CST). CST states that implementing "the right kind of computational structure suffices for the possession of a mind, and for the possession of a wide variety of mental properties" (p. 326).[10] Triviality results appear

---

[9] See Piccinini and Anderson (2018) for recent discussion.

[10] There are two important caveats here. One is that CST is not committed to the claim that computational structure determines every aspect of mentality: the content of mental states might be at least partially determined by external factors, and therefore does not supervene on computational

to challenge CST: if every physical object implements every automaton, then, if CST is true, every physical object implements the automaton that suffices for the possession of a mind.[11] Hence, every physical object is a cognitive system. In other words, if CST is true, then rocks, chairs, and hurricanes all possess the kind of mind that we do. But assuming that rocks, chairs, and hurricanes do not possess minds, triviality results indicate that CST, and theories of mind that assume CST, are simply false.

This argument can be put as follows:

3. Every physical object implements every automaton (triviality).
6. If CST (*the implementation of the right kind of automaton suffices for the possession of a mind*), then rocks, chairs, and hurricanes—as physical objects—possess minds (from 3).
7. Rocks, chairs, and hurricanes do not possess minds.[12]
8. CST is false (from 6 and 7).

Searle and Putnam reason along these lines when arguing against certain theories of mind and cognition that assume CST. Searle argues against "computationalism," which is the view that mental processes and states are *entirely* computational.[13] This strong form of computationalism is obviously committed to CST. Putnam's critique is aimed at the view known as *computational functionalism*—namely, the view that mental types are individuated by their functional role, which is determined by the causal relationship between mental states, inputs, and outputs (or "functional organization," in Putnam's words).[14] Computational functionalism identifies this functional organization with the computational structure of the system,[15] and therefore assumes CST.[16] Thus,

---

structure. The second caveat is that "sufficiency" in this context refers to *nomological* sufficiency, not metaphysical sufficiency. According to Chalmers, phenomenological properties supervene nomologically but not metaphysically on computational structure.

[11] I equate computational structures and automata here for the sake of argument (in Section 5.3, however, I will argue against this equivalence).

[12] Some panpsychists, however, might contend this premise, or at least insist that the rock's fundamental parts have mental properties (Goff, Seager, and Sean 2020).

[13] As Piccinini (2009) notes, however, most computationalists are not committed to the claim that mental processes are entirely computational.

[14] See Levin (2018) for an explication of this view. Broadly speaking, functionalism identifies kinds of mental states by their functional role.

[15] This view, also known as *machine functionalism*, was once put forward by Putnam (1967) and Fodor (1975). Only later did Putnam come around to criticizing it. See Shagrir (2005) for a critical review of the evolution of Putnam's view of computational functionalism.

[16] Computational functionalism can be seen as CST plus the view that types of mental states are identified by their functional, i.e., computational properties. Like CST, however, computational functionalism is not committed to the view that every aspect of mentality, such as mental content, is identified functionally. The identification of mental content with functional properties is known as functional-role semantics (Block 1986).

triviality results, if correct, undermine computational functionalism as well—because they imply that rocks, chairs, and hurricanes all implement a computational structure sufficient for possessing a mind.

Triviality results also threaten theories of mind and cognition that do not assume CST. Indeed, all the theories that state that some brain states are computational (without assuming that being computational is sufficient for mentality) would appear to be in jeopardy. If triviality results are true, then these theories are stating something quite trivial, since everything possesses a computational structure—in fact, every computational structure.

Searle says that triviality has even worse consequences for computation and cognition. It indicates that "syntax *is essentially an observer-relative notion*" (1992: 209) and thus that "*computation is not an intrinsic feature of the world. It is assigned relative to observers*" (p. 212). But if computation is not intrinsic to physics, Searle argues, then the claim that the brain is a computer "does not get up to the level of falsehood. It does not have a clear sense" (p. 225). The reason is that we make an empirical claim about the brain, but the decision as to whether or not something is a computer is a matter of assigning a computational interpretation. And further consequences follow for computational cognitive science. Searle argues that "*there is no way that computational cognitive science could ever be a natural science*" (p. 212). Computational cognitive science is in the business of investigating whether, what, and how the brain computes—but if it transpires that computation is not an intrinsic feature of the world, there will be nothing to discover.[17]

The conclusion is that triviality results have potentially devastating implications for the notions of implementation and computation, as well as for philosophical and empirical theories of mind and cognition that associate mentality with computation. How can these implications be blocked? The strategy taken by most critics is to avoid the triviality results themselves (i.e., premise (3)). This approach is entirely reasonable: even if you think that triviality results do not lead to devastating consequences about computers and minds, you might still want to salvage the important notion of implementation, given its immense theoretical and practical importance. Thus, some people—including Searle and Putnam themselves—qualify the claim about the extent of homomorphism (premise (1)): Searle requires a big enough wall to implement every program, while Putnam requires continuity and non-cyclical behavior, and focuses on finite-state automata. However, these qualifications alone do not appear to be sufficient to circumvent the consequences of triviality results. Most critics, then,

---

[17] Putnam, too, thinks that his argument has unsettling consequences for cognitive science, but he does not flesh them out in *Representation and Reality*. He does, however, present functionalism and cognitive science as complementary projects (1967: 434–435) and implies that cognitive science is no less than science fiction (1997, 1999: 118–119).

go after premise (2), and accuse Putnam and Searle of assuming an overly liberal notion of implementation, arguing that it takes more than simple homomorphism to implement an automaton.[18]

5.2 Avoiding Triviality

As we have just stated, most responses to Putnam and Searle argue that the triviality results rest on an overly liberal notion of implementation (known as the *simple mapping account* [Godfrey-Smith 2009]). They claim that it takes more than a simple mapping to implement an automaton, algorithm, or program.[19] These critics, however, disagree over the constraints on implementation that should be added to the simple mapping account. Some require the implementing physical states to be *causally* connected, meaning that the mirroring state-transition, $p \rightarrow q$, between two physical states is a causal relation.[20] Others require the implementation relation to have some *modal force*, meaning that it supports relevant *counterfactuals*.[21] Even more restrictions are placed by the *dispositional account*[22] and the *mechanistic account*.[23] Yet another condition that is sometimes invoked pertains to the *grouping* of physical states, which places some restrictions on the ways in which we lump together physical objects and properties into types of states.[24] Other people mention *pragmatic* constraints, which link the implementing automaton to a particular function of the object.[25]

---

[18] Searle himself admits that a more adequate, restrictive notion of realization does not lead to triviality results: "I think it is possible to block the result of universal realizability by tightening up our definition of computation" (1992: 209).
[19] But see Schweizer, who defends the simple mapping account (2014) and argues that it is consistent with a reasonable version of the computational theory of mind (2019b).
[20] See Chrisley (1994); Melnyk (1996); and Chalmers (1996, 2011). But, for criticism of the causal constraint, see also Copeland (1996); Bishop (2009); and Schweizer (2014, 2019a).
[21] See Chalmers (1996, 2011); Copeland (1996); and Scheutz (1999, 2001). But see also Schweizer (2014, 2019a) for criticism of the modal constraint. Copeland, e.g., says that while a sufficiently large wall might implement Wordstar in Searle's sense of implementation, it does not implement Wordstar in the adequate sense that involves counterfactual scenarios (pp. 347–350). Copeland thus redefines the notion of implementation. He says that e implements (executes) the algorithm α iff <e,L> is *an honest* model of SPEC. The wall is a non-standard (not-honest) model of the specification of Wordstar. An honest model is one in which the labeling is not specified ex post facto as labeling, and supports counterfactuals about the behavioral consequences of the implemented algorithm (pp. 350–351).
[22] See Klein (2008).
[23] See Piccinini (2007, 2015) and Miłkowski (2013).
[24] See Scheutz (1999, 2001, 2012) and Godfrey-Smith (2009). Copeland's requirement that "the labelling scheme must not be ex post facto" (p. 350) falls into this category and constitutes, together with the counterfactual (discussed previously), an *honest model* of the system.
[25] See Egan (2012); Fresco (2015); and Matthews and Dresner (2017). See also Millhouse (2019), who proposes a formal simplicity constraint.

One easy way to avoid triviality is to go the semantic route. Under this approach, we require that the implementation be an interpretation or representation function as well—even if only in a very minimal sense, whereby the implementing physical states represent the states of the implemented automaton.[26] Thus, if the states of the rock do not represent the states of an automaton S, then the rock does not implement S (of course, something has to be said about the representation function).[27] Now, if it turns out that the only way to avoid too strong a triviality is through the semantic route, then triviality results can serve as the basis of a formidable argument for a semantic account of computation. But many argue that we can minimize the magnitude of the triviality results without appealing to semantic factors. Let us examine one attempt to do just that.

5.2.1 Chalmers's Account of Implementation

I focus on Chalmers's (non-semantic) account of *implementation* (Chalmers 1994, 1996) for several reasons. First, it is perhaps the most fully developed and influential account of implementation. Second, it includes causal, modal, and grouping constraints, and is therefore a good representative of many other accounts of implementation. Moreover, other accounts rely on and build upon Chalmers's account. Lastly, it appears to have managed to circumvent Putnam's and Searle's triviality results.

Chalmers agrees with Putnam and Searle that implementation is a mapping or mirroring relation: "A physical system implements a given computation when the causal structure of the physical system mirrors the formal structure of the computation" (2011: 326). More specifically:

> A physical system implements a given computation when there exists a grouping of physical states of the system into state-types and a one-to-one mapping from formal states of the computation to physical state-types, such that formal states related by an abstract state-transition relation are mapped onto physical state-types related by a corresponding causal state-transition relation. (p. 326)

This passage highlights that the implementation of an abstract automaton ("computation") by a physical system involves a structure-preserving mapping

---

[26] See, among others, Rapaport (1999); Sprevak (2010); and Blackmon (2013).
[27] It should be noted that this semantic route is perfectly in accord with Putnam and Searle. Putnam and Searle do not urge us to embrace triviality. They use triviality for a *reductio* argument against certain views about the mind that assume that computation is non-semantic. They argue that this assumption leads to the absurd consequence that rocks and chairs possess minds.

(homomorphism) between the physical system and the automaton. But it also states that the mirroring physical states are *grouped* into state-types, and that the state-transition relations between these physical states are *causal* relations.

In his criticism of Putnam, Chalmers argues that the state-transition relations should also have some *modal* force. Indeed, this is his main criticism of Putnam's notion:

> The problem, I think, is that Putnam's system does not satisfy the right kind of state-transition conditionals. The conditionals involved in the definition of implementation are not ordinary material conditionals, saying that on all those occasions in which the system happens to be in state *p* in the given time period, state *q* follows. Rather, these conditionals have modal force, and in particular are required to support counterfactuals: *if* the system were to be in state *p*, then it would transit into state *q*. (1996: 312)

More specifically, the actual state-transitions of the implementing system fail to satisfy conditionals that are associated with both the exhibited and the unexhibited state-transitions of the implemented automaton. With regard to the exhibited states, Chalmers acknowledges that Putnam assumes that these transitions are meant to be causal; his criticism is that the implementing states are not *reliable*. The physical state-transitions from *p* to *q* are entirely independent on actual environmental conditions at the time of the run. They are not sensitive even to slight changes in the environmental conditions. They do not ensure that if the environmental conditions were slightly different (causing the system to still be in *p*), the behavior would be slightly different as well (and be counted as *q*).

The more interesting case pertains to unexhibited state-transitions (where these occur). The causal structure of the physical object should mirror all possible formal state-transitions of the implemented FSA. In Putnam's proof, however, the rock implements only a *single run* (the transition from *P* and *Q* and back), not other runs that might exist. If the FSA has other state-transitions (e.g., $P_1 \to P_2$ and $P_2 \to P_1$), these transitions should also be mirrored by the rock's dynamics.

Chalmers concludes that the notion of implementation should be formulated in terms of a stronger condition:

> A physical system implements an inputless FSA in a given time-period if there is a mapping *f* from physical states of the system onto formal states of the FSA such that: for every formal state-transition $P \to Q$ in the specification of the FSA, if the physical system is in a state *p* such that $f(p) = P$, this causes it to transit into a state *q* such that $f(q) = Q$. (1996: 315)

Chalmers argues that Putnam's proof does not meet this condition. At the same time, he notes that it can do so with only a slight revision. Ignoring environmental conditions helps to solve the first problem, as clocks are often insensitive to environmental variations. Having enough (different) physical states solves the second problem. This can be as simple as dials (e.g., various marks on a rock). Chalmers proves that every physical system containing a clock and a dial implements every input-less FSA (p. 317). This triviality result does not apply to every physical system, but it does apply to a large number of them.

Next, we will examine FSAs with inputs and outputs. Inputs and outputs (I/O) require the satisfaction of further dependencies—since a certain state with one input (i_1, P) leads to one state and an output (P_1, o_1), but the same state with a different input (i_2, P) might lead to a totally different result, e.g., (P_2, o_2). These dependencies can be accommodated if we add to the machine an *input memory* that records all possible inputs.[28] Such a machine (with a dial) would implement every FSA with I/O. This moves us well beyond universal realization, but it is still troublesome: input memories are not hard to instantiate, so a rock with some input-recording device implements an FSA for primality-testing.[29] This result also raises the specter of collapsing functionalism to behaviorism: if two systems have the appropriate I/O devices and satisfy the strong conditionals, they satisfy all FSAs with inputs and outputs (p. 324).

The final, yet key, measure put forward by Chalmers concerns the notion of *combinatorial state automaton* (CSA). Broadly speaking, a CSA is much like an FSA, except that it has a more complex, combinatorial internal structure. Each state is a combination of substates, and any state transition is sensitive to the combinatorial structure of the previous combined state (p. 324). Thus, whereas an internal state of an FSA is *monadic*, the internal state of a CSA is a vector of values or substates. In a Turing machine, for example, this vector ("configuration") includes the symbols written on the memory tape. The definition of implementing a CSA is as follows:

> A physical system P implements a CSA M if there is a vectorization of internal states of P into components $[s_1, s_2, \ldots]$, and a mapping f from the substates s_j into corresponding substates S_j of M, along with similar vectorizations and mappings for inputs and outputs, such that for every state-transition rule $([I_1, \ldots, I_k], [S_1, S_2, \ldots]) \to ([S'_1, S'_2, \ldots], [O_1, \ldots, O_l])$ of M: if P is in internal state $[s_1, s_2, \ldots]$ and receiving input $[i_1, \ldots, i_n]$ which map to formal state and input $[S_1, S_2, \ldots]$ and $[I_1, \ldots, I_k]$ respectively, this reliably causes it to enter an internal

---

[28] Godfrey-Smith (2009) proposes an even simpler alternative to resolve the I/O issue.
[29] In a response to Brown (2012), Chalmers (2012) adds the requirement that the input memory and dial should be causally and counterfactually connected to the outputs (p. 236).

state and produce an output that map to $[S'_1, S'_2, \ldots]$ and $[O_1, \ldots, O_l]$ respectively. (2011: 329)

These CSAs, according to Chalmers, are not at all easy to implement, as this requires a complex internal causal structure of the sort found in very few physical systems. While rocks might still implement simple FSAs, they do not implement more complex combinatorial state automata (CSA), which are more likely to be minds implemented by brains.

This concludes the main components of Chalmers's theory of implementation. In this way, he argues, we can settle on a non-semantic theory of implementation and computation that avoids the consequences of triviality results.[30] Let us now examine whether *implementation*, as Chalmers defines it, indeed escapes these consequences. In Section 5.3, we will examine whether Chalmers provides an adequate theory of *computation*.

5.2.2 Weak Triviality and Its (Non-)Consequences

Let us begin by distinguishing between *strong* and *weak* triviality. *Strong triviality* is what I referred to as *triviality* earlier (Premise 3)—namely, that every physical object implements every automaton. *Weak triviality* is the claim that a great many—perhaps all—physical objects implement some automaton.[31] We can think of strong and weak trivialities as lying at either end of a spectrum, with many options between them. Chalmers does not seek to provide an account of implementation that sidesteps triviality altogether: he acknowledges that, according to his account, even rocks might implement a very simple (e.g., one-state) automaton, and therefore he is willing to accommodate weak triviality. His aim is to block the dire consequences of strong triviality for implementation, computation, and cognition. In assessing Chalmers's account, then, we want to examine (a) whether weak triviality (or something very similar to it) blocks the dire consequences of strong triviality to the notions of implementation, computation and cognition, and (b) whether Chalmers's account manages to avoid overly strong triviality. I address the first issue in this section, and the second issue in Section 5.2.3.

Let us assume, then, that every physical object implements an automaton. Objects such as rocks and chairs implement very simple automata, perhaps

---

[30] Chalmers writes: "It will be noted that nothing in my account of computation and implementation invokes any semantic considerations, such as the representational content of internal states. This is precisely as it should be: computations are specified syntactically, not semantically" (2011: 334).

[31] If implementation is equated with computation, then this distinction amounts to the distinction between limited ("weak") and unlimited ("strong") pancomputationalism ("triviality").

even one-state automata. Other objects, such as minds, presumably implement automata that are more complex. What are the implications of this thesis of weak triviality for implementation, computation, and cognition? Does it have the dire consequences implied by strong triviality? I agree with Chalmers that the answer is *no*: weak triviality does not have devastating implications for implementation, computation, or cognition.[32]

Let us start with the notion of implementation. Strong triviality implies that implementation does not distinguish between rocks and chairs on the one hand, and laptops and minds on the other. They all implement the same set of automata—namely, the set of all automata. But weak triviality has no such consequences: (Chalmers's) implementation does distinguish between physical objects. A rock implements (say) a simple one-state automaton, whereas a laptop implements a far more complex one. Computer scientists—theoreticians and practitioners alike—would quite happily admit that very simple objects implement simple automata, such as the automaton whose only state-transition is $P \to P$, or the one whose sole state-transition is $P \to Q$. The force of the notion of implementation is not in denying this harmless result, but in distinguishing between rocks, chairs, laptops, and minds. Weak triviality is perfectly compatible with these distinctions.

What about computation? Strong triviality has some undesired results for the notion of *computational equivalence*, because it implies that the automata implemented by a system play no role in determining the system's type of computational identity (and thereby in determining whether or not two systems are computationally equivalent). Once again, this is because every physical object implements every automaton—so implementation cannot play a role in classifying computing systems. Moreover, assuming that computation *is* an implementation of an automaton, strong triviality implies that every physical object performs every computation (unlimited pancomputationalism). Weak triviality has no such consequences, but rather only implies that different objects implement different automata. Minds might implement one automaton, rocks another. If so, the notion of implementation might play a role in classifying rocks and minds into different computational types, and we do not reach unlimited pancomputationalism.

Finally, let us consider mind and cognition. Strong triviality, when supplemented with the sufficiency thesis, implies that rocks and humans have the same mentality. Weak triviality has no such consequences, since it does not entail that rocks and humans implement the same kind of computational structures— e.g., that rocks implement the complex automaton necessary to qualify as

---

[32] See also Chrisley (1994), who thinks that weak triviality is harmless, and that computation is never vacuous, even if it is universally realizable.

possessing a mind. Moreover, weak triviality bears none of the consequences that Searle mentions with respect to cognitive science. Weak triviality is consistent with the claim that cognitive science is an empirical science that is in the business of discovering computational properties. Cognitive science might not discover that the brain computes. However, cognitive science is still in the business of discovering which automata are implemented in the brain, and which of them are relevant to cognition. Relatedly, weak triviality is consistent with the claim that implementation and computation are objective and intrinsic to physics. The claim that everything implements an automaton does not mean that implementation and computation are not objective—just as the fact that every physical object has a mass does not mean that mass is not an "objective" and "intrinsic" feature of the physical world (see the discussion in Chapter 1).

5.2.3 Does Chalmers's Account Avoid the Consequences of Strong Triviality?

As we have seen, reducing triviality from strong to weak is enough to avoid the dire implications of strong triviality. The next question is whether Chalmers's account indeed avoids the results of overly strong triviality. According to some, it does not: without further constraints on the grouping of physical types, they argue, we have not improved much on the results of Putnam and Searle.[33] There are two types of further constraints mentioned by the critics. One is that each variable (substate) of the automaton maps to an independent element of the implementing physical system.[34] Another is that the disjoint physical states that are grouped together to form a single physical type (which in turn maps to a formal state) must be *similar*[35] in some non-trivial sense.[36] Thus, Scheutz (2012), for example, argues that "without a notion of 'legitimate grouping of physical states' all sorts of physical systems would implement unintended computations" (p. 75). Scheutz (2001) also suggests that implementation should be defined in the context of a fixed canonical physical theory (such as circuit theory), in which the grouping into physical types is already given. Miłkowski (2011) requires that the grouped physical states belong to the relevant causal structure, which he identifies with the isolated level of a mechanism. Godfrey-Smith (2009) argues

---

[33] See Brown (2012); Scheutz (2012); and Sprevak (2012).
[34] See Godfrey-Smith (2009) and Sprevak (2012).
[35] As Godfrey-Smith rightly remarks, we need not suppose that the disjoint physical states that are grouped together are identical.
[36] Copeland (1996) deals with this problem by ruling out ex post facto labeling systems.

that the substates that are grouped into coarse-grained categories should be physically similar.[37]

In his response to this criticism, Chalmers associates the "independent elements of the physical system" with spatially separable components of the system[38]—but admits that this requirement may be too strong, as it rules out certain good implementations.[39] This addresses the quest for the first constraint. As for the grouping of physical types, Chalmers tentatively mentions other pertinent constraints, such as a *naturalness constraint* on physical state-types or a *uniformity constraint* on physical state-transitions. The naturalness constraint suggests that the grouped physical types are somehow natural; the uniformity constraint signifies that the causal mechanism is uniform. However, Chalmers also states that it is not obvious how to formulate the required constraints in a clear and precise way.[40] An alternative way to strengthen the notion of implementation is to place more restrictions on inputs and outputs. Chalmers says that "it is generally useful to put restrictions on the way that inputs and outputs to the system map onto inputs and outputs of the FSA" (2011: 329). The question is how to specify I/O and where to locate them (i.e., proximal or distal); this is by no means easy to resolve.[41]

I leave to the reader to decide whether Chalmers's responses are satisfactory. My view is that Chalmers has gone a long way toward avoiding overly strong triviality. He appears to show that rocks and walls (as described by Putnam and Searle) do not implement every automaton, and, apparently, do not implement complex automata at all. At the very least, Chalmers demonstrates that the proofs for these effects are not valid when the notion of implementation is supplemented with additional constraints. If we impose certain causal and counterfactual constraints on the implementing system, we can rule out certain trivialities. If we add even further constraints regarding the spatial separation of the implementing substate components and the grouping of physical states (e.g., naturalness and uniformity), we rule out other trivialities. Thus, those who insist that rocks and walls indeed implement every automaton should come up with modified proofs—and this, to the best of my knowledge, has not yet been done.

---

[37] Godfrey-Smith (2009) also requires that the inputs and outputs be of the right physical kind. But this requirement is more suitable to functionalist theories of mind (which have their problems too, as Putnam points out), not to implementation in general. Godfrey-Smith then argues that even this requirement does not discharge triviality altogether.

[38] Also, in a response to Scheutz (2012), Chalmers (2012) excludes a temporal individuation of physical states.

[39] See Chalmers 1996: 329–330. But see also criticism by Sprevak (2012) and a response by Chalmers (2012).

[40] See, e.g., Chalmers (1996: 312; 329).

[41] Broadly speaking, functional specification is subject to different schemes of individuation. Physical specification is too restrictive, and intentional specification undermines the main goal of CST, which is to account for minds in non-semantic and non-intentional terms.

If this is correct, then the consequences of strong triviality do not automatically follow from the modified notion of implementation.

5.3 From Implementation to Computation

I have suggested that Chalmers's (non-semantic) account of implementation appears to circumvent an overly strong triviality and its consequences. As such, I believe that this account (perhaps with some modifications) is a good basis for a theory of implementation. The next, and more pertinent, question to ask is whether this account of implementation can serve as an account of computation. Chalmers assumes that it can. He takes his account of implementation to be an account of physical computation. I would beg to differ: while I actually agree that implementation, as Chalmers defines it, is *necessary* for computation under certain qualifications (to be discussed in Section 5.3.1), I do not believe that it is sufficient for computation (Section 5.3.2).

My claims about implementation and computation rest on the premise that we can keep the notions of implementation and computation apart. That said, such separation is not of the essence at this juncture: if someone were to insist that computation *is* implementation, so be it.[42] My argument in this case would also imply that Chalmers's account is not good for implementation either.

For the rest of this chapter, when using the term *implementation* I will be referring to Chalmers's notion of it, and will argue that implementation is necessary, but not sufficient, for computation.

5.3.1 Is Implementation Necessary for Computation?

I think that the answer is a qualified *yes*. One qualification is that implementation is not confined to specific formalisms such as finite-state automata. The implemented formal structures can be Turing machines, (abstract) neural networks, continuous formalisms (as is the case in some analog computation), or any other formalism that describes physical dynamics of some sort. In short, I would agree that implementing some formal structure is necessary for computation—however, it is not necessary for the implemented structure to be a formalism related to the standard mathematical theories of automata, computability, and so forth.

---

[42] This view is explicitly expressed by Chalmers (2011); Piccinini (2017); Sprevak (2018); and many others.

I underline this point because some have argued that the combinatorial state automaton model cannot serve as a general account of implementation (or, as a consequence, as a necessary condition for computation). They say that the CSA model does not cover all sorts of computation, and even does not describe Turing machines.[43] Chalmers (1996, 2011) concedes that the translation from Turing machines and other automata to CSA has some difficulties, but maintains that the account could potentially be modified and extended to handle other models of computation. He declares that "computations are generally specified relative to some formalism, and there is a wide variety of formalisms: these include Turing machines, Pascal programs, cellular automata, and neural networks, among others" (2011: 326). Among the other formalisms are the ones found in dynamic systems, evolution, and artificial life, including continuous mathematics; he says that "the current framework can fairly easily be extended to deal with computation over continuous quantities such as real numbers" (2011: 347). When talking about subsymbolic computation and neural networks, Chalmers says:

> Note that the distinction between symbolic and subsymbolic computation does not coincide with the distinction between different computational formalisms, such as Turing machines and neural networks. Rather, the distinction divides the class of computations within each of these formalisms. Some Turing machines perform symbolic computation, and some perform subsymbolic computation; the same goes for neural networks. (2011: 352)

I leave to the reader to decide whether or not, and to what extent, Chalmers's account of implementation can be extended to cover other formalisms, or whether we have to develop another theory that will cover these formalisms. The important point is that computation is not confined to specific formalism. A computing physical system has to implement *some* formalism.

Another qualification concerns the computational properties (or descriptions) of the implementing physical systems. The qualification is that these properties cannot be architectural. As I argued in Chapter 4, architectural properties cannot constitute a necessary condition for computation. This qualification is related to the first one: those who propound an architectural account would often identify computation with the implementation of specific formalisms such as automata—and this would then be identified with specific architectural properties such as digital structures. My counterclaim is that *if* implementation can be reduced to certain architectural properties, then implementation is not necessary for computation.

---

[43] See Brown (2012); Klein (2012); Miłkowski (2011); and Sprevak (2012). Some of them also doubt that the CSA model is a good model of cognition.

Chalmers himself does not take the architectural route. Rather, he argues that computational properties are *organizational invariants*, regardless of their architecture. He first characterizes the *causal topology* of a system in terms of the abstract causal organization of the system—namely, "the pattern of interaction among parts of the system, abstracted away from the make-up of individual parts and from the way the causal connections are implemented" (2011: 337). He then defines an *organizationally invariant* property as one that preserves the causal topology of the system. For example, replacing biological neurons with silicon neurons might preserve the causal topology of the brain, assuming that the patterns of interactions (such as excitation and inhibition) between the neurons remain the same.[44] According to Chalmers, flying, digesting, and most other properties are not organizational invariants: replacing biological parts of the stomach with pieces of metal while preserving causal patterns would no longer be an instance of digestion, according to Chalmers.

Following Piccinini (2015), I will use the more general term *medium-independence*. While Chalmers characterizes medium-independence in terms of organizational invariance, Piccinini and others provide a slightly different characterization (to be discussed in Chapter 6). Both characterizations aim to capture two important (and related) features of computation. One is that computational properties are abstract in the strong sense—non-physical or even formal or mathematical. Chalmers talks about *abstract causal organization*, which I take to be very close to functional organization. The other feature is that the same abstract causal organization can be found in systems with very different physical, chemical, or even biological properties. Chalmers uses the term *organizational invariance* to highlight this feature; others talk about *multiple realization*.

What is the difference between medium-independence and architectural profile? They are similar in that both refer to the functional organization (or abstract causal structure or architecture). But there is an important difference as well: architectural accounts refer to *specific kinds* of functional organizations—for example, those that are step-satisfaction. Accordingly, they exclude systems that lack these specific kinds of functional organization—for example, those that are more continuous and do not proceed in steps. In contrast, medium-independence embraces all sorts of functional organizations

---

[44] Chalmers enumerates the changes that preserve topological organizations as follows:

(a) moving the system in space; (b) stretching, distorting, expanding and contracting the system; (c) replacing sufficiently small parts of the system with parts that perform the same local function (e.g. replacing a neuron with a silicon chip with the same I/O properties); (d) replacing the causal links between parts of a system with other links that preserve the same pattern of dependencies (e.g., we might replace a mechanical link in a telephone exchange with an electrical link); and (e) any other changes that do not alter the pattern of causal interaction among parts of the system. (2011: 337–338)

and does not associate computation with specific organizations. Medium-independence excludes processes that do not exploit this functional organization, such as flying and digesting. Different accounts of medium-independence provide different understandings of what *exploitation* means in this context. According to Chalmers, flying and digesting are not computing processes, since they do not proceed by virtue of their organizationally invariant properties—that is, their organizationally invariant properties are not causally relevant to flying and digesting. When put in terms of implementation, we can say that flying and digesting do not occur *qua* implementing formalisms.

I agree with Chalmers that organizational invariance (or medium-independence) is a necessary condition of computation. In the following section, however, I argue that it is not sufficient: the stomach and other systems of that ilk do not compute, even *qua* implementing formalisms.

5.3.2 Is Implementation Sufficient for Computation?

The main argument against the sufficiency of implementation to computation is from *limited pancomputationalism*—namely, the claim that every physical system performs some computation. I argue that this claim misplaces the problem with Chalmers's account: the real problem is not with the claim that, under some description, rocks, hurricanes, and stomachs compute. Rather, the problem is that these systems do not compute even under the description that they implement some type of formalism.

Let us start with the argument from limited pancomputationalism. Chalmers's notion of implementation accommodates weak triviality—namely, the claim that every physical system implements at least one automaton, even if only of the one-state variety. But if we assume that implementing an automaton is sufficient for computing, the conclusion is that *every* physical system computes (limited pancomputationalism). Many have preferred to avoid limited pancomputationalism; they see it as a flaw, even if not a disastrous one. The flaw is that limited pancomputationalism violates the classification criterion: rocks, hurricanes, and stomachs do not compute. The way to avoid limited pancomputationalism is to reject the assumption that implementation is sufficient for computation.[45]

---

[45] Versions of this argument appear in Miłkowski (2013) and Piccinini (2015).

The argument then goes like this:

1. Every physical system implements at least one automaton (weak triviality).
2. Some physical systems do not compute (denying limited pancomputationalism).

Conclusion: Implementation is insufficient for computation (from 1 and 2).

As noted earlier, Chalmers is not put off by limited pancomputationalism. In his view, we can concede that rocks, chairs, stomachs, and hurricanes have computational (i.e., organizationally invariant) properties. We can even admit that when describing these computational properties, stomachs (etc.) compute (hence, he rejects Premise 2). But, he continues, limited pancomputationalism is consistent with the claim that digestion is not computation: stomachs also have non-computational—perhaps physical, chemical, and biological—properties. Assuming that digestion relies—even if only in part—on some of the non-computational properties, we have no reason to refer to digestion as computation. To put it another way: while one might say that the stomach computes *qua* (or in virtue of) implementing some formalism, it does not digest *qua* implementing this formalism. Implementing a formalism, hence computing, plays no role in digestion. The same goes for rocks, chairs, hurricanes, and most physical systems. While these systems might compute under some description, they do not compute under their "normal" description.

I think that Chalmers's reply is quite reasonable, and that he shows nicely that limited pancomputationalism is consistent with the claim that digestion is not computation. The problem with Chalmers's account lies elsewhere: stomachs, rocks, and hurricanes are not taken to compute even when described as implementing. Chalmers can insist that stomachs do not digest *qua* implementing—but he cannot say that stomachs do not compute *qua* implementing. If stomachs do not compute *qua* implementing—if they do not compute when described as implementing—then Chalmers must relinquish his claim that implementing is sufficient for computing.

To make the point more clearly, let me modify Premise 2 in the argument against the sufficiency of implementing for computing. Instead of saying that stomachs do not compute, we would settle on the weaker claim that stomachs do not compute even when described as implementing. In that case, the modified argument looks like this:

1. Every physical system implements at least one automaton (weak triviality).
2* Some physical systems do not compute, even when described as implementing.

Conclusion: Implementation is insufficient for computation.

But why should we accept Premise 2*? The main observation behind 2* is that scientists often describe stomachs, rocks, and so on in terms of mathematical formalisms (which the systems implement)—and yet no one describes such systems as computing, even under these implementation descriptions. Consider our neural n-queens network. I agree with Chalmers that when a system is described as computing, we are referring to its causal topology—namely, the fact that it consists of $n \times n$ units, that the units are interconnected, that the connections are symmetrical, that units on the same row inhibit each other, and so on. Moreover, we describe the dynamics of the network in terms of an energy-function formalism, and show that with these topological properties, the system converges to fixed points ("attractors") that are global minima points of the energy function. These fixed points are solutions to the n-queens problem. So far, so good—as this means that both implementation and computation are present here.

However, consider a spin-glass magnetic system, which is a lattice of particles whose abstract causal structure is identical. As I noted in Chapter 4, the field of attractor neural networks was developed by physicists who imported the formalisms developed in statistical mechanics into the world of neural networks. Noting that the causal topology of spin-glass systems is similar to that of interconnected networks, they applied the mathematical principles in statistical mechanics to that of fully interconnected networks. In particular, the organizationally invariant properties of magnetic systems—their interconnections, symmetry, and so on—are also found in attractor neural networks, and the energy function that describes the stabilization of magnetic system on minima points through a gradient descent process ("reducing temperature") also describes the dynamics of the networks.

Let us focus on the abstract causal structure of the spin-glass system: its components, organization, and dynamics. If Chalmers is correct, under this description the spin-glass system is a computing system. But it is not—or at least, physicists do not refer to its dynamics as computation. We do not view the lattice, even when referring to its organizationally invariant properties, as computing. While I agree with Chalmers that the lattice implements the pertinent network/energy-function formalism, we do not take such a system as computing, even *qua* its causal topology. The very same scientists (Hopfield 1982; Amit 1989) who describe the neural networks as computing do not describe the spin-glass systems as such. Again, I agree with Chalmers that when we describe the process of reducing the temperature of the spin-glass system at its points of equilibrium, we are not describing it as computing (much as we do not describe digestion as such). My point is that we do not describe such a system as computing even when we abstract from its physical details and focus only on its organizationally invariant properties. The same goes for other physical systems: merely describing the causal topology of stomachs, hurricanes, or rocks does not make them computing systems.

Note that I do not deny that the magnetic system implements a formalism of some sort. Nor do I deny that under certain circumstances, the lattice *can* be used (or described) as computing. My point is that the departure from implementation to computing requires further conditions (such as representing the numbers) that the magnetic system may or may not satisfy.

Of course, one is always free to insist that the magnetic system computes. My claim is that considering the lattice (*qua* its causal topology) as computing reflects neither the practices of scientists, nor how most of us refer to computing. Therefore, if there is an alternative account that does conform to these practices, it would be preferable to Chalmers's account. What kind of account would this be? The easy route is to add one or more constraint(s) to the account that rocks, stomachs, and hurricanes do not satisfy. As noted earlier, many have proposed strengthening Chalmers's notion with other constraints. Most of those, however, will not help in excluding our lattice from the computational domain. The individuation of the states of lattice is no less "natural," uniform, or appealing to spatially separable components of the system.

Another candidate for an additional constraint is an architectural property. Copeland (1996), for example, characterizes implementation of an algorithm partly in terms of step-by-step applications of a certain propagation rule. This step-by-step constraint, he argues, rules out many physical systems as computing: "According to an account that takes the notion of an algorithm and its supporting architecture seriously, the solar system does not compute" (p. 351). This architectural route is not viable, however. As argued in Chapter 4, the architectural property cannot differentiate between the neural network and the spin-glass system, as they both have the same architecture.[46] At one point, Chalmers suggests that we can solve the problem of limited pancomputationalism by stating that computation is the implementation of sufficiently complex automata (Chalmers 1994: 400; 2012: 242). According to this proposal, rocks, chairs, and many other physical objects presumably have too-simple architecture—because they only implement (e.g., one-state automata)—and therefore do not compute. But this proposal excludes only very simple objects. It does not exclude hurricanes, stomachs, or spin-glass systems, which presumably implement more complex formalisms.

A more recent contender is a teleological constraint (Piccinini 2015; Coelho Mollo 2018). According to this proposal, the network computes because it has the (teleological) function of computing—whereas the spin-glass system does

---

[46] See also Campbell and Yang (2019), who advance a similar claim while specifically targeting Copeland's account.

not compute because it lacks such a function. We will consider this proposal in Chapter 6. Finally, proponents of the semantic view are happy to point out that the missing constraint has something to do with the content of the states. Thus, the n-queens network computes because its states carry certain content regarding the location of the queens on the chessboard. Conversely, the lattice does not compute, because its states carry no content. When we assign certain content to the particles (numbers, queens, etc.), the lattice might be seen as computing. But when we do not assign it content, it cannot be said to compute. We address this semantic view in Chapter 7.

5.4 Summary

In this chapter, we considered the thesis that computing is implementing, with a particular focus on Chalmers's account. Starting with the much-discussed triviality results, we saw that it is possible to avoid overly strong triviality results. If we impose certain causal and counterfactual constraints on the implementing relation, we can rule out certain trivialities. If we add even further constraints on the spatial separation of the implementing substate components and on the grouping of physical states such as naturalness and uniformity, we rule out yet more trivialities. We are still left with some trivialities, but these trivialities no longer endanger the notions of implementation and computation—or not compellingly so. Moreover, as Sprevak (2018) remarks, the remaining weaker trivialities impose important constraints on the adequacy of accounts of computation. In the final section of this chapter, I cited one kind of weak triviality—that every physical system implements an automaton of some sort—to argue that implementation, at least of the sort proposed by Chalmers, is not sufficient for computation. In Chapter 8, I will invoke another kind of triviality (that some physical systems simultaneously implement more than one automaton) in order to provide another argument for the insufficiency claim. These results, with some further considerations, will be taken to prompt a semantic view of computation. But before turning to the semantic view, let us consider the view that computation is a mechanism.

6
Computation as Mechanism

The mechanistic account of computation has evolved into a formidable theory of physical computation. Its main advocate is Gualtiero Piccinini, who began developing this approach in his doctoral dissertation (Piccinini 2003b), then further articulated and shaped it in a series of papers (in particular, Piccinini 2007, 2008b), culminating in an overarching and comprehensive account of computation in his book *Physical Computation: A Mechanistic Account* (Piccinini 2015). Another mechanistic account has been put forward by Marcin Miłkowski, who advances it in the context of mind and cognition (Miłkowski 2013). Yet another account is proposed by Nir Fresco (2014), who seeks to promote an account of concrete digital computation as a foundation for cognitive science. David Kaplan (2011; Kaplan and Craver 2011) provides a mechanistic account in the framework of computational models in neuroscience.[1] Boone and Piccinini (2016), Dewhurst (2018a), Coelho Mollo (2018), and others provide more recent articulations of the account. The mechanistic account is apparently the dominant view about computation today.

This chapter focuses on Piccinini's account, which is the most comprehensive and detailed account of physical computation to date.[2] After presenting and reviewing the account (Sections 6.1–6.2), I discuss what I believe to be its two main shortcomings (Sections 6.3–6.4). My conclusion is that, despite its salient virtues, the mechanistic account falls short of satisfying the key classification and explanation desiderata of an account of computation.

6.1 An Outline of the Mechanistic Account

The term *mechanism* has various uses and meanings. The mechanistic account of computation, however, should be understood in the context of the recent

---

[1] Kaplan, however, focuses on the computational model, rather than on the computing mechanism. His account analyzes the relationships between the computational properties of the model (e.g., being a computer simulation) and the properties of the target (modeled) system. Our interest is in the accounts that analyze the target system as a computing system; see the discussion in Chapter 1, and also the pertinent discussion in Rusanen and Lappi (2016).

[2] I review Miłkowski's work in more detail elsewhere (Shagrir 2014).

mechanistic wave in philosophy of science.[3] As Piccinini puts it: "The mechanistic account begins by adapting a mechanistic framework from the philosophy of science. This gives us identity conditions for mechanisms in terms of their components, their functions, and their organization, without invoking the notion of computation" (2015: 3). This framework emphasizes the centrality of the so-called mechanistic explanations in the sciences, especially in biology and neuroscience. Computation and computational explanations should be accounted for within this framework. In Piccinini's words:

> The central idea is to explicate computing mechanisms as systems subject to mechanistic explanation. By mechanistic explanation of a system X, I mean a description of X in terms of spatiotemporal components of X, their functions, and their organization, to the effect that X possesses its capacities because of how X's components and their functions are organized. To distinguish systems whose capacities are subject to mechanistic explanation in the present sense from other systems, I call them mechanisms. To identify the components, functions, and organization of a system, I defer to the relevant community of scientists. (Piccinini 2007: 506)

According to this proposal, computational explanations are mechanistic explanations, computational processes are mechanisms, and computational properties are mechanistic properties. However, clearly not all mechanistic explanations and properties are computational. Computational explanations, properties, and mechanisms have their distinctive mark. As Piccinini puts it in the introduction to *Physical Computation*:

> A mechanistic account of computation must add criteria for what counts as computationally relevant mechanistic properties. I do this by adapting the notion of a string of letters, taken from logic and computability theory, and generalizing it to the notion of a system of vehicles that are defined solely based on differences between different portions of the vehicles. Any system whose function is to manipulate such vehicles in accordance with a rule, where the rule is defined in terms of the vehicles themselves, is a computing system. (2015: 3)

In the course of the book, Piccinini sets out the three main elements that identify computational mechanisms. One is a teleological function. Another is the

---

[3] The *loci classici* are Bechtel and Richardson (1993); Machamer, Darden, and Craver (2000); Glennan (2002); and Craver (2007). For more recent expositions, see Illari and Williamson (2012) and Andersen (2014a, 2014b).

manipulation of vehicles; Piccinini refers to this feature as *medium-independence*. The third is a rule. He concludes *Physical Computation* with a concise characterization of computation:

> A physical system is a computing system just in case it has the following characteristics:
> - It is a functional mechanism.
> - One of its functions is to manipulate vehicles based solely on differences between different portions of the vehicles according to a rule defined over the vehicles. (2015: 274)

It should be noted that Piccinini's account has undergone a number of modifications over the years. Most importantly, it began as a version of an architectural account, identifying computation with some form of digital mechanism. As such, in 2008 he defined *computation* as follows:

> A *computation* ... is the generation of output strings of digits from input strings of digits in accordance with a general rule that depends on the properties of the strings and (possibly) on the internal state of the system. Finally, a *string of digits* is an ordered sequence of discrete elements of finitely many types, where each type is individuated by the different effects it has on the mechanism that manipulates the strings. Under this account, strings of digits are entities that are individuated in terms of their functional properties within a mechanistic explanation of a system. (2008b: 34)

This definition, however, suffers from the same difficulty faced by other architectural accounts (see Chapter 4)—which is that, according to this account, computing machines, such as analog computers, are not computing mechanisms:

> Given the generality of the mechanistic account, it may be surprising that it excludes so called analog computers (in the sense of Pour-el [1974]). Analog computers do not manipulate strings of digits. Rather, they manipulate real (i.e., continuous) variables. Hence, they are left out of the present account. But analog computers can be given their own mechanistic account in terms of their components, functions, and organization. (2007: 519–520)

In a similar vein, connectionist systems and neural networks compute to the extent that they manipulate strings of digits according to a rule. Piccinini argues that most connectionist systems perform computation in this sense, but not all of them do (2007: 518). In his paper "Some Neural Networks Compute, Others

Don't" (2008c), Piccinini makes a similar claim about neural networks in general. The term *compute* in the title refers to the manipulation of *strings of digits* according to a given rule. Most neural networks compute in this sense of computation, but others don't.

At some point, however, Piccinini extends his account to other forms of computation. He introduces the very general notion of *generic computation*, which includes both analog and digital computation. His definition of computation in terms of manipulations of strings of digits becomes the definition of digital computation (2015: 177–178), and his new definition of generic computation is put in terms of *vehicles*, which are either variables or specific values of a variable (2015: 121). In this new, extended definition, all neural networks compute, although some perform non-digital computations (2015: 221–223). Physical computation is now defined in terms of *medium-independence* more generally, regardless of whether computation is digital or not: "Concrete computations and their vehicles can be defined independently of the physical media that implement them" (2015: 122).

In this chapter, I discuss the more recent account of generic computation. Piccinini argues that this account fulfills the required desiderata set for an account of computation, objectivity, explanation, classification ("the right things compute," "the wrong things don't compute"), miscomputation, and taxonomy. I focus on the classification and the explanation criteria, and raise two kinds of criticism against the account. One is that there are computational explanations that do not satisfy the norms of mechanistic explanations (Section 6.3). The other is a set of critical comments about the elements of the account, and in particular against the proposed criterion of teleological function (Section 6.4). Before that, I clarify the role of the mechanistic approach in the proposed account of physical computation (Section 6.2).

6.2 What Is "Mechanistic" in the Mechanistic Account?

A major advantage of the mechanistic account is that it does not rely on theoretical notions from logic and computer science such as *algorithm*, *program*, *effective procedure*, *automata*, *formal proof*, and others. Instead, it appeals to the mechanistic framework in the philosophy of science. One might wonder, however, to what extent the proposed mechanistic account really fits in with the mechanistic framework. It appears that *mechanistic* plays a different role in Piccinini's account than that played by *semantic, mapping, architectural*, and other notions in rival accounts. Semantic, mapping, and architectural properties play a *classificatory* role in accounts of computation. They are used to *exclude* certain non-computing processes; they contribute to meeting the *wrong-things-don't-compute* desideratum. According to semantic accounts, processes that do not involve semantic properties do not compute. According to mapping accounts, processes that do not "implement" (i.e., bear mapping relationships to) an abstract structure (e.g., an automaton) do not compute. According to architectural accounts, processes that do not possess the required architectural profile (e.g., step-satisfaction) do not compute. The term *mechanism*, however, is not used to exclude systems that do not compute; it plays no role in meeting the *wrong-things-don't-compute* desideratum. Of course, in mechanistic accounts all computations must be mechanistic, just as in semantic accounts all computations must be semantic (etc.); in that sense, being mechanistic is a prerequisite of computation, much as properties such as being semantic (or step-satisfaction, etc.) are meant to be prerequisites of computation. But in Piccinini's account, being mechanistic (unlike being semantic, etc.) does not do any classificatory work— for *non-computing systems are also mechanisms*. Indeed, according to Piccinini, the mechanistic account of computation aims to distinguish between *computing and non-computing mechanisms*:

> The main challenge for the mechanistic account is to specify properties that distinguish computing mechanisms from other (non-computing) mechanisms—and corresponding to those, features that distinguish computational explanations from other (non-computational) mechanistic explanations. (2015: 120)

What are the properties that distinguish computing mechanisms from other, non-computing mechanisms? If we refer to the definition, three are apparent: the *teleological function*, which excludes planetary systems, the weather, and many other systems (p. 145); *medium-independence*, which excludes cooking and cleaning (p. 122) as well as digestive processes (pp. 146–147); and the governing *rule*, which excludes random-number generators (p. 147). But what is also apparent is that none of these properties—teleological function, medium-independence, or rule—bears a special relationship to the mechanistic framework: they can be, and indeed have been, adopted in non-mechanistic accounts of computation. According to Fodor (1994), computational (i.e., syntactic) properties are conceived as high-order physical properties, and in this sense are medium-independent.[4] Hardcastle (1995) discusses computation in certain

---

[4] See also Haugeland (1978), who talks about medium-independence in the context of automatic formal systems, and Edelman (2008: 7) and Chalmers (2011), who talk about organizational invariance (see also Chapter 5).

teleological terms. Copeland (1996) and many others associate computation with a rule (e.g., algorithm).

Another key feature of the mechanistic account is that it is non-semantic:

> At the origin of the mechanistic account are two central theses. First, computation does not presuppose representation. Unlike most accounts in the philosophical literature, the mechanistic account does not appeal to semantic properties to individuate computing mechanisms and the functions they compute. In other words, the mechanistic account keeps the question whether something is a computing mechanism and what it computes separate from the question whether something has semantic content and what it represents. (2007: 502)[5]

But, again, being non-semantic is not a distinctive feature of mechanistic accounts. There are many accounts of computation that are neither mechanistic nor semantic—such as mapping and syntactic accounts. Moreover, it seems that an account of computation can be both mechanistic and semantic (Miłkowski 2017). For example, in Piccinini's definition of computation, we can replace the teleological function with a *semantic function* (or at least need a reason to refrain from such a replacement). Another, more tangible example would be the computational analysis of the navigational capacities of rats in terms of the representational functions of cells ("place cells") in the hippocampus (O'Keefe and Nadel 1978). This analysis is arguably both mechanistic and semantic.[6]

None of this undermines the adequacy of Piccinini's definition. The upshot, rather, is that central features in the account—teleological function, medium-independence, rule, and being non-semantic—are not in and of themselves tied to the mechanistic framework. There are mechanistic analyses that lack these features, and non-mechanistic analyses that do have them. In other words, we could do without the term *mechanism* altogether, and replace it with the term *process* or *dynamics* in the definition of computation. We could say, for example, that a physical system is a computing system/process/dynamics in the case that its teleological function is to manipulate vehicles based solely on differences between different portions of the vehicles according to a rule defined over the

---

[5] See also Piccinini (2008a, 2015); Miłkowski (2013); and Fresco (2014).

[6] Why, then, contrast the two? My guess is that the answer is something like the following: Advocates of the mechanistic accounts seek to block the appeal of semantic accounts, which arises from the failure of other non-semantic accounts—such as causal (Copeland 1996; Chalmers 2011) and syntactic (Fodor 1980; Stich 1983) accounts—to deal with certain problems, in particular the problem of computational implementation, which leads to triviality results and the risk of pancomputationalism (see Chapter 5). They see the mechanistic account as adequately addressing the computational implementation problem, thereby salvaging the non-semantic route. Indeed, this articulation is quite explicit in Fresco (2014) and Piccinini (2015, chs. 2 and 3; 2017).

vehicles (the same point applies to Piccinini's earlier account, where the vehicles are strings of digits).

So, in what sense is the proposed account mechanistic? Piccinini writes:

> The present account is *mechanistic* because it deems computing systems a kind of functional mechanism—mechanism with teleological functions. Computational explanation—the explanation of a mechanism's capacities in terms of the computations it performs—is a species of mechanistic explanation. (2015: 118)

As I understand it, the claim that computation is a mechanism is designed not so much to *exclude* non-computing systems (which are also mechanisms), but rather to *include* computations within the set of mechanisms. This is a contentious goal, given that many researchers have contrasted computations with mechanisms and computational explanations with mechanistic explanations. But once this goal is achieved—which is the purpose of Piccinini and Craver (2011)—it is natural to analyze computation and computational explanations from within the mechanistic explanatory framework. The mechanistic framework would then provide the tools to explicate computational explanation, as well as the features (medium-independence, teleological function, etc.) that define the computing mechanism in general. It would provide the tools needed to distinguish computational explanations from other, non-computational mechanistic explanations, and, accordingly, between computing and non-computing mechanisms. In other words, Piccinini's account is mechanistic in the sense that computing systems (much like non-computing systems) are mechanisms, and the mechanistic framework naturally provides the means to account for computing mechanisms and their (mechanistic) explanations.

If this is correct, the assessment of the mechanistic account of computation can be divided into two parts. One has to do with the assertion that computational explanations are mechanistic (Section 6.3), and the other with the claim that computing mechanisms are characterized by teleological function, medium-independence, and rules (Section 6.4).

6.3 Computational and Mechanistic Explanations

The upshot of the previous section is that the success of the mechanistic account of computation largely depends on the claim that "computational explanation— the explanation of a mechanism's capacities in terms of the computations it performs— is a species of mechanistic explanation." This is in no way an obvious assertion. Many have argued that there is even a tension between the mechanistic framework and computational explanations. As a teaser, it would be interesting to note that in his influential book *Explaining the Brain*, Carl Craver (2007) does not even mention the central role of computational approaches in the study of brain and cognitive sciences.[7] This might certainly give the impression that computational approaches do not fit squarely within the mechanistic framework. Beyond this somewhat anecdotal point, however, there are many who view, and even contrast, computational explanations with mechanistic explanations. Thus, some philosophers view computational explanations as types of functional analyses, and argue that the latter are autonomous and distinct from mechanistic explanations.[8] Functional analyses specify functional properties, whereas mechanistic explanations specify structural properties and components that realize the functions.[9] Cognitive neuroscientists also often contrast computational and mechanistic explanations; as Chirimuuta (2014) remarks, this distinction is commonplace in neuroscience.[10]

It is not surprising that the mechanists (Piccinini and Craver 2011; Miłkowski 2013; Piccinini 2015, chap. 5) address this concern directly by disputing the distinction between functional analysis and mechanistic explanations. Both functional analysis and mechanistic explanations are taken to be decompositional and constitutive. They explain certain capacities (e.g., an input-output function) by showing how these capacities are constituted of more basic capacities (including their functions, behaviors, or activities) organized together. Some functional analyses also locate the subcapacities in subcomponents of the system, which are individuated functionally. Mechanistic explanations, however, always specify these components; most importantly, they specify the structural properties of the components, including their location, trajectory, size, and shape. Another way to put this is that functional analyses specify functional properties, whereas mechanistic explanations specify structural properties that realize those functions.

Piccinini and Craver (2011), however, resist the view put forward by Fodor (1968), Cummins (1983, 2000), and others that functional analyses are explanations that are autonomous and distinct from mechanistic explanations.

---

[7] The term *computation* is mentioned only once, in the context of an information-processing task; see Levy (2009), who comments on this in his review of the book.

[8] See also the discussion in Chapter 4 about functional analysis, and how it relates to computation.

[9] See, e.g., Cummins (1983, 2000) and Fodor (1968). Cummins, e.g., says:

> It is therefore important to keep functional analysis and componential analysis [i.e., mechanistic explanation] conceptually distinct. Componential analysis of computers, and probably brains, will typically yield components with capacities that do not figure in the analysis of capacities of the whole system. (2000: 125)

[10] See also Chirimuuta's interpretation of Carandini and Heeger (2012): "It is unlikely that a single mechanistic explanation [for normalization phenomena] will hold across all systems and species: what seems to be common is not necessarily the biophysical mechanism but rather the computation" (p. 141).

On the contrary, they argue that functional analyses "gain their explanatory force by describing mechanisms (even approximately and with idealization) and, conversely, that they lack explanatory force to the extent that they fail to describe mechanisms" (p. 284). More specifically, they claim:

> Functional analyses are *sketches of mechanisms*, in which some structural aspects of a mechanistic explanation are omitted. Once the missing aspects are filled in, a functional analysis turns into a full-blown mechanistic explanation. By this process, functional analyses are seamlessly integrated with multilevel mechanistic explanations. (p. 284)

What about computational explanations? Piccinini and Craver (2011) might have given the impression that computational explanations are sketches of mechanisms.[11] This is a reasonable interpretation if we understand computational explanations to be types of functional analyses (which are described as sketches). Moreover, Piccinini and Craver classify Marr's computational and algorithmic levels—the only example of computational explanations in their paper—as sketches.

In *Physical Computation*, however, Piccinini (2015) states explicitly that computational explanations can be full-blown mechanistic explanations: "Computational explanations count as full-blown mechanistic explanations, where structural and functional properties are inextricably mixed" (p. 124). True, computational explanations specify medium-independent properties. Nevertheless, Piccinini argues that

> such an explanation is still mechanistic: it specifies the type of vehicle being processed (digital, analog, or what have you) as well as the structural components that do the processing, their organization, and the functions they compute. So computational explanations are mechanistic too. (p. 98)

One might wonder how computational properties can be both medium-independent and structural (e.g., implementational) properties at the same time. Piccinini's answer is that computational, medium-independent properties "place structural constraints on the media that realize them and the mechanisms that operate on them" (p. 124). Elsewhere, he adds that "structural and functional properties are not neatly separable within a mechanism. There is no such thing as a purely functional component, or purely functional property" (p. 99). The

---

[11] This interpretation is attributed to them by Chirimuuta (2014); Rusanen and Lappi (2016); and Shagrir (2016).

bottom line, then, is that computational explanations are full-blown to the extent that they refer to relevant functional and structural properties.

How well do computational explanations fit within the mechanistic framework? Do they satisfy the constraints imposed on mechanistic explanations? There are four kinds of objections to the premise that computational explanations are mechanistic, all aimed at showing that at least some computational explanations do not entirely conform to the norms of mechanistic explanations. The first objection is that some computational explanations are not *decompositional* (Section 6.3.1). The second objection is that at least some full-blown computational explanations do not refer to *structural properties* (Section 6.3.2). The third objection is that the computational level does not integrate squarely within the *mechanistic-implementational hierarchy* (Section 6.3.3). The last, and perhaps most interesting, objection is that some aspects of computational explanations are not in the business of revealing *causal structure* (Section 6.3.4). As we shall see, some of these objections concern the scope of mechanistic explanations more generally, but I will keep the discussion closer to computational explanations.[12]

6.3.1 Computational and Decompositional Explanations

Mechanistic explanations are decompositional: they explain a phenomenon by breaking down the phenomenon into subcapacities and/or subcomponents whose activities and organization constitute the phenomenon. Scholars have noted that some explanations do not involve componential analysis, hence are not mechanistic. They do not break down the explanandum capacities into subcomponents and their organization. Rathkopf (2018) argues that mechanistic explanations apply to *nearly decomposable systems* (Simon 1962), where nodes in a network have more and perhaps stronger connections with each other than with nodes outside the module. Many network models, however, provide non-decompositional explanations for non-decomposable systems where part-whole decomposition is not possible. For example, a network model that accounts for patterns of traffic in a road network explains the amount of traffic in each road based on dependence relationships that span the entire network. Thus, the reason that a certain road connecting two edges has lighter traffic depends on

---

[12] Some might argue that computations are not mechanisms as understood within the mechanistic framework, and therefore computational explanations are not mechanistic. Others might argue that even if computing processes are mechanisms, their (computational) explanations are often not mechanistic. Rather than distinguish between these claims, I shall simply consider what they have in common—that not all computational explanations are mechanistic explanations.

the structure and organization of the entire network, and cannot be explained by decomposing the network into separate components and their organization. In a similar vein, Weiskopf (2011) notes the existence of non-componential models in cognitive science; Huneman (2010) argues that in some cases the explanation appeals not to causal structure, but rather to the topological or network properties of the system; and Levy (2013) claims that decomposition (and localization) plays a lesser role in population genetics, ecology, and other macro-biological populational disciplines.

One can extend this point to the analysis of computing systems as well (Rathkopf 2018). Consider the attractor neural network for the n-queens problem discussed in Chapter 4. The network, as we recall, converges to a solution, which is a configuration in which exactly n queens are located on the $n \times n$ board, and no two queens are on the same row, column, or diagonal. This means that in every fixed point, exactly n cells are activated, and no two cells among them are on the same row, column, or diagonal. This network is in no way a nearly decomposable system: each cell in the network contributes to the activation of any other cell in the network. In other words, the dependence ("synaptic") relationships span the entire network. Moreover, there are no modules of cells in which the connections between cells are stronger than others. There are, of course, strong inhibitory connections between cells that are "on the same row," but each cell in this row has also strong inhibitory connections with a different set of cells (that are "on the same column"). The set of cells that are on a specific row, coupled with their same-column cells, comprises the entirety of the cells in the network.

How does the network converge to a solution for the n-queens problem? Is the analysis (explanation) of its behavior decompositional? In my view, it is not—at least not entirely. One can certainly decompose the system into its subcomponents (cells), their activity (which is exactly the same for every cell), the relevant (i.e., inhibitory or excitatory) relationships between the cells, and their organization (which cells activate other cells). Moreover, you can use this decomposition to explain certain features: one can tell how any given cell is about to behave by observing the current activity of all other cells and their relationship with that cell. One can also tell if a certain configuration of the network (where some cells are active and other not) will be followed by another configuration.

Nonetheless, this explanation falls far short of accounting for the main features we want to explain. It does not explain why the network relaxes at all (and, after all, many networks never relax); it does not clarify why, when starting from some (arbitrary) initial configuration, the network gradually—over many iterations—arrives at a fixed point (attractor). Most notably, it does not account for why the attractors of the networks are precisely the solutions for the n-queen problem. In other words, it does not explain why the attractors are configurations in which exactly n cells are activated, but there is no pair of activated cells on the same row, column, or diagonal.

As we saw in Chapter 4, in order to explain these features, one must know something about the topological structure of the network. Relaxation is explained by certain topological features (e.g., that the weights between cells are symmetrical), while solutions are explained by other topological features of the network, such as strong inhibitory relationships. When one knows something about these features, it is easier to explain the behavior of the network. Some mathematical theorems that refer to the topological properties explain the relaxation of the network on *some* fixed point. They also show that these fixed points are "minima points" of the energy landscape. When one looks at the energy function (eq. 1 in Chapter 4), one can further understand why these minima points are the solution to the problem: the minima points of the summations are zero points, and it is not hard to see that these zero points are achieved exactly when n cells ("queens") are activated, and none of them are on the same row, column, or diagonal. This insight is achieved not through decompositional analysis, but through an analysis of the energy function, which is completely blind to the contribution of each specific cell. The analysis examines the relationships between the values of the energy function, whereby each value represents the *total activity* ("energy") of the system.

Craver (2016) argues that the non-decomposable networks considered by Rathkopf are causal networks composed of nodes and interactions. This is also true of the queens model that consists of $n \times n$ cells ("components") and their (inhibitory and excitatory) relationships. The non-decomposable networks appear to be explanatory because they refer to this "base level"—namely, a set of causally organized parts, or a mechanism. In the case of the queens model, it is certainly true that the explanation refers in part to the cells, their activity, and their interaction. Some part of the explanation also refers to the fact that the minima points are those in which no two cells on the "same row" are activated. But it is also true that referring to the cells, their activity, and their interaction alone fails to explain the behavior of the network that we want to explain—namely, its relaxation on solutions. As we have just seen, the explanation of this feature must refer to the topological structure of the system. One might insist that the mere reference to the activity of the cells, their activity, and their interaction classifies them as mechanistic. This may be fine: the debate here is not over labeling. I am willing to concede that the analysis of the network is mechanistic, in the sense that it includes some decompositional analysis. The point is that another part of the analysis, which is not decompositional, is nevertheless computational. And if I am right about this, then it becomes less attractive to identify computational explanation solely with decompositional—and hence mechanistic—analysis.

The upshot is that there is (at least occasionally) a gap between computational explanations and decompositional analysis. This gap is exemplified in the explanation of (some) computing networks. The queens network can definitely be decomposed. This decomposition reflects the basic causal structure of the network and contributes to the explanation of its behavior. The point is that at least a substantial part of the computational explanation relies on a state-space ("global") analysis of its energy function. This analysis does not get into the activation values of each cell, the relationships between pairs of cells, and so on, but rather takes into account the topological features of the network and the relationships between the "total" states (configurations) of the network.

6.3.2 Abstract Explanations and Structural Properties

It has been argued that at least some computational explanations can be full-blown explanations, even if they do not refer to structural properties and therefore are not mechanistic. One premise of this argument is that mechanistic explanations essentially refer to structural properties of components—such as their location, size, direction, mass, and so forth. This criticism pertains to abstract explanations more generally: abstract explanations can be full-blown (i.e., not sketches), even if they make no reference to structural properties. Some of the claims refer to functional analyses, which belies the claim that functional analyses are sketches (Weiskopf 2011; Levy and Bechtel 2013; Barrett 2014; Shapiro 2017), whereas others specifically discuss computational explanations (Chirimuuta 2014; Egan 2017). These authors do not deny that some abstract explanations track causal structure. Rather, they argue that these abstract explanations can be full-blown explanations without referring to structural properties of the causal structure in question. Some relate these explanations to *multiple realization*, pointing out that the same abstract explanation applies to systems with different structural properties (Haimovici 2013; Barrett 2014; Chirimuuta 2014). Others do not link these abstract explanations to multiple realization (Shapiro 2016; Egan 2017).[13]

A word about abstraction is in order at this point. *Abstract explanations* can refer to explanations that omit certain details of the described causal structure; these details can be structural, functional, or of other types (Weisberg 2013).[14] Some have interpreted the mechanists as expounding the *"more details, the*

---

[13] Shapiro is famously skeptical about the scope and significance of multiple realization (Shapiro 2000; Polger and Shapiro 2016). He says that "even if functional properties are not multiply realizable, functional analysis can be autonomous from mechanistic explanation, and psychological explanation can be autonomous from neuroscientific explanation" (2017: 1057). Egan (2017) emphasizes the normative aspect of computational explanations.

[14] We assume that computational explanations are abstractions in a stronger sense—namely, that they are formal, e.g., mathematical descriptions. Whether these formal descriptions really refer to

better" (MDB) premise—which states that the explanatory force of the mechanistic analysis is proportional to the amount of detail that it provides about the mechanism (Levy and Bechtel 2013, Chirimuuta 2014). Thus, they view the mechanists as downplaying the explanatory force of any kind of abstract explanation. Piccinini (2015) clarifies, however, that he rejects the MDB premise: mechanistic and computational explanations—even ideally complete ones—can be, and often are, abstract. They aim to specify as many *relevant* properties as possible—namely, the features that are relevant to produce the explanandum phenomenon. The mechanists only insist that some of the relevant properties must be structural.[15]

The more interesting criticism, however, is that abstract explanations can be full-blown explanations without referring to structural (e.g., implementational) properties at all. Thus, Shapiro (2017), for example, argues that Sternberg's task-analysis explanation of the recall process of strings of numerals is full-blown, even though it ignores any implementational details of the task. Shapiro does not deny that functional properties place structural constraints on any mechanism that implements them (and vice versa). He claims that placing structural constraints does not render the constraining properties structural or the explanation mechanistic. Every abstract explanation is constrained to some extent by implementational details (Shapiro 2017). Moreover, while implementational properties can serve as evidence in support of candidate explanatory models and of distinctions between them, this does not make them an integral part of the explanation. The norms of confirmation and explanations are not the same (Weiskopf 2011; Shapiro 2017). Thus, Shapiro (2017) concludes that "the boldness of the claim that all explanations in the cognitive sciences must be mechanistic depends on being able to show that alternative forms of explanation contain at least tacit commitments to mechanisms. But these commitments, even if tacit, should be substantive (p. 1054). Otherwise, says Shapiro, "the dispute between the mechanistic hegemonists and the functionalists threatens to descend into one over labelling" (2017: 1056)

As previously noted, Piccinini agrees that abstract explanations can be full-blown mechanistic explanations, insofar as they specify the relevant properties of the explanandum phenomenon. In particular, computational explanations are full-blown mechanistic to the extent that they specify relevant medium-independent properties of the mechanism. He argues that some of these medium-independent properties are structural, or at least have structural

---

abstract (e.g., medium-independent) properties or objects depends on one's view about the ontology of abstract entities in physical systems.

[15] See also Boone and Piccinini (2016); Craver (2016); Kaplan (2017); and Craver and Kaplan (2020).

aspects. In fact, he dismisses the functional/structural distinction altogether. Thus, computational explanations can be full-blown mechanistic even though they do not specify other structural properties; in fact, they may not specify implementational ("medium-dependent") properties at all. The same line of reasoning applies to functional analyses more generally: these can be full-blown mechanistic even without specifying implementational, medium-dependent properties, and they are full-blown to the extent that they specify the relevant properties of the mechanism. The specification must include some structural properties—but these properties need not be implementational, that is, medium-dependent. If this understanding is correct, then the claim is not that functional analyses *must* be sketches, but rather that many—even all—available explanations that are described as functional analyses are in fact sketches.

This reply is somewhat vague with respect to the substantive commitments of mechanistic explanations. Can Sternberg's task analysis be considered a mechanistic explanation? It certainly does not refer to implementational, medium-dependent properties. But it does arguably capture the actual causal structure of memory. I think that we are still in the dark with respect to what counts as a medium-independent structural property. Piccinini says that a computational explanation is mechanistic to the extent that "it specifies the type of vehicle being processed (digital, analog, or what have you) as well as the structural components that do the processing, their organization, and the functions they compute" (2015: 98). But many analyses appear to satisfy the requirement of being full-blown, despite the fact that they are described by Piccinini as "sketches." Piccinini refers to Marr's computational and algorithmic levels as sketches, although they appear to satisfy the requirement of being full-blown mechanistic. The computational theory of edge detection states that the elements in the visual systems that perform edge detection are retinal ganglion cells, as well as LGN cells and the pyramidal cells in V1, and that their relevant activity is the activation of these cells. It also specifies the organization of the cell (e.g., feed-forward) and its relevance to the computation. The algorithmic level specifies the type of analog-to-digital vehicle and how this structure is relevant to the computation. As such, it appears to provide a complete account (in the abstract) of the operations of the mechanism.

Another concern with Piccinini's response is that his dismissal of the functional/structural distinction revives the argument that computational explanations are distinct and autonomous. Although computational explanations can be full-blown mechanistic, they are nevertheless distinct from implementational mechanistic explanations: the former refer to medium-independent (functional and structural) properties, whereas the latter refers to medium-dependent, implementational, properties. In other words, we can reformulate the distinctness thesis around the medium-independent/medium-dependent distinction, rather around the dismissed functional/structural distinction. Computational and implementational explanations—be they mechanistic or not—are distinct, as they specify different properties of the mechanism. To clarify, the suggestion is not to return to the anachronistic picture, which Piccinini and Craver rightly reject, in which the computational level (e.g., computational psychology) and the implementational level (e.g., neuroscience) are completely detached from each other. We have seen a great deal of conversation between disciplines over the past few decades, not to mention the key role played by cognitive neuroscience and computational neuroscience in brain research today. Rather, the claim is that the two explanations are distinct in the sense that they specify very different kind of properties, that is, medium-independent (computational) and medium-dependent (implementational).

To summarize, it seems that there is some tension between computational explanations (which are abstract) and mechanistic ones. Some computational explanations can be full-blown abstract without appealing to structural properties, such that they do not satisfy the norm of referring to structural properties. In response, Piccinini says that computational explanations can refer to medium-independent properties that are structural: the more important distinction is between medium-independent (computational) properties and medium-dependent (implementational) ones. One concern with this reply is that it blurs the distinction between sketches and full-blown mechanistic explanations. Another is that it reintroduces the divide between computational explanations and implementational ones. Even if both are (full-blown) mechanistic, computational explanations are still distinct and (perhaps) autonomous from implementational explanations. In the next section, we discuss another facet of this computational/implementational divide.

6.3.3 Computational and Implementational Hierarchies

According to the mechanistic view, mechanistic explanations are hierarchical.[16] This means that there is a hierarchy of mechanistic explanations whereby each component in an explanation is itself explained mechanistically. This claim raises the following question about the mechanistic view of computation: how are computational and implementational explanations related? Piccinini points out the problem when referring to Marr's renowned tri-level framework: "His 'levels' are not levels of mechanisms because they do not describe component/

---

[16] This section relies on Elber-Dorozko and Shagrir (2019); see also Harbecke (2020), who also raises the question of integration between the hierarchies.

subcomponent relationships. The algorithm is not a component of the computation, and the implementation is not a component of the algorithm" (2015: 98). Indeed, the realization relationship—of medium-independent properties by some implementational, medium-dependent properties—is *not* a relationship of part and whole. The 0s and 1s in the digital computer might be implemented by certain specific voltages, but the realizing voltages are in no way parts of the 0s and 1s; the two are perhaps correlated, or even identical. So how are the computational and the implementational levels related within the mechanistic framework? Moreover, assuming that we can have computational and implementational hierarchies, how are these hierarchies integrated or related to each other?[17]

To see the point more vividly, consider the figure from Botvinick, Niv, and Barto (2009), in which they describe models of reinforcement learning in the context of decision making (Figure 6.1).[18] On the left side we see an actor-critic computational ("abstract") model in which both the action strengths and the value function are learned through an interaction with the environment. On the right side we see an implementation model ("neural correlate") of the implementing neural structures.[19]

As we can see, the implementational model is not a lower mechanistic level of the computational model. The implementational, medium-dependent components are not parts of the computational, medium-independent parts. This means that the two models cannot be two separate levels of a single mechanistic hierarchy. So how should we understand the relations between the models within the mechanistic hierarchy? Moreover, each model can be thought of as a level within a part-whole relation hierarchy. It can be argued that the components of the computational model can be further analyzed in terms of computational subcomponents and their relations,[20] whereas the implementational components can be further analyzed in terms of implementational subcomponents and their

---

[17] According to the mechanistic framework, a complete explanation at each level would include all (and ideally only) the causally relevant relationships and activities that constitute the explanandum phenomenon.

[18] I will not enter here into the details of the model. See Elber-Dorozko and Shagrir (2019), where we describe some models of reinforcement learning from computational neuroscience.

[19] Botvinick, Niv, and Barto talk about "the computational and neural underpinnings of . . . behavior" (p. 262).

[20] Piccinini (2015) describes the computational hierarchy in computers as follows:

> Computing systems, such as calculators and computers, consist of component parts (processors, memory units, input devices, and output devices), their functions, and their organization. Those components also consist of component parts (e.g., registers and circuits), their functions, and their organization. Those, in turn, consist of primitive computing components (paradigmatically, logic gates), their functions, and their organization. (pp. 118–119)

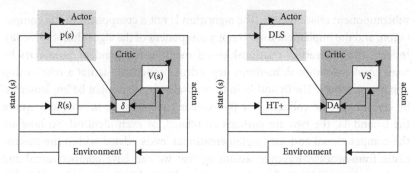

Figure 6.1 Computational and implementational models of reinforcement learning, side by side. R(s): reward function; V(s): value function; δ: reward prediction error; π(s): policy (action-selection function); DA: dopamine; DLS: dorsolateral striatum; HT+: hypothalamus and other structures; VS: ventral striatum (From Botvinick, Matthew M., Yael Niv, and Andrew G. Barto. 2009. "Hierarchically Organized Behavior and Its Neural Foundations: A Reinforcement Learning Perspective." *Cognition* 113: pp. 262–280. Reproduced with permission from Elsevier).

relations.[21] We can then ask: how do these two hierarchies, the computational and the implementational, relate to each other?

There are two ways that the mechanist can address these questions. One is lumping together the implementational and the computational models. This means that the relevant computational properties are lumped together with their implementational properties. In this picture we do not really have two separate levels (and hierarchies), but only one: the relevant computational properties are brought together with their implementational properties on the same level(s) of explanation. This simple solution suggests that computational and implementational properties figure together in the same explanation and in the same level(s) of the mechanistic hierarchy. This solution fits in quite nicely with the picture in which computational explanations are sketches of mechanisms. In that picture, the computational sketches become full-fledged mechanistic explanations only when we complement the sketches with the same-level implementational properties. When both kinds of properties are mentioned, we have a full-fledged mechanistic explanation—and hence a level of mechanism. The mechanistic hierarchy simply embeds within it a subhierarchy of computational sketches. Some argue, however, that this view is inconsistent with scientific practice, which often appeals to computational explanations as full-fledged

---

[21] See also Botvinick, Niv, and Barto (2009) and Elber-Dorozko and Shagrir (2019), who describe two hierarchical models, computational and implementational, in the context of reinforcement learning.

ones (Haimovici 2013). As noted, Piccinini (2015) also refrains from the view that computational explanations are essentially sketches.

A second option is to keep the two models apart. The two models comprise complete mechanistic explanations that are related through the implementation relationship: each computational component of the computational model is mapped to (implemented by) an implementational component of the implementational model. The same goes for the two hierarchies. Each level in each hierarchy is a complete explanation of the phenomenon at the level above it. In addition, the computational properties in the computational hierarchy are implemented by implementational properties in the implementational hierarchy (in reality, there may not be a perfect match between the two hierarchies, and computational properties at the same level may be implemented by implementational properties at different levels). This solution can more readily accommodate the notion that there is a multiple realization of cognitive functions, since the same computational hierarchy can be related to (i.e., implemented in) different implementational hierarchies.

This picture fits in quite nicely with the functional view of computational explanation, according to which computational explanations are full-fledged functional (yet non-mechanistic) explanations. According to this functional picture, computational explanations are distinct and autonomous from mechanistic explanations (Cummins 1983; Fodor 1968), which fits in with the solution in which the two hierarchies are distinct. Computational and implementational properties do not figure together in the decompositional explanation of the same capacities. Instead, only computational properties are part of the decomposition of computations. Implementational properties can still figure in explanations of computations, but these explanations would not be mechanistic, because there is no part-whole relationship between the explanans and explanandum.

What about the view that computational explanations are both abstract and full-fledged mechanistic explanations? It would be difficult to see how the first solution could be consistent with this view. If computational explanations are complete mechanistic explanations, why do they require additional implementation details of the same mechanistic level of explanation? The second solution is not necessarily inconsistent with the view that computational explanations are both abstract and full-fledged mechanistic explanations. For example, if one understands computational states and properties to have causal powers, one can view the computational hierarchy as a hierarchy of complete mechanistic explanations. However, the role of the implementational hierarchy has yet to be explicated. One possible way to elucidate this complex picture is to maintain that the implementation relation is part of the computational explanation; its role is to explain how the more abstract (functional) hierarchy is implemented (Kaplan 2017; Coelho Mollo 2018). But we would still note that this implementational explanation of abstract capacities cannot be mechanistic, since the implementation is not a part-whole relation.

The upshot is that there is a tension between computational explanations and the idea that mechanistic explanations are hierarchical. The mechanist can choose the one-level (and one-hierarchy) picture, at the price that computational explanations essentially become sketches of the mechanism. Alternatively, he or she might opt for the level-apart (and two-hierarchy) picture, but then this picture would fit better with the view that computational explanations are functional.

6.3.4 Information Processing and Causal Structure

The most pressing objection to the mechanistic account, in my view, is that it downplays the central role of informational or representational aspects in the cognitive and neural sciences. In those contexts, it is difficult to understand the relevance of computation when isolated from its informational or representational context.[22] In Chapter 9, I argue at some length that an important chunk of computational theory in cognitive science and neuroscience is devoted to addressing certain *why* questions whose explanations do not seem to involve causal mechanisms. These explanations (models) do refer to causal structure. The point is that they do not aim to track (only) causal relationships, but rather aspects related to the fact that the described system is information-processing. In this section I will review this criticism in brief.

Chirimuuta (2014) locates these *why* questions in the so-called *interpretative models* (Dayan and Abbott 2001). Dayan and Abbott note that in addition to phenomenal (descriptive) and mechanistic models, theoretical neuroscience also invokes interpretational models. These models "use computational and information-theoretic principles to explore the behavioral and cognitive significance of various aspects of nervous system function, addressing the question of why nervous systems operate as they do" (2001: xiii).[23] Chirimuuta argues that answering these *why* questions involves explanations that typically make reference to efficient coding principles. Her chief example is the normalization equation, which models cross-orientation suppression of simple cell responses in

---

[22] Rescorla (2016) emphasized this point in his review of Piccinini (2015).
[23] Chirimuuta (2014) distinguishes between *A-minimal models* and *B-minimal models*. The models described in Section 6.3.2 are A-minimal models. These models are abstract yet "causal-mechanical" explanations in the sense that they track the causal structure of the system. The B-minimal models are abstract explanations, yet they are not causal-mechanical. According to Chirimuuta, at least some computational models, e.g., the interpretative models, are B-minimal models.

the primary visual cortex and other systems (Carandini and Heeger 1994, 2012; Heeger 1992).

Very briefly, while cells in V1 were found to selectively respond to bar-shaped stimuli in a preferred orientation (Hubel and Wiesel 1962), it turns out that this response is significantly reduced ("suppressed") if stimuli with a non preferred orientation are superimposed on the preferred stimuli. Heeger (1992) proposed the normalization model to account for this phenomenon. The idea is that in addition to the excitatory input from LGN, each V1 cell also receives inhibitory inputs from its neighboring V1 cells, which are sensitive to bars at different angles. As Chirimuuta emphasizes, this normalization equation—which quantitatively describes the cells' responses—has subsequently been found in other parts of the nervous system (Carandini and Heeger 2012). This raises the question "Why should so many systems exhibit behavior described by a normalization equation?"—to which the answer is that "for many instances of neural processing individual neurons are able to transmit more information if their firing rate is suppressed by the population average firing rate" (Chirimuuta 2014: 143). This answer, it seems, makes reference to computational principles (in this case, a certain analysis from information theory), rather than to causal structure: "My key claim is that the use of the term 'normalization' in neuroscience retains much of its original mathematical-engineering sense. It indicates a mathematical operation—a computation—not a biological mechanism" (Chirimuuta 2014: 142). Of course, no one doubts that the normalization function is implemented within some neural structure, and that the implementation is important to the overall understanding of the functioning of the nervous system.[24] The point is that some aspects of the explanation—the ones associated with computational principles—are not mechanistic.[25] As Chirimuuta puts it, the explanation of why the normalization function is useful for the organism "departs fully from the model-to-mechanism mapping framework that has been proposed as the criterion for explanatory success" (p. 129).[26]

---

[24] See Kaplan (2017), who shows how the normalization equation is implemented differently in different species.

[25] Chirimuuta concludes with an endorsement of a claim for the distinct nature of computational explanation in neuroscience. She argues that while Piccinini and Craver (2011) and Kaplan (2011) correctly reject an anachronistic version of the distinctness thesis by some philosophers of mind, they fail to notice that many computational neuroscientists justifiably and clearly "distinguish between mechanistic and computational explanations, and that this distinction is characterised by efficient coding explanations, rather than generic functional explanations" (Chirimuuta 2014: 147).

[26] Chirimuuta refers to Kaplan's *model-to-mechanism mapping* (3M) requirement (Kaplan 2011; Kaplan and Craver 2011). According to this requirement: "(a) the variables in the model correspond to components, activities, properties, and organizational features of the target mechanism that produces, maintains, or underlies the phenomenon, and (b) the (perhaps mathematical) dependencies posited among these variables in the model correspond to the (perhaps quantifiable) causal relations among the components of the target mechanism" (Kaplan and Craver 2011: 611).

William Bechtel and I have put forward a somewhat similar claim, while focusing on Marr's computational-level theories (Shagrir 2010; Bechtel and Shagrir 2015; Shagrir and Bechtel 2017).[27] We argue that computational-level theories link the computed mathematical function and the explanandum information-processing task. They aim to explain why the computed mathematical function (e.g., derivation) is appropriate to the explanandum information-processing task (e.g., edge detection). This explanation, we suggest, has to do with the system-environment relationships, and not with an internal mechanism.[28] The upshot is that mechanistic accounts focus on how the mathematical operations are implemented and performed. Computational explanations, however, *also* aim to account for the relationship between those operations and the information-processing task at hand. We return to this claim in Chapter 9.

6.3.5 Summary

One of the promises of the mechanistic view of computation is to provide an overarching explanatory framework from which we can understand and account for the explanatory role of computational explanations. When we view computational explanations as mechanistic explanations, we can understand their explanatory role, as well as what distinguishes them from non-computational mechanistic explanations. In this section, I have challenged this claim. I agree that at least some computational explanations satisfy at least some of the norms of mechanistic explanations. I have argued, however, that computational explanations do not sit squarely with the mechanistic framework. Some computational explanations are seemingly non-decompositional; some computational explanations do not clearly refer to structural properties; computational explanations do not naturally integrate within the mechanistic-implementational hierarchy; and some aspects of computational explanations do not aim to track causal structure, but rather to answer certain questions about information processing. Although these claims are controversial, I believe that when they are taken together, they indicate that the mechanistic framework is not *the* natural place to account for computational explanations of physical systems.

---

[27] Rusanen and Lappi (2016) also associate the *why* questions with Marr's computational level theories and argue that computational theories provide explanations that express formal, non-causal dependencies.

[28] Rusanen and Lappi (2016) and Egan (2017) argue that Marr's computational-level theories provide explanations that express formal, non-causal dependencies.

6.4 Rules, Medium-Independence, and Teleological Functions

Piccinini's definition of computation includes three main elements—namely, *rule*, *medium-independence*, and *teleological function*. He invokes them in order to exclude non-computing systems. As noted in Section 6.2, these conditions are not tied to the mechanistic approach, and so are not affected by the criticism thereof. It is therefore essential to examine whether they constitute an adequate account of computation.

Rules. The appeal to governing rules plays a relatively minor role in the mechanistic account. It is chiefly invoked to exclude random-number generators from the domain of physical computing systems. The requirement of rules is fairly modest: a rule "is a map from inputs (and possibly internal states) to outputs" (Piccinini 2015: 121). Like Cummins, Piccinini does not require the system to represent the rule. A system that acts in accordance with dynamic equations, for example, satisfies the requirement. The meaning of *inputs* and *outputs* is not specified in detail. When discussing stomachs, Piccinini says that their "inputs" and "outputs" might not be of the same kind, but he admits that this is not a decisive objection to the rule requirement (p. 147). I would relax the requirement even further, as I am not sure that inputs and/or outputs are required at all (see Chapter 4).

Now consider random-number generators. Obviously, some computing systems include some stochastic or probabilistic elements ("randomness") and run probabilistic algorithms (p. 147). Moreover, the notion of a probabilistic Turing machine is pivotal in computability theory. Thus, some randomness must comply with the rule requirement. The degree of randomness that designates a system as non-computing is left open. A genuine random generator, however, does not compute: "There is no rule for specifying which digit it will produce at which time" (p. 147).

I agree that genuine random-number generators do not compute. But I am not sure that the rule requirement helps much in their exclusion. In some sense, random-number generators act according to rules. They might receive an input (e.g., pressing a button) and they produce outputs (i.e., strings of digits). Moreover, genuine random-number generators, if they exist at all, are carefully crafted to generate real randomness, and they act in accordance with the laws ("rules") of nature. It is true that these rules do not specify what digits are produced at what time, but these rules certainly specify which outputs they produce—namely, *digits*.

Ultimately, the outcome of this discussion is that the rule requirement is not crucial. Computation proceeds according to certain rules—but so do almost all (or even all) other physical systems. Piccinini says that the rule requirement helps to exclude random-number generators—but it is not clear that even this is true, as genuine random-number generators also follow certain rules.

Medium-independence. A more fundamental requirement is that computing processes are *medium-independent*. Piccinini attributes medium-independence to the vehicles of computation:

> A vehicle is medium-independent just in case the rule (i.e., the input-output map) that defines a computation is sensitive only to differences between portions (i.e., spatiotemporal parts) of the vehicles along specific dimensions of variation—it is insensitive to any other physical properties of the vehicles. (p. 122)

Piccinini also states that "the rules are functions of state variables associated with certain degrees of freedom" (p. 122). Coelho Mollo (2018, 2019) also uses the notion of *degrees of freedom*—characterizing them as "dimensions of variation of physical variables: for instance, a rigid robot that can only move forward, backward, left, and right, has two degrees of freedom, insofar as its position can vary only along two spatial dimensions" (2019: 436). Coelho Mollo remarks that this characterization of medium-independence is closely related to Chalmers's notion of *causal invariance* (discussed in Chapter 5). He says that insofar as causal-invariant properties "are individuated in a way that fully abstracts away from the physical constitution of their realisers, they are individuated in medium-independent terms" (2019: 447).

As noted in Chapter 5, medium-independence is related to two key features of computation: that computing processes are abstract, and that they are multiply realizable. What links these two features together is implementation ("realization"): a physical computation implements an abstract structure (e.g., an automaton), but "a given computation can be implemented in multiple physical media" (Piccinini 2015: 122). Computations are therefore medium-independent in that they "can be defined independently of the physical media that implement them" (p. 122). Notably, medium-independence is stronger than multiple realization. While the former entails the latter, the opposite is not true: a process can be multiply realizable without being medium-independent (Piccinini 2015: 122–123; Coelho Mollo 2019). Computing systems have the highest degree of multiple realizability: their individuation places no constraints on the physical medium that implements them—only on their degrees of freedom (Coelho Mollo 2019).

According to this account, medium-independence plays a significant role in classifying computing and non-computing systems and processes. Detecting edges (by desktops and brains) is defined in terms of the medium-independent properties, and, as such, they satisfy the requirement for computing. Cooking, cleaning, exploding, and so on are not medium-independent, since they "are defined in terms of specific physical alterations of specific substances" (Piccinini 2015: 122). The same goes for digestive processes, which are defined "in terms of specific chemical changes to specific families of molecules" (p. 147) and are therefore medium-dependent. Coelho Mollo makes a similar point, noting that

> weather systems are not medium-independently characterized—to be a weather system involves being composed of large amounts of gas molecules of certain kinds (depending on atmospheric composition) and having causal powers that depend on their intrinsic physical properties (e.g. density, temperature). (p. 438).

I agree that medium-independence is a necessary condition for computation; I also agree that medium-independence is the source of the abstract nature and multiple realizability of computation. But, notably, medium-independence alone does not rule out non-computing processes such as digesting, cleaning, and cooking (etc.) as computing. As noted in Chapter 5, these processes might have medium-independent (organizationally invariant) properties too. If every physical process (system) implements some type of formal structure, then every physical process—such as digesting, cleaning, or cooking—has medium-independent properties. In this respect, non-computing systems are no different from computing systems: conceivably, there is a description by which the stomach possesses degrees of freedom no less than other computing systems. What makes the difference is that digestive processes, *qua* digestion, are medium-dependent and, as such, do not compute. The same applies to cooking and cleaning. Cognitive processes—detecting edges, recognizing faces, multiplying numbers, and so forth—are (arguably) cognitive by virtue of processing medium-independent properties and, therefore might be deemed to compute.

To be sure (and as emphasized in Chapter 5), I do not claim that focusing on the medium-independent properties of the stomach renders it a computing system. On the contrary: we can abstract from the medium-dependent properties of the stomach and describe its processes in terms of rules that are sensitive to degrees of freedom alone. But this description does not render the stomach—or any other physical system—a computing system. According to the mechanistic account, what excludes the stomach and other physical systems from the computational domain is the teleological function: the stomach, like many other physical systems, lacks the teleological function to carry out medium-independent processes—namely, processes that are sensitive to degrees of freedom alone. In other words, even if the stomach carries out these medium-independent processes under some description, it lacks the teleological function to carry them out, and hence it does not compute.

If all this is correct, then medium-independence plays a lesser role in deeming physical systems non-computing. The stomach, the weather, and other physical systems are not computing, but not because they lack medium-independent processes. Actually, every physical system carries out such medium-independent processes, at least under some description. These systems do not compute because they lack the appropriate teleological function. And this means that it is the teleological function, not the medium-independence, that does the job of excluding these systems from the domain of computing systems.

One could argue that medium-independence does the job of distinguishing digestive and medium-independent processes within the stomach, and in this sense it deems a digestive process as non-computing. Medium-independence might also do the job of distinguishing between computing and non-computing processes within a *computing* system. It might deem, for example, some neural, medium-dependent processes within our visual system as non-computing. This might well be correct. I do not want to undermine the importance of medium-independence. My point is that the mechanistic account puts a heavy burden on the teleological function: in most cases, you must conjoin medium-independence with the teleological function to exclude non-computing systems. It is time to examine whether or not the teleological function can carry the burden.

Teleological functions. Piccinini views computation as a mechanism with *teleological functions*. Most importantly, one of its functions, according to his account, is to perform computation—namely, "to manipulate vehicles based solely on differences between different portions of the vehicles according to a rule defined over the vehicles" (p. 274). What is a teleological function? Piccinini devotes a lengthy discussion to this question (2015: chap. 6; Maley and Piccinini 2017) and adopts a goal-directed (dispositional) approach:

> A *teleological function* (generalized) is a stable contribution to a goal (either objective or subjective) of organisms by either a trait or an artifact of the organisms. (p. 116)

This characterization encompasses both biological systems (e.g., brains) and artifacts (e.g., laptops and smartphones). Possible goals include survival and reproduction, among others.

Teleological functions play a major role in the classification of computing and non-computing systems. According to Piccinini, planetary systems, the weather, and many other systems do not compute because they have no teleological function (p. 145; Coelho Mollo 2019). Planetary systems (and the like) satisfy the rule and medium-independence conditions, at least under some description: they manipulate vehicles based solely on differences between various parts of the vehicles according to a defined rule. However, they do not compute because, even under this description, they fulfill no teleological function whatsoever. What about cooking, cleaning, and digesting? Much like planetary systems, these processes satisfy the rule and medium-independence conditions, at least under some description. However, unlike planetary systems, they do have some teleological functions—and yet stomachs do not compute, because they lack the *right kind* of teleological function. They do not have the teleological function to manipulate vehicles based solely on differences between different portions of the vehicles according to a defined rule.

An example of a computing process is our visual system. Our early visual processes detect visual edges in the retinal images. These visual edges represent "physical edges" in the perceiver's visual environment (visual field), such as object boundaries. According to some computational theories (Marr 1982), our visual system detects edges by computing the zero-crossings of second-derivative operations (this theory is discussed at greater length in Chapter 9). According to Piccinini, the system does not only compute this mathematical function because the detection is achieved through medium-independent processes. It computes because the teleological function of the visual processes is to carry out these medium-independent processes.

While I agree with Piccinini that computing is related to a task or goal of some kind, I am more skeptical about the need for teleological functions. Teleological functions are not a natural fit with computation (Dewhurst 2016). Piccinini himself rules out two approaches to teleological functions that cannot be used to account for computation. He excludes etiological or historical theories (e.g., Millikan 1984; Neander 1991, 2017) because he thinks that computation is grounded in the current causal powers of the system. In particular, these theories cannot account for spontaneous computations: the visual system computes differentiation, according to Piccinini, even if this computation has no historical roots. Perspectival theories (e.g., Hardcastle 1999; Craver 2013) are also ruled out, because they introduce a dimension of observer-relativity that Piccinini aims to avoid.

Another difficulty with teleological functions is their alleged tension with medium-independence. Medium-independence states that the identity conditions of computation are not tied to any physical medium, whereas teleological functions are defined in terms of specific causal powers of physical systems (Coelho Mollo 2019). According to Coelho Mollo, resolving this tension requires adopting some version of functionalism.

My main criticism, however, is that Piccinini's account does not show in sufficient detail how the teleological function correctly classifies computing and non-computing systems. The account of teleological functions is suggestive and detailed—but is also very general, and seldom refers to computing systems. When applied to computing systems, important details are left unspecified—such as the goals of a computing system, or how these goals constrain the individuation of medium-independent processes. Instead, the account assumes that such constraints are imposed when the system is computing, and are not imposed when the system is not. But this simply amounts to assuming that the teleological function fulfills its purpose, rather than demonstrating that it does.[29]

In fact, I would argue that when we analyze the applicability of the proposed teleological functions to physical computing systems, we see that these functions do not fulfill their classification task very well. Let us start with non-computing systems. We do not typically attribute goals to galaxies and planetary systems, but if survival is a goal, then we should be told why not-collapsing ("surviving") cannot be a goal of a galaxy. Unlike gas leaks (discussed by Maley and Piccinini 2017), galaxies also "reproduce" new stars and "pursue [their own] inclusive fitness" in the sense that they extract energy from their environment in order to maintain their internal stability. Presumably, some topological and geometrical ("medium-independent") properties contribute to the survival of galaxies and planetary systems. Remove the geometrical relationships between the planets and the sun—such as those described by Kepler's laws—and the planetary system would vanish. And yet planetary systems do not compute.[30]

Next, consider stomachs. It is not controversial that stomachs have teleological functions and that digestion (which is medium-dependent) is a stable contribution to the survival of organisms. But what about the medium-independent processes that take place within the stomach? Do they also contribute to the survival (or other goals) of organisms? Perhaps they do and perhaps they do not— we certainly cannot rule out the possibility that they do. Assume, for the sake of argument, that they do: let us say that there are certain topological properties (such as points of equilibrium) in the stomach that are important for the organism's well-being. Would we deem the stomach to be a computing system in that case? I think we would not (see also Chapter 5). But even if I am wrong about this, it is the task of the mechanistic account to demonstrate why we would view the medium-independent processes in the stomach as computing.

Let us turn to computing processes such as edge detection. According to computational theories of vision, it is agreed that the early visual processes compute; they compute the zero-crossings of second-derivative operations. It is also agreed that this computation contributes to the well-being of biological organisms and artifacts. But what is the teleological function of medium-independent visual processes? Many would agree that the visual process has the teleological function of detecting edges—namely, producing representations of physical edges

---

[29] See also Dewhurst (2018b) for another criticism along these lines.
[30] This and perhaps the other difficulties mentioned later might be dealt with by a different, e.g., etiological, account of functions.

(outputs) from representations of light intensities (inputs). Achieving this task further contributes to the survival of organisms in their environment. Piccinini, however, refrains from a semantic characterization of computation; his account is non-semantic. He says that "a physical computing system is a mechanism whose teleological function is computing mathematical function f" (p. 121). In the case of edge detection, the system not only computes second-derivative operations (as we agreed). Its teleological function, according to Piccinini, is to compute this (non-semantic) mathematical function. But why think that the teleological function that defines computation is the mathematical function? In what sense does it contribute to the survival of the organism more than the mathematical functions performed by the stomach? And why isn't the semantic task sufficient for the survival of the visual system?[31]

Another case in point is the immune system. There is a rich literature on the medium-independent ("computational") properties of immune systems (Jerne 1974); some have even compared them to the topology of neural networks (Dasgupta 1997). These topological properties contribute to the ability of immune systems to attack invaders. In some cases, immune networks are not characterized as computing systems (Hoffmann 2008), even though they have the kind of teleological function that should result in them being considered as such. In other cases, natural and artificial immune networks are characterized as computing, especially when they are viewed as information-processing.[32] Here, too, it is far from obvious that it is the mathematical and not the semantic function that makes the immune network a computing one (or not).

Piccinini provides a philosophical analysis of the nature of teleological functions, but as far as I can tell, this analysis falls short of showing that the teleological function that is relevant to computing systems is mathematical and not semantic. In Chapter 8, I provide an argument for the claim that the semantic task is crucial to the individuation of computation. If I am right about this, then characterizing computation solely in terms of non-semantic teleological functions is inadequate.

In summary, the three stated conditions for computing have their virtues, but also their limitations. It is doubtful that the *rule* requirement fulfills its limited role of excluding random-number generators. The *medium-independence* requirement has its merits, but plays a lesser role in distinguishing computing from non-computing systems. The bulk of the account falls on the shoulders of the

---

[31] It should be noted that the semantic (representational) task itself is often defined in terms of a teleological function. In fact, according to Piccinini (Morgan and Piccinini 2018), the "shared conception" is that a representation is an internal entity that "has the function of responding to, or tracking, the distal entity" (p. 10).

[32] Dasgupta writes: "The natural immune system is a subject of great research interest because of its powerful information processing capabilities. In particular, it performs many complex computations in a highly parallel and distributed fashion" (1993: 5).

teleological function requirement. However, it has yet to be shown that this requirement adequately distinguishes computing from non-computing systems.

6.5 Summary

This chapter addressed the mechanistic account, particularly as set out in Piccinini's *Physical Computation*. I raised two kinds of criticism. The first was that some computational explanations do not satisfy the norms of mechanistic explanations (Section 6.3). The second was that the main criteria of the account—rules, medium-independence, and teleological functions—do not appear to constitute an adequate characterization of physical computation (Section 6.4).

Nonetheless, the mechanistic account has many advantages in its favor. It is the most systematic and detailed account of physical computation to date. It is the first account that clearly disengages physical computation from logic and computability theory, thereby sidestepping many of the pitfalls of earlier accounts. In its most recent iteration, it abandons the architectural approach and appropriately characterizes computation in terms of medium-independence (including a detailed and sound characterization of medium-independence). Last but not least, it recognizes that computation cannot be characterized solely in terms of medium-independence, and that another crucial element is missing. It proposes that teleological functions are the missing element. However, as I have shown, there are reasons to doubt that these functions are up to this task. Instead, I argue that the missing element is in fact the semantic properties of computational states and processes. This is the focus of the final part of this book.

7
The Semantic View of Computation

A semantic view of computation asserts that semantic properties are an essential aspect of the nature of physical computing systems. A primary motivation in favor of the semantic view is that it arguably meets the classification criterion of distinguishing computing from non-computing physical systems. Computing systems such as desktops and brains appear to involve representations, be they derivative or natural. Many instances of non-computing systems—such as stomachs, hurricanes, and rocks—do not involve semantic properties, and therefore cannot be deemed to be computing. If this is correct, then the semantic view is superior to the existing non-semantic views reviewed in previous chapters. Another argument ("the master argument") for the semantic view is presented in Chapter 8. In this chapter, my aim is twofold: to explain what is meant by the semantic view (Section 7.1), and to defend it in the face of a long list of objections that have been raised against it (Section 7.2). I will further develop my own account in the next two chapters.

7.1 What Is a Semantic View of Computation?

A semantic view of computation states that semantic properties are somehow *essential* to the nature of computation. Philosophers have used terms such as *involve*, *bear upon*, *inform about*, *is relevant to*, and *have* (semantic properties) to capture the tight linkage between semantic properties and computation.[1] But what do they mean? After all, everyone agrees that computation often involves information or representation, whether in manufactured systems or in natural ones—yet this is perfectly consistent with non-semantic views.[2] In order to

---

[1] Sprevak identifies the semantic view with the claim that "computation essentially involves representational content" (2010: 261). Rescorla writes that "on the semantic view, *all* physical computational systems have semantic or representational properties" (2014: 1298). Piccinini says: "I call any view that computational states are representations that have their content essentially a *semantic account of computation*" (2015: 27).

[2] Thus, Frances Egan (who holds a non-semantic view) writes that "computational theories treat human cognitive processes as a species of information processing" (1995: 181). Piccinini, who also argues against the semantic view, writes: "In our everyday life, we usually employ computations to process meaningful symbols, to extract useful information from them" (2015: 26).

answer this and other questions, we must characterize the semantic view in more detail.

7.1.1 Essential Involvement

Those who discuss the semantic view understand the locutions of the terms *involve*, *have*, and so forth in terms of *individuation*. Thus, both proponents and critics of the semantic view describe it as a claim about the individuation, taxonomy, or identity conditions of computation—whereby the individuated entities can be systems (Piccinini, Rescorla), processes (Sprevak, Dewhurst), states (Piccinini, Dewhurst), or events. The claim is that the individuation of these systems (etc.) takes into account their semantic properties—namely, that these play a role in determining whether a given system is computing or not; whether two systems are computationally similar or computationally different; whether or not changes in semantic properties alter computational identity; and so on.

I have associated the locution *involve* (etc.) with individuating computation. Next, we ask what is meant by *essential* in the phrase "essentially involve." The simple answer is that *essential* means *always*, whereby *always* refers to any computation, actual or possible.[3] So the semantic view asserts that semantic properties always impact the individuation of computation. The demand is not that the individuation takes into account all semantic properties, or only semantic properties (Piccinini 2015: 27), but rather that some semantic properties, perhaps with certain non-semantic properties, always affect (or "impact," "play a role in," or "enter into") the individuation of computation.

7.1.2 Non-Semantic Views

The semantic view asserts that semantic properties *always* affect computational individuation. This assertion gives rise to a variety of views that are not semantic. A *non-semantic view* asserts that semantic properties *never* affect computational individuation. As previously noted, such a view is consistent with the claim that computation often involves semantic properties, insofar as it asserts that computational *individuation* never considers these semantic properties (see, e.g., Egan 2010; Piccinini 2015; Dewhurst 2018a).

---

[3] We might ask about the scope of the possible—namely, if it refers to any computation that is physically, metaphysically, or even logically or conceptually possible. Given that our focus is physical systems, we can be satisfied with *physically possible* computations.

An important brand of non-semantic views associates computational individuation with certain semantic terms (e.g., *intentionality, representation, information*). These theories do not necessarily make substantial claims about content—such as that it can be naturalized, eliminated, and so on (discussed later). Rather, they simply deny that the so-called semantic terms are about semantic properties. Miłkowski (2013), for example, argues that computation is information-processing, but that the term *information* is not a semantic entity: "Information processed by a computer need not refer to, or be about, anything in order to be the inputs or outputs of a computation" (p. 48). He further states that the inputs and outputs always carry information in the trivial sense of causal nexus, but need not be representational or carry information "in the normal sense" (p. 48). Miłkowski does not deny that computing systems ever carry informational content in the normal sense, or that informational content is real. Rather, he claims that a computing system is information-processing even if the system does not carry any informational content in the normal sense. Fresco (2014) identifies concrete digital computation with the processing of instructional information—but, like Miłkowski, he does not appear to require that this information have content. Stich (1983), who is an eliminativist about intentional content, suggests that intentional terms such as *belief* and *desire* refer to syntactic properties.

Some philosophers argue that computational individuation takes into account semantic properties in some cases, but not in others. We might call this view *neither semantic nor non-semantic* (NSNNS). Rescorla (2013; 2017) explicitly argues in favor of NSNNS. He suggests that the semantic view is prevalent in computational cognitive science, and that the non-semantic view dominates computer science. Burge appears to subscribe to this view as well: on the one hand, he famously argues that visual content affects the individuation of computational-cognitive states (1986; 2010); on the other, he also says, in relation to a specific computation, that "here we have computation *without* representation" (Burge 2010: 424). Thus, both Rescorla and Burge do not uphold the semantic view. But they also seem to think that in the context of cognitive science, where computation involves cognitive representation, the content of the representation affects computational individuation—and in that regard, they do not support the non-semantic view.[4]

---

[4] Lee (2021) proposes a pluralistic view of computational identity that is similar to NSNNS in its claim that semantics affects computational individuation in some cases but not in others.

7.1.3 Variants of the Semantic View

As it turns out, there are at least two different variants of the semantic view that relate to computational individuation. According to one version, semantic properties play a role in distinguishing between computing systems (such as brains and desktops) and non-computing systems (such as stomachs and hurricanes). According to the other version, semantic properties play a role in distinguishing between different kinds of computing systems—for example, between brains and desktops. In both views, semantic properties are a factor in the identity conditions, or the individuation, of computation. In the former version, semantic properties are part of the identity of a system as a computing system, and as such are essential to rendering it a computing system. In the latter version, semantic properties determine computational equivalence—namely, the classification of (computing) systems, processes, events, and states into computational types, or varieties.[5]

The former version of the semantic view asserts that semantic properties are essential to meeting the classification desideratum—namely, to distinguishing between computing and non-computing systems. Let us call this version the *C-semantic* view (whereby C stands for *classification*). The second version asserts that semantic properties are essential to meeting the taxonomy desideratum—namely, to distinguishing between different types of computation. In this version, semantic properties play an essential role in computational equivalence, as they determine (perhaps in conjunction with other factors) whether or not two instances (tokens) of computation belong to the same type of computation. Let us call this version the *E-semantic* view (whereby E stands for *equivalence*).

On the face of it, there is no reason to uphold one version and reject the other. But there are those who do. Fodor adopts the C-semantic view. He says more than once that there is "no computation without representation" (Fodor 1975: 34; 1980: 122) and "no representations, no computations" (1975: 31). According to Crane, Fodor adopts this criterion in order to distinguish between computing and non-computing systems:

> According to reductionists like Fodor . . . what distinguishes systems that are merely describable as computing functions (such as the solar system) from systems that genuinely do compute functions (such as an adding machine) is that the latter contain and process representations—no computation without representation. (Crane 2016: 154)[6]

---

[5] This distinction is also emphasized by Sprevak (2018) and Lee (2021).
[6] Fodor himself says that "the solar system is not a computational system, but you and I, for all we now know, may be" (1975: 74 n. 15).

Fodor, however, rejects the E-semantic view (see, e.g., 1994: 7–16). According to him, computational kinds are individuated by their syntactic properties, whereas syntactic individuation does not appeal to semantic properties. I will leave aside the question of whether this claim—adopting the C-semantic view but rejecting the E-semantic view—is consistent.

In principle, one could hold the E-semantic view without adopting the C-semantic view. Arguably, this is indeed Burge's view: he believes that semantic properties are essential to the individuation of computational-cognitive states while conceding that, in other cases, "we have computation *without* representation." But, in fact, there is not much sense in adopting the E-semantic view without adopting the C-semantic view. If semantic properties *always* affect the individuation of computational kinds (E-semantic view), this implies that there is no computation without representation. And if there is no computation without representation, this would strongly suggest that the semantic properties of representations play a role in the identity condition of computing versus non-computing systems (C-semantic view). It would therefore make more sense to attribute a variant of the NSNNS view to Burge (as we did earlier), whereby the individuation of computational kinds (equivalence) takes semantic properties into account when computation operates on representations, but does not do so when computation involves no representation. I myself would argue here for both the C-semantic and E-semantic views.

7.1.4 Semantics

The semantic view refers to semantic features, usually properties. But what is meant by *semantic*? This is a notoriously hard question when considering semantics in general. The short answer, however, is that when confined to computing systems, semantic properties refer to representational or informational content. The semantic view claims that the individuation of computational systems (etc.) makes an essential reference to the content of the states of the system.

What is this content? Most would agree that content involves *aboutness*. The states of computing systems that have content denote or refer to certain other objects, events, properties, and so on. These entities can be located in the environment of the (computing) system; in distant, counterfactual, or abstract domains; or within the system itself. Some identify content with the referents themselves, while others identify it with certain perspectives of these referents (such as "senses"). All agree, however, that *aboutness* implies directionality: while objects with content refer to certain other entities, those entities might not refer back to them, nor might they have semantic properties at all.

Next, we ask what kinds of content play a role in computing systems. Following others, I suggest taking a pluralistic stance. One aspect of this pluralism is that different computing systems might operate on different kinds of semantic properties (Piccinini and Scarantino 2011; Piccinini 2015: chap. 14). Some computations operate on so-called representations whose content is interpretative—in the sense that it is derived by an observer, designer, or user (usually external). Laptops operate on semantic properties that are defined, at least in part, by the user of the machine. Other computations might operate on representations whose content is non-derivative. Most people would argue that the content of the computations that take place in our brain, for example, is not defined by the interpretation of an external observer.[7]

Another aspect of pluralism pertains to the factors that determine content.[8] In classical cognitive science, the content of a computing system might receive a functional, model-based treatment.[9] Others—specifically neural computations—might operate on representation or information whose content is at least in part causally based, or so it is often assumed in cognitive neuroscience.[10] It may turn out that there is a single account of the content of computing systems.[11] The semantic view is perfectly consistent with this scenario. But it is also consistent with the far more reasonable scenario that computation allows for various kinds of semantic properties.

Sprevak (2010) suggests that the semantic properties involved with computing systems might not always have a particularly complex structure (*minimalism*). Some computations operate on propositional and compositional representational systems, while many others appear to operate on representational systems that lack these rich structures. (Cells in V1, for instance, do not appear to have a complex propositional structure.) Another facet of minimalism pertains to internal representations: some computations involve internal representations, but others do not. Two-layer feed-forward networks map input representations to

---

[7] Another motivation for the pluralistic stance is the fact that computation is associated with different semantic *terms*, such as *representation, information, content, coding,* and *encoding*—as well as *symbols, signs, signals, denotations,* and *data structures*, to name only the most popular terms.

[8] Some philosophers, e.g., Cummins (1989), further distinguish between the identity conditions of representations (i.e., what facts make something a representation—namely, having *some* content) and the identity conditions of content and/or information (i.e., what facts determine the *specific* content of, or the information carried by, that entity). Fodor (1987) answers the first question in functional-computational terms and the second in causal-based terms; see also Ramsey (2016). I will not get into this distinction. Instead, when I talk about semantic properties, I refer to representational or informational *content*.

[9] See Cummins (1989) and Ramsey (2007).

[10] David Marr (1982), e.g., writes that "the apocryphal grandmother cell" (p. 15) is "a cell that fires only when one's grandmother comes into view" (p. 15n.); a comprehensive treatment of the causal view (augmented with a teleological component) is provided, e.g., by Dretske (1981, 1988).

[11] Some philosophers, including Cummins (1989) and Ramsey (2007), argue that computation forces us to identify representation in only one way—usually some form of functional-based account.

output representations, but the mapping involves no internal ("hidden-layer") representations. Finally, some computations, while operating on internal complex propositional structures, do not have mental content.

Some philosophers think that the semantic properties involved in computation have a normative aspect (Cummins 1989; Ramsey 2007). Normativity implies that the representation may be right or wrong, and that there is a possibility of misrepresentation. My representation of *cow*, R, might be tokened by a horse under some darkish conditions. This does not necessarily mean that R is a representation of *cow-or-horse*. R is still a representation of *cow*; in the described case, R misrepresents the horse as a cow.[12] The normativity of semantic properties is usually associated with a certain function, in the sense of a goal or purpose. In the computers we design, such as desktops, the purpose or function of representation is derived from us, the designers or users. In some natural systems, however, the function may evolve from an adaptive process, such as evolution (Millikan 1984) or learning (Dretske 1988). My view is that the representational or informational content in computing systems is *always* normative—in the sense that there is an issue of correctness in their application. I will not argue for this claim explicitly, although I do think that it follows from the argument in favor of the semantic view that will be presented in Chapter 8.

To sum up, by *semantic* we refer to the informational or representational content of computation (states, systems, processes, etc.), which means that these states come with directional reference (*aboutness*). We do not require there to be a single account of the content of all computing systems: there might well be different kinds of content, as well as different determinants of content (*pluralism*). The relevant representation or information might not be mental, propositional, or compositional—nor even internal (*minimalism*). According to some philosophers, content always involves the possibility of misrepresentation (*normativity*).

7.1.5 Non-Semantic Accounts of Semantic Properties

Before moving on, I should say something about the so-called non-semantic accounts of *semantic properties* (as opposed to non-semantic views of *computation*). How should we treat these accounts with respect to the semantic view

---

[12] Dretske (1988) distinguishes between semantic properties that have a normative aspect ("representation") and semantic properties that do not ("information"). Hence, according to Dretske, there is no misinformation: R carries the information that there is a horse in front of me, even under the darkish conditions in which I misrepresent this horse as a cow. Other scholars do not accept this distinction and introduce instances of misinformation (Piccinini and Scarantino 2011).

of computation? If these accounts abolish semantics, do they not nullify the semantic view of computation? I will distinguish between three sorts of accounts.

Naturalistic theories. Naturalistic, or reductive, accounts aim to recharacterize semantic properties (such as content) in non-intentional and non-semantic terms. They aspire to reduce content to some other more natural property; Fodor famously captures this goal in the slogan "*If aboutness is real, it must really be something else*" (1987: 97). Many philosophers have developed naturalistic accounts of content.[13] Some identify content with functional role (Block 1986), or with isomorphism-based structures (Cummins 1989; Ramsey 2007). Others have turned to causal-based accounts (Fodor 1987; 1990), and some have augmented it with an adaptive (learning or evolution) component (Dretske 1988; Millikan 1994). These accounts are often taken to compete with each other with respect to mental content. But, as previously noted, when it comes to the context of computation, different accounts might describe different kinds of computing systems (Floridi 2011; Piccinini and Scarantino 2011).

At first blush, these theories of content, if successful, appear to blur the distinction between semantic and non-semantic accounts of computation, since the features that make the semantic accounts semantic are really not semantic (they are "something else"). However, I do not think that the semantic view is under any real pressure. First, the semantic view of computation, though consistent with a naturalistic approach to content, is not committed to naturalism (and, given pluralism about content, it is certainly not committed to one specific account of content). The semantic view is amenable to the possibility that the content of at least some computing systems cannot be naturalized at all. Second, the idea that computational content can be naturalized is rather hypothetical. It is very doubtful that we can naturalize the (apparently derivative) content in the systems that we design. It is also far from certain that mental content can be naturalized. None of the competing theories of mental content has provided a fully satisfactory account of mental content. So, even if computational content can be naturalized—something that is very much in doubt[14]—we are not there yet.

Third, and most importantly, the debate between semantic and non-semantic views of computation is about the nature of computation, not about the nature of content. The debate is about the features that play a role in the individuation of computation. The semantic view says that the individuation of computation always takes content into account; a view that is not semantic (e.g., non-semantic views) denies this. Whether content itself can be identified in non-semantic terms is an important but separate issue. To compare: Consider the debate on

---

[13] In the context of information, these accounts are known as *semantic theories*, as they target the informational content or meaning of information.

[14] Thus, Sprevak notes that "many contemporary philosophers suspect that representation simply cannot be naturalized" (2013: 547).

whether or not computation is sensitive to content. As far as I can tell, this issue is orthogonal to the debate between naturalists and their opponents. Computation would be sensitive to content (or not) regardless of whether content can be naturalized. The same applies to computational individuation. Content would (or would not) affect computational individuation regardless of whether content can be naturalized. Thus, the semantic view is viable, irrespective of whether content can be naturalized.[15]

Formal theories. Formal accounts characterize in formal (i.e., logical, mathematical, or statistical) terms the ways that semantic properties are communicated, are transformed, relate to each other, are composed, and so forth. The most well-known theory of this kind is Shannon's theory of communication, which introduced a "non-semantic" notion of information (Shannon 1948; Wiener 1948; Shannon and Weaver 1949). It does not appear to recharacterize information or informational content in non-semantic terms—rather, it takes informational content as given: "Frequently the messages have meaning; that is they refer to or are correlated according to some system with certain physical or conceptual entities. These semantic aspects of communication are irrelevant to the engineering problem" (Shannon 1948: 379). The engineering problem concerns the channels of information; more specifically, its aim is to account for the amount of information that can be transmitted in those channels under various circumstances.[16] There are other formal theories that belong to this sort of non-semantic account. Algorithmic information theory, for example, concerns the amount of information ("complexity") encoded in string or other data structures.[17] Operational semantics provides axiomatic systems to prove certain properties of computer programs, such as correctness and validity.[18] Denotational semantics constructs mathematical objects ("denotations") that represent the operations of computer programs.[19] And there are others.

All these theories aim to give a formal treatment of aspects that are related to semantic properties. There are two ways to interpret these theories. In one interpretation, these theories aim to formally characterize certain non-semantic (e.g., syntactic) aspects of the structures that carry informational or representational content. This interpretation may be more faithful to the first two theories of information. Apparently for this reason, these theories are said to introduce a non-semantic notion of information. In another interpretation, the mathematical

---

[15] It is nevertheless true that if one is a naturalist about computation (or even just mental computation), then one's motivation to adopt the semantic view would depend on whether or not we successfully naturalize mental content.

[16] Fresco (2014: 135–136); Piccinini (2015: 226–229).

[17] Algorithmic information theory was developed independently by Kolmogorov (1965); Chaitin (1977); and others.

[18] See Plotkin (2004).

[19] See Scott and Strachey (1971).

theory provides the meaning of the computer programs. Such an interpretation might be more faithful to denotational semantics and other theories of formal semantics.

In any event, the formal theories do not appear to undermine the viability of the semantic view of computation. In the first interpretation, they simply characterize some non-semantic features, leaving the semantic properties intact. In the second interpretation, they provide a certain non-semantic (e.g., formal) account of semantic theories, and as such fall under the category of naturalistic accounts.[20] Under neither interpretation do the formal theories deny the existence of semantic properties or their role in computational individuation. Thus, formal theories and the semantic view of computation can happily coexist.

Eliminativist theories. Some accounts take an eliminative strategy: they deny that the defining semantic properties "are real" (as Fodor puts it), in the sense that they really exist or conform to real kinds. Quine (1960) famously denied that meanings and intentional states actually exist. The Churchlands deny that beliefs and desires, with their intentional content, are real (Paul Churchland 1981; Patricia Churchland 1986). Stich (1983) is an eliminativist with respect to intentional content, arguing that intentional states, such as beliefs and desires, should be defined in syntactic terms. Ramsey (2007) argues that there are no real non-classical (e.g., neuroscientific) representations. Such eliminativist theories have certain affinities with instrumentalist theories that are similarly not committed to the existence of semantic properties. Dennett (1987), for example, is famous for taking this stance with respect to intentional states and content. The main difference is that instrumentalists still think that semantic properties can be useful for explanatory or predictive purposes, whereas eliminativists tend to deny even that.

The eliminativist theories, if successful, pose a threat to the semantic theories of computation. If semantic properties are not real, then semantic properties, it seems, cannot play an individuative role in computational theories. But I see no real threat here to the semantic view. First, the nullification of the semantic view depends on the success of the eliminativist theories, which are very much in dispute. Second, the specific theories that are on the market do not eliminate every semantic property, only some of them. Quine and the Churchlands eliminate intentional content, but offer alternatives. Quine talks about *stimulus meaning*. The Churchlands opt for neuroscientific representational theories that posit computational-neural states whose content is defined in isomorphism-based functional terms (Churchland 2007). Ramsey is an eliminativist with respect to

---

[20] A similar point is made about Tarski's *theory of truth*, which can be interpreted as a formal semantic characterization of the semantic notion of truth, or as a "non-semantic" theory about how the bearers of truth (i.e., sentences) relate to each other with respect to preserving truth (see Sher 1991, 1996).

non-classical theories, but opts for classical theories that posit computational-representational states, whose content is also defined in isomorphism-based functional terms (Ramsey 2007). These claims are in accord with the semantic view. The proponent of the semantic view can say that in the domains where there are no real semantic properties, there is no real computation; computation occurs only in the domains where there are real semantic properties. Yet another option is to adopt a computational (instrumentalist) stance—namely, that computation occurs only in domains where there are useful, albeit not real, semantic properties. This computational stance is, as far as I can tell, also in accord with the semantic view of computation.

To recap, none of the reviewed theories appears to put real pressure on the semantic view of computation. Naturalistic theories aim to reduce semantic properties to non-semantic ones. The semantic view is not committed to naturalism about content, but is consistent with it. The semantic view is the claim that semantic properties play a part in the individuation of computational states, whether or not they are naturalized. Formal theories can be interpreted as accounting for some of the non-semantic properties of computation (and perhaps other processes). As such, they are perfectly consistent with the semantic view. They can also be interpreted as formal theories of content, and, as such, can be considered a species of naturalistic theories. Eliminativist theories are very controversial, and usually do not target every kind of content. They are therefore consistent with the semantic view that wherever we have real content, that content affects computational individuation.

7.1.6 What the Semantic View Is Not

It is important to distinguish the semantic view from other, closely related approaches. First, the semantic view of computation is consistent with a non-semantic view of *implementation*. A non-semantic view of implementation asserts that the relation of implementing formalism (e.g., an automaton) by a physical system does not involve semantic properties. A semantic view of computation is consistent with this assertion. Of course, if you think that computation *is* nothing but the implementation of a formalism, then you cannot hold a semantic view of computation *and* a non-semantic view of implementation at the same time. But the semantic view of computation is not committed to the identification of computation with implementation. Proponents of the semantic view of computation are free to maintain that the implementation of formalisms is non-semantic, but that counting the implementing physical system as computing essentially involves semantic properties. I return to this issue later, in my reply to Objection 1.

The semantic view can also be distinguished from the view that *computational descriptions* (such as those in theories and explanations) make explicit reference to semantic properties. Proponents of the semantic view are free to maintain that computational descriptions are themselves formulated in formal (e.g., mathematical) terms and do not make explicit reference to semantic properties. The semantic view is the claim that considering the described system as computing essentially involves semantic properties. I discuss this issue in greater detail later, in my reply to Objection 2.

The semantic view of computation is distinct from the view that computing processes are insensitive to semantic properties.[21] One widespread view is that the computing processes in my laptop operate on symbols or bits—and yet these processes are completely "blind" (as opposed to "sensitive") to how we interpret the symbols, that is, to their semantic properties. In that regard, computing processes are not sensitive to semantic properties. There are those who invoke this blindness when objecting to the semantic view, on the assumption that the claims about individuation and sensitivity are closely related. I shall address this claim in my reply to Objection 9. For now, suffice it to say that the semantic view is a claim about the sort of properties that matter to the individuation of computation, rather than about the properties to which computation is sensitive.

Finally, the semantic view differs from externalism about computation—the claim that the individuation of computation essentially takes into account features that are external to (that is, located outside) the computing system. Both of these approaches are concerned with individuation. However, externalism is not committed to the claim that the external features are semantic, whereas the semantic view is neutral about the semantic internalism/externalism debate.[22]

7.1.7 The Gist of My Account

Since I will gradually develop my own account of computation in Chapters 8 and 9, it seems advisable to highlight its distinctive features at this point. My account differs from other semantic accounts (e.g., Churchland and Sejnowski 1992; Ladyman 2009; Sprevak 2010) in one or more of the following aspects. First, I distinguish between implementation, which I take to be a non-semantic relation (see Chapter 5), and computation, which I take to be semantic. Moreover,

---

[21] See also Piccinini (2008a); Sprevak (2010); and Rescorla (2012).
[22] Authors who argue for externalism, without committing (or even objecting) to the semantic view, include Bontly (1998); Horowitz (2007); Piccinini (2008, 2015); and Shea (2013). We can also distinguish externalism about computation from computational externalism. Computational (or wide) externalism is a claim about the location of the vehicles of computation (e.g., Wilson 1994), whereas externalism about computation is a claim about what individuates computational states, regardless of where they are located.

I take it that while a physical system typically implements at any time more than one formalism, only one of these formalisms typically serves to identify the computational structure (or vehicle) of the system in a given context. Take, for example, a device that outputs three physical properties, L, M, and H (the detailed examples are provided in Chapter 8). Ignoring other properties, we can group these properties in different ways: {L,M,H}, {L+M,H}, {L,M+H}, and so on. Each grouping might be mapped to ("implement") a different formalism. Typically, however, only one of them is relevant to the computational structure of the system in a given context.

Second, my view is semantic in that the content of physical states or properties determines their computational individuation. If, for example, the physical properties L and M have $CONTENT_1$ and H has $CONTENT_2$, then the relevant grouping, for the purposes of computational individuation, is {L+M,H}. We will say that the device has two computational, "abstract" properties: $COMP_1$ and $COMP_2$. $COMP_1$ is associated with the physical properties L+M and with $CONTENT_1$. $COMP_2$ is associated with the physical property H and with $CONTENT_2$. One can say (correctly) that there is no computational difference between {L+M,H} and {L,M+H}, as the latter also leads to two computational types. However, in Chapter 8, we will see that when also considering inputs and internal states, these two groupings can yield very different computational structures.

Third, I do not think that every change in content alters the individuation of computational structure. Assuming that the physical properties L and M have $CONTENT_3$ and H has $CONTENT_4$, the relevant grouping, for the purposes of computational individuation, is still {L+M,H}. If, however, L has $CONTENT_1$ and M and H have $CONTENT_2$, then the relevant groupings, for the purposes of computational individuation, will change to {L,M+H}. In Chapter 8, I will show that these alterations in groupings can also result in different computational structures. The important point, however, is that in *all* these cases, the sameness and differences of the content of physical properties play a role in the formation of computational types.

Fourth, I take it that computational descriptions, explanations, and theories are all formal in that, as a general rule, they do not explicitly mention the contents of computational states, but only their medium-independent properties. Nevertheless, these descriptions (etc.) are computational, rather than merely mathematical, only if they refer to medium-independent structures that were grouped (individuated) via their contents (as just discussed). These medium-independent structures—the ones that are grouped (individuated) via their contents—are the computational structures or vehicles of the system. This will be further clarified in Chapter 8.

Last, I take it that another important element of computation is a modeling component, which I discuss in Chapter 9. This component helps to exclude representational systems that are non-computing, and is also key to understanding the distinctive features of computational explanations.

7.1.8 Supporting the Semantic View

There are several arguments in favor of the semantic view.[23] Two of them are more central than the others. The *standard argument* (mentioned at the outset of the chapter) is that semantic properties are enormously helpful in distinguishing computing from non-computing systems.[24] The standard argument supports the C-semantic view. It goes like this: The semantic view helps to satisfy *the-right-things-compute* part of the classification criterion (Premise 1). This is because the paradigm cases of computing systems carry informational or representational content—they are information-processing systems. The paradigm cases include cognitive, neural, and perhaps other natural systems, as well as artificial computing systems (artifacts) such as chess machines, air traffic controllers, word processors, and smartphones and laptops more generally. No less importantly, the semantic view helps to satisfy *the-wrong-things-don't-compute* part of the classification criterion (Premise 2). This is so because semantic properties exclude many systems that do not carry informational or representational content, such as stomachs, hurricanes, solar systems, rocks, and many others. Lastly, advocates of the semantic view point out that the non-semantic accounts face serious difficulties in meeting the classification criterion, as I have pointed out in the previous chapters (Premise 3). If all these three premises are correct, then the semantic view has an edge on its non-semantic counterparts (conclusion).

The opponents of the semantic view might challenge each of those three premises. Regarding the first premise, they might argue that there are still many representational systems that do not compute. But this objection does not immediately undermine the semantic view. The semantic view asserts that semantic properties are *necessary* for the individuation of computation. It is not committed to the claim that semantic properties are *sufficient* for the individuation of computation. It would be nice, of course, to see a semantic account that excludes the alleged representational systems (I return to this task in Chapter 9).

---

[23] See Sprevak (2010), who lists some of them.
[24] A version of this argument is put forward, e.g., by Crane, who concludes: "What distinguishes systems that are merely describable as computing functions (such as the solar system) from systems that genuinely do compute functions (such as an adding machine) is that the latter contain and process representations—no computation without representation" (2016: 154).

One could also say that the semantic condition is empty—because every physical system is, in some sense, representational. I discuss this contention in the reply to Objection 3. Many have challenged the second premise on the grounds that there are computations without representations, which are wrongly excluded in the semantic view. I address this challenge in my replies to Objections 1 and 2. Finally, one can still debate the third premise, on the grounds that an adequate non-semantic account of computation can be, or has already been, found.

The *master argument* for the semantic view aims to show that semantic properties essentially affect the classification of physical systems into computational types. Allegedly, it shows that semantic properties play a role in determining whether any two physical systems are computationally the same or different. This argument directly supports the E-semantic view, as it shows that semantics matters when it comes to computational equivalence. Given that there is little reason, if any, to embrace the E-semantic view and reject the C-semantic view, the master argument also supports the C-semantic view. Chapter 8 is devoted to the master argument. Together, the standard argument and the master argument provide a solid support for both versions of the semantic view.

Supporting the semantic view also requires removing some powerful objections to it. In the following section, I will address what I take to be the most pressing objections.

7.2 Objections to the Semantic View

In this section, I will reply to nine objections to the semantic view. These objections deserve more discussion than provided here; my aim is to say in brief how these objections can be addressed by proponents of the semantic view.[25]

Objection 1: There Are Computations Without Representations
The most common objection to the semantic view is raised in examples of computations that involve no semantic properties whatsoever. Rescorla relies on Block's *vending automaton* (Block 1978; Godfrey-Smith 2009) as an example:

> As a counter-example to the semantic view, consider a simple, finite-state vending machine discussed by Godfrey-Smith (2009). The machine has two inputs (I_1 = 5 cents, I_2 = 10 cents), three outputs (O_1 = null, O_2 = Coke, O_3 = Coke & 5 cents), and three internal states S_1, S_2, and S_3, governed by the transition table [not presented here]. Call this machine "VEND." The implementation condition for VEND does not seem to involve meaning, representational

---

[25] Some of the objections receive a more detailed treatment by Sprevak (2010).

content, or "aboutness." A physical system can implement VEND even if its states lack any semantic interpretation. Of course, one might impose representational talk upon the system. For instance, one might say that a system entering into state S_2 thereby "represents" that five cents more are required for a Coke. At best, such representational attributions reflect a Dennettian "stance" towards the system (Dennett [1987]), not a genuine constraint the system must satisfy to implement VEND. Nothing about VEND itself seems to require that we attribute representational import to states S_1, S_2, and S_3. Nothing about VEND's transition table assigns any essential role to semantics, representation, or content. (2013: 684)

In another paper, Rescorla (2014) says the same thing about an automaton dubbed ELEV, which, when implemented, can be used to operate elevators.

Reply: Rescorla talks here about the semantic view of *computational implementation*, thus identifying computation with implementation. As I noted earlier, the semantic view of computation is not committed to the semantic view of computational implementation. It is in fact consistent with a non-semantic view of implementation. I actually agree with Rescorla that the implementation conditions of VEND (and ELEV) do not involve semantic properties such as "meaning, representational content, or 'aboutness.'" As I noted in Chapter 5, I agree that the notion of implementing an automaton (and a formalism more generally) is non-semantic.[26] Implementation, however, is different from, and an insufficient condition of, computation, as demonstrated by Rescorla's example. As noted in Chapter 5, virtually every physical system—rocks, hurricanes, stomachs, and ventilators—implements some automaton of that sort, perhaps even more than one, and yet we do not treat them as computing systems. In fact, we do not deem old-style vending machines to compute, even though they implement VEND: they emit cans and coins upon receiving the correct amount of money, but they compute nothing. The same goes for elevators that implement ELEV. If we treat all these systems as computing, the notion of computing becomes useless and trivial, adding nothing to the notion of a physical process. Computing starts when we "*impose* representational talk upon the system." But as long as we do not impose semantic properties on the vending machine (and I agree with Rescorla that we don't), the machine, though it implements VEND, is not computing.

---

[26] Subscribing to an NSNNS view, Rescorla thinks that in some cases the implementation conditions involve semantics. I think that implementation is never semantic.

Objection 2: Computer Science and Its Branches Individuate Computation Non-Semantically

Many suggest that we should examine how computer science treats computational individuation. When we do, we see that computation is individuated non-semantically. Piccinini, for example, writes that "many readers, especially those familiar with computer science and computability theory, will readily agree that in those disciplines, computational states are individuated by their formal or syntactic properties" (2008a: 208), and that "the whole mathematical theory of computation can be formulated without assigning any interpretation to the strings of symbols being computed" (2008a: 212). Thus, in their chapter on Turing machines, Lewis and Papadimitriou (1981) talk about a function *f* from symbolic configurations (strings) to symbolic configurations regardless of any interpretation, and then say that a Turing machine, *M*, *computes* this function (p. 175ff.).

Reply: A proponent of the semantic view might note that some branches of computer science do actually involve semantics, and to a significant degree (Turner 2013). Others would suggest that we put aside computer science in this matter, as it deals with mathematical objects, not physical ones (Sprevak 2010). My strategy is a bit different. I actually agree that many theories of computing systems describe only formal (e.g., syntactic) properties, and not semantic ones. However, my reply is that it does not follow that computational individuation is non-semantic.[27] But I will make two comments in advance.

First, I agree that there are formal theories that individuate the states of computing systems non-semantically. In fact, in my reply to Objection 1, I even insisted that a theory of implementation selects the implemented automaton without appealing to semantic properties. The fact that some theories (whether or not they are referred to as computational) focus on the non-semantic properties of computation implies nothing about whether or not the individuation of computation is semantic. Think of the formal theories of information (such as Shannon's information theory) discussed earlier. These can be interpreted as analyzing certain non-semantic properties of information processing. As such, they individuate informational states without appealing to the informational content of the states. But we cannot conclude from this that we can individuate informational states, *qua* informational, without appealing to informational content. Similarly, a proponent of the semantic view might concede that, at least occasionally, automata theory describes and classifies Turing machines without appealing to semantics by focusing on the strings of symbols, regardless of their interpretations.[28] This is because the Turing machines have very interesting

---

[27] See also Crane (1990).
[28] However, there are other instances where we individuate the strings by appealing to their representing numbers (Boolos and Jeffrey 1989).

properties, such as halting, that are not related to the interpretation of the strings. This, however, does not imply (without further argument) that computational states, *qua* computational, are individuated non-semantically.

Second, I also agree that *computational* theories provide formal descriptions of computing systems. They will, for example, describe the formalism that a physical system implements. Yet it does not follow that this formal structure is individuated non-semantically. The aim of the *master argument* for the semantic view (see Chapter 8) is to illustrate just this: it shows that semantic properties can determine which of the formal properties implemented by the system are selected by *computational* theories and explanations. Thus, the semantic view is at least consistent with the formal nature of computational theories.

A semanticist can thus reply to the objection as follows: Computational theories apply to systems that have, or at least can have, semantic properties (and as long as they don't have semantic properties, they don't compute). The talk about the computation of non-semantic functions is just a derivative of the semantic talk about computation. Thus, to return to Lewis and Papadimitriou, their talk about the computation of string-theoretic function is a derivative of their discussion of "Turing-computable functions from natural numbers to natural numbers" (1981: 177), where they introduce an interpretation function from symbols to numbers. It is true that computational theories often provide formal descriptions of computing systems without explicitly mentioning semantic properties. It is also true that there are formal theories (still under the heading of computer science) that study computing systems regardless of their semantic properties; they focus on the non-semantic properties of computing systems. But, as just mentioned, all this is consistent with the semantic view of computation.

Objection 3: The Semantic View Is Consistent with Limited Pancomputationalism

The semantic view is consistent with the claim that every physical system—including stomachs, hurricanes, and rocks—computes. This is certainly true if one assumes that every physical system carries information.[29] But even without making this assumption, the semantic view is arguably consistent with pancomputationalism—simply because such systems, according to a certain interpretation, are representational systems. For example, I can assign certain content (such as numbers) to the states of the stomach. Under this assignment, the stomach transforms representations of numbers into representations of numbers. The same goes for virtually every physical system. Nothing stops me

---

[29] See Piccinini (2015, 2017), who points out that this assumption, together with others, leads to pancomputationalism.

from assigning content to the states of hurricanes, rocks, and chairs. Thus, the semantic view is not helpful in excluding non-computing systems. It is consistent with pancomputationalism, with all of its pitfalls.

Reply: I do not think that every physical system carries information (at least, not in the manner that we characterized computational content in Section 7.1.4). But let us assume, for the sake of argument, that it does. Let us even assume, for the moment, a very liberal version of the semantic view, whereby carrying information is a sufficient criterion for computation. In that case, the proponent of the semantic view would indeed embrace limited pancomputationalism. But limited pancomputationalism does not imply that the semantic view has no role in excluding non-computing processes. As Chalmers has noted, one can subscribe to limited pancomputationalism and still deny that digestion is computation: the fact that the stomach computes does not mean that its digestion is an instance of computation (see Chapter 5). Following this reasoning, the proponent of the semantic view can say that the stomach carries information (and therefore computes), but that digestion does not proceed *in virtue of* information-processing, and therefore it does not qualify as computation. The same goes for other non-computing processes.

As we have just noted, however, the more reasonable stance is that stomachs, hurricanes, and rocks represent nothing. Their states convey no content whatsoever. In that case, according to the semantic view, these systems do not compute. Thus, the semantic view is actually *at odds* with limited pancomputationalism. What is true is that the semantic view is consistent with *very limited pancomputationalism*. This even weaker thesis states that every physical system *could*, under certain circumstances, be a computing system. If we assign content, such as numbers, to the states of the stomach, we could perhaps use it to compute the solution of certain specific equations. Of course, we would still require that the other conditions of computation be met for the stomach to be deemed computing; if those conditions were met, then the stomach would compute.

I see nothing wrong with this result, however. On the one hand, the semantic view correctly classifies stomachs, hurricanes, rocks, and many other non-representing systems as non-computing. On the other hand, it does not rule out such systems as computing when we turn them into representational systems. Whether or not these systems really do compute when they are turned into representational systems depends on other conditions of computation (as previously noted, the semantic view does not rule out additional, non-semantic conditions of computation): if those conditions are met, then such systems will be considered computing, and if they are not, then they will not be (and whether or not these conditions are met has little to do with the semantic view).

Objection 4: The Semantic View Is Inconsistent with the Objectivity of Computation

Computation is objective: "That my laptop is performing a computation seems to be an objective fact (as opposed to a fact that depends on how an observer chooses to interpret my laptop)" (Piccinini 2015: 34). However, at least some of the contents of computation appear to be interpreted; that is, they are derived from an observer, designer, or user. Thus, the semantic view is inconsistent with the objectivity of computation, and is therefore false.

Reply: As noted in Chapter 1, we can understand the objectivity requirement in (at least) two different ways. On one understanding, objectivity is contrasted with "free interpretations"—for example, that the observer can view the system as implementing every formalism (triviality). But as noted in Chapter 5, the semantic view is in accord with the denial of triviality. In fact, the semantic view is consistent with the claim that there are thick constraints (to use Coelho Mollo's term) on the assignment of computational descriptions to physical systems.

On another understanding, objectivity is contrasted with observer-dependence. About this I argued that there is no reason to adopt too strong an objectivity constraint, as *partial objectivity* can suffice. Partial objectivity is the conjunction of two claims: the claim that every computational property of some computing physical system (such as brains) is objective (PO1), and the claim that some computational properties of every computing physical system (such as laptops) are objective (PO2). The semantic view is consistent with PO2. Laptops might have other, non-semantic computational features that are objective. Implementation, for example, is entirely objective, in my view. Thus, whether a system is a universal computer or simply implements a given formal structure is objective. Other features of computers—such as their components and organization—are objective as well. The semantic view is also consistent with PO1. It may well be the case that the contents of other computing systems, such as brains, are objective (and naturalized). If this is indeed the case—that the content of cognitive and neural computational states is objective—then we have an important subclass of computations that are entirely objective (i.e., not observer-dependent).

I also note that, on Piccinini's mechanistic account, the teleological function that turns the laptop into a computer *is* dependent on a designer or user (as far as it contributes to the designer's or user's goals). But Piccinini apparently distinguishes between the designer/user and the observer. This distinction is made more clearly by Coelho Mollo (forthcoming), who says that computing systems are not objective if they are observer-dependent—namely, if they depend "on explanatory perspectives that observers take toward physical systems" (this view is characterized as "perspectivalism"). However, Coelho Mollo also argues that computing systems can be both mind-dependent and objective. For

Coelho Mollo, the teleological function of laptops and other computing systems is dependent on their designers and/or users and yet objective, as an external observer has no choice but to assign computation (or not) to the physical system. I will leave aside the distinction between designers/users and observers,[30] and note that the semantic view is consistent with the claim that the contents of the laptop's states are derived from, or determined by, the designers/users and not by external observers. Thus, laptops might also be observer-independent (in this sense), according to the semantic view.

Objection 5: The Semantic View Is Inconsistent with the Naturalistic Project of the Mind

There are various versions of this objection. In one version, the semantic view of computation is inconsistent with the claim that there are theories of content (and/or mentality more generally) that are both naturalistic and computational. A theory is *naturalistic* if all its explanans are non-mental and non-semantic properties or terms; it is *computational* if some of its explanans are computational properties or terms. But if computational properties are semantic, even in part, they cannot play a role in a naturalistic theory. Here is one formulation of this objection:

> One problem with naturalistic theories of content that appeal to computational properties of mechanisms is that, when conjoined with the semantic view of computational individuation, they become circular. For such theories explain content (at least in part) in terms of computation, and according to the semantic view, computational states are individuated (at least in part) by contents. (Piccinini 2008a: 222)

The circularity to which Piccinini refers implies that if naturalistic theories provide explanations in terms of the semantic properties that define the pertinent computation, then these theories are not truly naturalistic; hence, naturalistic-computational theories are inconsistent with the semantic view of computation.

A somewhat stronger version of the argument states, in addition to the inconsistency claim, that computational theories in the cognitive and brain sciences aim to provide a naturalistic account of mentality. This is supposed to give us more reason to adopt the claim that there are adequate theories of the mental that are both naturalistic and computational—and therefore to reject the semantic view. Here is a formulation of that argument:

---

[30] But see Hemmo and Shenker (2019), who argue, to the contrary, that computing systems can be both observer-dependent and objective (the observers being *measuring devices*).

The ultimate aim of cognitive science is to offer, not just any explanation of mental phenomena, but a naturalistic explanation of the mind. The objective is to explain how a system can be mental in terms that do not already presuppose mental life.... To a first approximation, cognitive science's strategy for achieving this goal is to explain mental life in terms of computations implemented by the brain. If this strategy is to work, then explanation in terms of implementation of computation had better be explanation in non-mental terms. The alternative would be incompatible with the naturalistic project. It would mean that, rather than explaining mental life in non-mental terms—in terms of computations implemented by the brain—cognitive science would ultimately be explaining mental life in terms of, inter alia, other mental properties. If it turns out that computational implementation itself needs to be explained in terms of mental properties like our beliefs, interests, attitudes, then the naturalistic aim of cognitive science—explaining mental life in non-mental terms via the notion of computation—is doomed to failure. (Sprevak 2012: 11)[31]

A third version of the argument states very explicitly that there are already adequate theories of mentality that are both naturalistic and computational. This claim is put forward by proponents of the philosophical theories that state that the mind is partly or entirely computational, and hence is explained by computational properties (these theories are the computational theory of mind, computationalism, and computational functionalism).[32] Add this to the inconsistency claim, and the conclusion is that the semantic view of computation is false.

Reply: It appears that the proponents of the semantic view are forced to deny that there are, or even can be, theories of content and mentality that are both naturalistic and computational. But this is not exactly the case: a theory can be both naturalistic and computational, according to the semanticist, provided that the theory *also naturalizes* the semantic properties of the relevant computational states. What you cannot have is a full-fledged naturalistic theory of mentality that does not further naturalize the semantic properties of computational states. The mistake, according to the semanticist, is to assume that the appeal to computational properties buys you naturalism for free. If your naturalistic theory appeals to computational properties, then it must also include an account of these computational properties in non-semantic and non-mental terms.[33] Otherwise, as Piccinini notes, you get vicious circularity.

---

[31] Sprevak himself is not committed to this argument. He presents the naturalistic constraint as a desideratum of Chalmers's account.
[32] See Piccinini (2009) and Rescorla (2015) for critical surveys of these theories.
[33] This point is neatly put by Crane:

> The notion of computation depends on the notion of representation.... The aim, then, is to explain representation: we must have a reductive theory of representation if we are going to vindicate our computational theory of cognition in accordance with the naturalistic assumptions. (2016: 154)

Now, it is true that the proponents of some naturalistic philosophical theories of content and mentality (such as the computational theory of mind) assume that computational properties need no further naturalization. They also often assume that these *philosophical* theories somehow reflect the practices and methodological aims of the *empirical* computational theories of cognitive and brain sciences. But I think that these assumptions are mistaken. Many scientists—both in the classical (e.g., Pylyshyn 1984) and non-classical (e.g., Churchland, Koch, and Sejnowski 1990) camps—tend to characterize computation semantically. It is therefore very doubtful that computational theories in the cognitive and brain sciences aim to account for content and mentality in naturalistic, non-mental, and non-semantic terms. I do not deny, of course, that computational theories in the cognitive and brain sciences account for mental and cognitive capacities. I will suggest, however (in Chapter 9), that these theories provide explanations by individuating input and output states at least partly in semantic terms.

Objection 6: Ascribing Content to Computational States Presupposes a Non-Semantic Individuation of These States

Piccinini (2004b) argues that any theory that ascribes mental content to the computational states of the mind/brain is committed to non-semantic individuation of computational states. The reason is that these theories are arguably inapplicable to ordinary computers (e.g., laptops), and so fail to show what computing minds and computing laptops have in common. Thus, these theories of content, when conjoined with the semantic view, "find themselves in a position from which they cannot tell what minds have in common with ordinary computing mechanisms." The only way to tell is to adopt "a non-semantic way to individuate computing mechanisms and their states" (p. 399).

Reply: The semantic view actually shows that the computational states of minds and laptops have a common feature: they are all individuated with reference to their content. The semantic view is not committed, however, to the claim that the content of computational states must be individuated by the same standards. As said previously, it is actually more reasonable to adopt a pluralistic approach to computational content. Informational and/or teleological theories might apply to the computational content of the mind/brain, whereas other theories might apply to the computational content of laptops. Thus, the semantic view is consistent with the claim that computing systems are all individuated by content. In fact, an advantage of the semantic view is that it can account for an important difference between computing minds and computing laptops: they are different kinds of computing systems because their content is individuated by different standards. Thus, without a further argument that shows why the semantic view is committed to a single kind of content, the conclusion of the objection—that ascribing computational content presupposes a non-semantic view—does not follow.

Objection 7: A Change in Content Does Not Change Computational Identity

One familiar intuition is that changing content does not alter the computational identity of the system. In my calculator, the digits 0 and 1 represent the numbers zero and one. But assigning the contents of *apple* to 0 and *orange* to 1 would not change computational identity. More generally, we would still say that two computing systems that implement the same formal (e.g., syntactic) structure are computationally equivalent. This demonstrates that content does not affect computational individuation—and hence that the semantic view is false.

Reply: Some would insist that a change in content comes with a change in computational identity. I do not. I agree that a change of content often does not change computational identity. In particular, changing the content of 0 and 1 in the example just given would not alter computational identity. But this does not falsify the semantic view. As I will show in Chapter 8, other changes in content might lead to a change in computational identity; some changes in content alter the formal/syntactic structure (vehicle) that carries the content. These cases give us reasons to reject the non-semantic view (according to which content *never* alters computational identity). As I will also explain in Chapter 8, these cases, in which changes in content alter computational identity, indicate that content *always* affects computational individuation.

Objection 8: The Semantic View Does Not Allow for Important Environment-Independent Generalizations of Computational Theories

Egan (1994) observes that if computational theories of cognition are environment-dependent, then some important environment-independent generalizations are lost. If we want to retain these generalizations, we must abandon the semantic view. Egan assumes (correctly) that if computational generalizations are affected by broad content, then they are environment-dependent.

Reply: This objection has certain affinities to Objection 7. Both aim to highlight some intolerable consequences of the claim that computational identity (or theories) is altered when the content (or environment) is changed. My reply is therefore similar to the one I gave to Objection 7. I agree with Egan that we must retain environment-independent computational generalizations, but I believe that retaining them is fully in line with the semantic view. The semantic view need not assert that an environmental (or content) change is accompanied by a change in computational identity. The semantic view thus does not imply that computational theories cannot generalize across different environments (or different contents). The semantic view only implies that computational theories cannot generalize across certain changes in the environment (or in content). Indeed, in such instances, it is actually good that we avoid the generalization—or so I will argue in Chapter 8.

Objection 9: Computational Processes Are Not Sensitive to Semantic Properties

Computational individuation takes into account causally relevant properties. Semantic properties, however, are causally irrelevant to the computational process—and therefore we can conclude that semantic properties are not taken into account by computational individuation. Why are semantic properties causally irrelevant? Because computational processes are "formal": even if they operate on representations, they are "sensitive" to formal/syntactic properties of the representations, not to their semantic properties.[34] My pocket calculator is sensitive to the shapes (as it were) of the 0s and 1s, not to their interpretations: we could change the interpretation, and the computational process would be just the same. In fact, the processes would be exactly the same without any interpretation at all (or indeed, without representation). This demonstrates that semantic properties do no causal work, and therefore are causally irrelevant to the computation.

Reply: One possible reply to this objection is to insist that computation is sensitive to semantic properties.[35] Another is to keep sensitivity and individuation apart: computational processes are not sensitive to semantic properties, and yet computational individuation can take semantic properties into account. These are simply different issues. Non-sensitivity is a claim about the causal dynamics of computational processes—namely, that the dynamics is not dependent on semantic properties. The semantic view is a claim about *individuation*—namely, that semantic properties play an essential role in the classification of token processes, states, and the like into computational types. As others have already noted, sensitivity need not affect individuation (Piccinini 2008a; Sprevak 2010;

---

[34] Fodor (using the term "apply to") famously writes: "I take it that computational processes are both *symbolic* and *formal*. They are symbolic because they are defined over representations, and they are formal because they apply to representations in virtue of (roughly) the *syntax* of the representations.... What makes syntactic operations a species of formal operations is that being syntactic is a way of *not* being semantic" (1980: 64).

Following Fodor, Egan says that computational operations "are sensitive only to *formal* (i.e. *non-semantic*) properties of the representations over which they are defined, not to their content" (2010: 254).

[35] See Block (1990); Peacocke (1994); O'Brien and Opie (2006); Figdor (2009); and Burge (2010: 95–98). Rescorla (2012) argues that computation is mostly insensitive to semantics but aims to show how it could be.

Rescorla 2012). The claim that it must affect individuation rests on a notion of causal relevance that raises a host of philosophical problems and is notoriously hard to explicate (Sprevak 2010).

Of these two approaches, I favor the latter: I think that computation is insensitive to semantic properties, but that this does not undermine the semantic view. However, I would like to propose a somewhat different explanation of why this is so. As previously noted, those who believe in non-sensitivity maintain that changing the content of computational states, S_1, S_2, etc., of the system does not affect the causal dynamics. They therefore conclude that changing the content of these states would not affect computational individuation: we would still identify the computational states of the system with S_1, S_2, etc. I agree with this assertion. But this reasoning assumes that we ascribe content to the *very same* states S_1, S_2, etc. In the next chapter, I will show that there are other cases in which ascribing different content would result in a different grouping of physical states into computational states. Under this ascription, the computational states would no longer be S_1, S_2, etc.—and therefore the computational identity of the system would change as well. These cases show that change in content alters computational individuation, even though computational operations are not sensitive to this content.

7.3 Summary

The first part of this chapter focused on the explication of the semantic view of computation (Section 7.1): what it states (Section 7.1.1), how it contrasts with its non-semantic counterparts (Section 7.1.2), and two of its variants (Section 7.1.3). I then sought to clarify what *semantics* means in computational contexts (Section 7.1.4) and examined the impact of non-semantic theories of content on the semantic view (Section 7.1.5). Finally, I distinguished the semantic view from other, closely related views (Section 7.1.6), highlighted the distinctive features of my account (Section 7.1.7), and briefly reviewed the main arguments in its favor (Section 7.1.8). The second part of the chapter focused on nine objections to the semantic view, and how—when more refined distinctions are made—the semantic view can overcome these objections.

8
An Argument for the Semantic View

In this chapter, I develop and defend an argument for the semantic individuation of computational states. Shagrir (2001, 2012a), Sprevak (2010), Rescorla (2013), and others have introduced versions of this argument. Its first and central premise is the simultaneous implementation of automata by physical systems. In the following sections, I describe the phenomenon of *simultaneous implementation* (Section 8.1), followed by the argument for the semantic view (Section 8.2), and a reply to two types of objections that have been raised in the literature (Sections 8.3 and 8.4).

8.1 Simultaneous Implementation

Simultaneous implementation is the phenomenon whereby a physical system implements multiple formal structures at the same time, at the same location, and even with the very same physical properties.[1] The formal structures in question are automata, but it is apparent that this phenomenon also extends to other formalisms. Many describe automata in terms of *total states* (e.g., Chalmers 1996). I describe them in terms of *gates* (or "neural cells"), which form the basis of real digital computing. The two descriptions are equivalent.[2]

Consider a physical system P, which is a tri-stable flip detector. P works as follows: It emits 7–10 volts if it receives voltages greater than 7 from each of the two input channels, 0–3 volts if it receives under 3 volts from each input channel, and 4–6 volts otherwise. I will use the symbols H, M, and L (high, medium, and low) to signify these three different *physical* properties (7–10 volts, 4–6 volts, and 0–3 volts)—which implies that the argument does not depend on these specific physical properties. The behavior of P is summarized in Table 8.1.

We next classify an input/output as L+M if it receives/emits under 6 volts (in the range of either 0–3 volts or 4–6 volts) and H if it receives/emits 7–10 volts; we assign the digit 0 to the emission/reception of L+M and 1 to the emission/reception

---

[1] Fresco, Copeland, and Wolf (forthcoming) offer a systematic study of this phenomenon; they call it the "indeterminacy of computation."

[2] See Minsky (1967), who invokes both descriptions (he presents gates as McCulloch and Pitts "cells") and demonstrates the equivalence relations between them (the proof is on pp. 55–58).

Table 8.1 Physical gate P: The gate maps voltages from two input channels, Input 1 and Input 2, into one output channel.

Input 1	Input 2	Output
7–10 V (H)	7–10 V (H)	7–10 V (H)
7–10 V (H)	4–6 V (M)	4–6 V (M)
7–10 V (H)	0–3 V (L)	4–6 V (M)
4–6 V (M)	7–10 V (H)	4–6 V (M)
4–6 V (M)	4–6 V (M)	4–6 V (M)
4–6 V (M)	0–3 V (L)	4–6 V (M)
0–3 V (L)	7–10 V (H)	4–6 V (M)
0–3 V (L)	4–6 V (M)	4–6 V (M)
0–3 V (L)	0–3 V (L)	0–3 V (L)

Table 8.2 P implements *AND*: Under the assignment of 1 to H and 0 to L+M, P implements *AND*.

Input 1	Input 2	Output
1	1	1
1	0	0
0	1	0
0	0	0

of H. Under this scheme, P implements an *AND-gate* (Table 8.2). When we classify inputs and outputs based on emission/reception of L and of M+H, and assign the symbol 0 to the emission/reception of L and 1 to the emission/reception of M+H, P implements an *OR-gate* (Table 8.3). As a result, the same physical system P simultaneously implements two distinct logic gates.

Let me contrast simultaneous implementation with Putnam's and Searle's *triviality results*. Roughly speaking, Putnam and Searle assert that almost every physical system implements every automaton (Chapter 5). As we noted earlier, Putnam and Searle are accused of adopting an excessively liberal notion

Table 8.3 P implements *OR*: Under the assignment of 1 to M+H and 0 to L, P implements *OR*.

Input 1	Input 2	Output
1	1	1
1	0	1
0	1	1
0	0	0

of implementation. Subsequently, it has been suggested that if we place causal, modal, or grouping constraints on the notion of *implementation*, we avoid such triviality results.

Simultaneous implementation is the much weaker claim that some physical systems simultaneously implement more than one automaton. The main difference from Putnam and Searle's constructions is that the proposed constructions are based on the very same physical properties of P—namely, its voltages. Another difference is that in the cases provided by Putnam and Searle, there is an arbitrary mapping from different physical states to the same states of the automaton, whereas in our case, the *same* physical properties of P are associated with the same inputs/outputs of the gate. The different implementations result from associating different voltages (across implementations, not within implementations) with the same inputs/outputs. Within implementations, however, relative to the initial assignment, each logical gate reflects the causal structure of P. In this respect, we have a standard implementation of logical gates, which satisfies the conditions set by Chalmers and others for implementation. It is easy to see that implementation also accords with the suggestions to group together physical states that are spatially located, physically similar, and "natural." Indeed, the proposal appears to accord with the standard ways in which we implement bits of 0s and 1s in physical systems.

There is nothing special about P; we could create other dual gates in a similar fashion. Consider, for example, a physical system Q that emits 7–10 volts (H) if it receives voltages higher than 7 volts (H) from exactly one input channel, 0–3 volts (L) if it receives under 3 volts (L) from each input channel, and 4–6 volts (M) otherwise (Table 8.4). Under the assignment of 0 to emission/reception of L+M and 1 to emission/reception of H, Q implements an *XOR-gate* (Table 8.5). Assigning the digit 0 to emission/reception of L and 1 to emission/reception of M+H, Q implements an *OR-gate* (Table 8.6).

Table 8.4 The physical gate **Q**.

Input 1	Input 2	Output
H	H	M
H	M	H
H	L	H
M	H	H
M	M	M
M	L	M
L	H	H
L	M	M
L	L	L

Table 8.5 **Q** implements *XOR*: Under the assignment of 1 to H and 0 to L+M, **Q** implements *XOR*.

Input 1	Input 2	Output
1	1	0
1	0	1
0	1	1
0	0	0

Table 8.6 **Q** implements *OR*: Under the assignment of 1 to M+H and 0 to L, **Q** implements *OR*.

Input 1	Input 2	Output
1	1	1
1	0	1
0	1	1
0	0	0

Table 8.7 Physical gate R. Under the assignment of 1 to H and 0 to L+M, R implements *AND*. Under the assignment of 1 to M+H and 0 to L, R implements *AND* (again).

Input 1	Input 2	Output
H	H	H
H	M	M
H	L	L
M	H	M
M	M	M
M	L	L
L	H	L
L	M	L
L	L	L

Another example is of a physical system R that emits 7–10 volts (H) if it receives voltages higher than 7 from each input channel, 0–3 volts (L) if it receives under 3 volts from at least one input channel, and 4–6 volts (M) otherwise (Table 8.7).

Grouping the inputs and outputs based on emission/reception of L+M and of H, and assigning 0 to emission/reception of L+M and 1 to emission/reception of H, R implements an *AND-gate*. Grouping the inputs and outputs around emission/reception of L and M+H, and assigning 0 to emission/reception of L and 1 to emission/reception of M+H, R implements an *AND-gate* again! This R system is interesting, in that the two implemented *AND-gates* may be in different "total states" under the same physical conditions. If, for example, the inputs are M from both input channels (making the output M), then this very same *physical run* simultaneously implements the (0,0) → 0 mapping under the first implemented *AND-gate*, but the (1,1) → 1 mapping under the second implemented *AND-gate*.

Importantly, I do not claim that each simple physical system implements every logical gate—nor, indeed, that they implement every complex combinatorial state automaton. Nevertheless, one can use these *physical* gates (i.e., P, Q, R, and the like) as the building blocks of physical systems that simultaneously implement more complex automata. As an illustration, let us construct Ned Block's

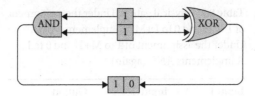

Figure 8.1 The automaton S$_{ADD}$ for computing (two-digit) addition (from Block 1990).

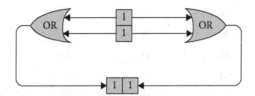

Figure 8.2 The automaton S$_{OR-OR}$ that consists of two *OR* gates.

device for addition (Block 1990), which consists of the two (syntactic) gates *AND* and *XOR* (Figure 8.1). Giving the strings of 1s and 0s a binary (semantic) interpretation, we can use this device to compute the addition of two-digit numbers. Using our physical gates **P** and **Q** as our building blocks, the very same device also implements, simultaneously, a very different automaton—namely, one that consists of two *OR* gates (Figure 8.2).

The more general point is that should you want to implement a zillion-gate automaton, you could use these and other tri-stable gates to implement another automaton of "the same degree"—where "the same degree" means something like "the same number of logical gates." Automata that include finite and infinite memory (as in Turing machines) are no obstacle to this result: we can individuate the 0s and 1s in the memory, just as we individuate the inputs and outputs. Thus, "the same degree" also means the "the same amount of memory." Note that I do not claim that this ability is shared by every physical system. We can assume that physical systems often implement only one automaton of the same degree. It might also be the case that, technologically speaking, it would be ineffective and very costly to use these tri-stable gates. The point is rather philosophical: Given some combinatorial state automaton (CSA), we can construct a physical system that simultaneously implements this CSA and another CSA of the same degree. These constructions are not just a logical or metaphysical possibility. They are nomological possibilities; in all likelihood, they are also technologically feasible.[3]

---

[3] One might think that the different Boolean function might ultimately combine into logically equivalent functions. Fresco, Copeland, and Wolf (forthcoming) prove, however, that the likelihood

8.2 The Master Argument: From Simultaneous Implementation to the Semantic Individuation of Computational States

What follows from simultaneous implementation? Not much, in my view, for the notion of implementation. In particular, I do not think that simultaneous implementation gives us a reason to adopt a semantic conception of implementation. I believe that a theory of implementation (such as in Chalmers 2011) can and does tolerate simultaneous implementation—or at least I do not assume otherwise. I do think, however, that this seemingly innocuous claim about simultaneous implementation has far-reaching consequences for the theories of computation and cognition. Elsewhere (Shagrir 2012b), I use simultaneous implementation to support a premise in an argument against the computational sufficiency thesis.[4] Here, I use it as a premise for an argument for the semantic individuation of computational states. I have advanced an earlier version of this argument in the past (Shagrir 2001). In this section I shall present a more refined version of the argument, and then address objections that have been raised to the argument (Sections 8.3 and 8.4).[5]

The argument has the following form:

(1) A physical system might simultaneously implement different automata $S_1, S_2, S_3, \ldots$

(2) A computational taxonomy of a physical system might take into account one automaton, Si, in one context, and another automaton, Sj, in another context.

(3) There must be a constraint that determines which automaton is taken into account by the computational taxonomy in a given context (from 1 and 2).

(4) This constraint is (at least in part) a semantic feature—namely, the contents of the states of the physical system.

Therefore: A computational taxonomy (individuation) of a physical system takes into account semantic features, that is, the contents of the system's states.

---

of this scenario is close to nil for a physical system that implements a function on five or more dual gates like this.

[4] Maudlin (1989) and Bishop (2009) advance arguments toward similar conclusions, while also assuming a premise weaker than strong triviality. Hemmo and Shenker (2019) argue that simultaneous implementation undermines the validity of Landauer's principle in physics.

[5] Note that the conclusion—of semantic individuation of computational states—does not automatically undermine CST. Supporters of CST can modify the thesis to state that implementing the right kind of automaton (as opposed to a computational structure) is sufficient for possessing a mind. Given that implementation is non-semantic, CST might still be the basis of a non-semantic theory of cognition.

Some comments: First, I assume that implementing a formal structure of some kind is necessary for computing. This might be an algorithm, a graph, a network, or even formal dynamics. For simplicity, my focus here is on automata; thus, the presumption in Premise 2 is that a computational taxonomy takes into account at least one automaton. The contentious claim of Premise 2 is that a computational taxonomy might not take into account all implemented automata. Typically, the taxonomy would take into account only one automaton in a given context—but I do not deny that there might be instances where it would take into account more than one. Second, I do *not* assume that implementing a formal structure such as an automaton is a semantic relationship; in fact, I maintain that the implementation is a non-semantic relationship between a physical system and an automaton. My claim is that computation is more than (non-semantic) implementation.[6] Third, the argument is in favor of the *equivalence* (and seemingly stronger) version of the semantic view (the *E-semantic view*)—namely, that semantic properties essentially affect the process of individuation of systems, processes, states, and so forth into computational *types*.

Lastly, the argument may be seen as an assertion about *computational vehicles* (Rescorla 2015; Shea 2018)—that the identity conditions of computational vehicles inextricably involve semantic properties. This does not mean that we cannot individuate vehicles in non-semantic terms. A vehicle is an implemented formal structure (e.g., the automaton $S_{OR\text{-}OR}$), and, as such, it can be individuated non-semantically. The claim is that when we treat this implemented automaton as a *computational vehicle* (*structure*), semantics sneaks in. Semantics gets into the picture when we classify the implementing physical states into computational types.

Let us now turn to the argument. The first step, as already noted, is about simultaneous implementation. The second step is the claim that computational identity may vary across contexts:

(2) A computational taxonomy of a physical system might take into account one automaton, Si, in one context and another automaton, Sj, in another context.

Take the physical machine M, which consists of the pair of physical gates P and Q. M simultaneously implements the automata S_{ADD} (Figure 8.1) and $S_{OR\text{-}OR}$ (Figure 8.2). The claim is that a computational taxonomy would take into

---

[6] See also the discussion in Chapter 7. Note that I have no reason to argue *against* a semantic view of implementation: a semantic view of implementation would immediately entail a semantic view of computation, which is my conclusion anyway. My aim is to show that one can argue in favor of a semantic view of computation even assuming a non-semantic view of implementation.

account S_{ADD} in one context (e.g., when M is performing addition), but might take into account S_{OR-OR} in another context, when M performs some other task.[7]

To illustrate this point more vividly, let us compare M to another physical system, M*, which implements S_{ADD} but not S_{OR-OR}. Let us say that I use M and you use M* to compute addition. In this case, we surely want the two systems M and M* to be regarded as computationally equivalent, regardless of whether or not M* also happens to implement S_{OR-OR}. Assume further that at some point M will no longer be used to compute addition, but rather will be deployed for another task associated with S_{OR-OR}. We would still count the two physical systems as computationally equivalent with respect to performing addition—but we would no longer say that the machines are computationally equivalent with respect to their current (different) tasks. With respect to their current tasks, a computational taxonomy of M would take into account S_{OR-OR} instead of S_{ADD}.

Alternatively, consider a sensor that implements the automaton S_1, which is used for some visual task. We would agree that implementing S_1 affects the computational individuation of our sensor. Assume that it turns out that this sensor simultaneously implements another automaton S_2. Assume, further, that this sensor, which simultaneously implements S_1 and S_2, is removed to another environment, where it serves an auditory task.[8] However, there it transpires that the method used for this auditory task is no longer S_1, but S_2. I believe that it would be quite reasonable to say that in the new auditory context, a computational taxonomy would take into account S_2 and not S_1. To see this more clearly, consider another sensor that uses S_2 for the same auditory task, yet does not simultaneously implement S_1. A computational taxonomy would still count the two sensors as computationally equivalent, since they both implement S_2 with respect to the auditory task—even though only one of them also implements S_1.

The upshot is that computational taxonomies might take into account different automata in different contexts. The computational structure—the implemented automaton taken into account by a computational taxonomy—of a physical system is sensitive to the task that the system performs in a given context.

It follows from (1) and (2) that

(3) There must be a constraint that determines which automaton is taken into account by the computational taxonomy.

If a physical system might implement more than a single automaton (as per Premise 1), and if a computational taxonomy might take into account different

---

[7] A *task* can be thought of as some capacity or function that the system performs. For the sake of simplicity, we can assume that the task can be described as some mapping input-output function.

[8] This example is a variant on the *visex/audex* thought experiment (Davies 1991).

automata in different contexts (as per Premise 2), there must be an additional constraint that determines which of the implemented automata is relevant to the computational identity of the system. If, for example, the machine M simultaneously implements both S_{ADD} and S_{OR-OR}, and assuming that a computational taxonomy counts only S_{ADD} when M is used for performing addition, there must be another factor that determines that the computational structure of M is S_{ADD} and not S_{OR-OR}.

However, one may wonder if such a determining constraint is really needed. By way of comparison, a physical taxonomy might classify a piece of wood and a piece of metal under the same kind of *temperature* (assuming they have the same temperature). But such classification does not require a further factor ("constraint") that singles out temperature from the other physical properties of the systems where they might differ (i.e., wood versus metal): the physical property *temperature* is not dependent on factors other than the temperature itself. Why, then, should computational taxonomy be any different? Why must we insist that a computational type is defined by a further factor, above and beyond the implementation of an automaton?

I agree that classifying a piece of wood and a piece of metal under the same kind of temperature does not call for a further constraint, above and beyond their temperature. A physical taxonomy has no choice but to classify the piece of metal and the piece of wood under the same kind of temperature (assuming they have the same temperature). Similarly, the classification of two systems M and M* under the same *implementational* type S_{ADD} (etc.) requires no further features, other than implementing this automaton. An implementational taxonomy has no choice but to classify M and M* under the same implementational type, S_{ADD}. A *computational* taxonomy, however, need not do so—even if they implement the same automaton S_{ADD}: rather, it might classify M as a S_{OR-OR} type and M* as a S_{ADD} type. This shows that the fact that M implements S_{ADD} is not sufficient to define a *computational* type. There must be a further factor involved in computational individuation in addition to the implemented automaton.

What could this further constraint be? What would account for the fact that computational identity is conferred by one automaton rather than another? It is perhaps safe to say that intrinsic implementational properties—physical, biological, neural, and others—cannot play this determinative role.[9] If a physical system simultaneously implements several different automata—if its intrinsic physical (etc.) properties simultaneously implement different automata—then these intrinsic properties (such as voltages) alone cannot serve to account for the system's computational identity. They cannot tell one automaton from another.

---

[9] We can think about "intrinsic" properties as non-relational properties or as properties that a system has by virtue of itself.

What about extrinsic properties? I will defer the discussion of non-semantic extrinsic properties to the discussion of Objection 2 (Section 8.4). For now, I will contend that:

(4) The relevant determining constraint is (at least in part) a semantic feature, namely, the contents of the states of the physical system.

Consider our physical system P. We recall that when assigning 0 to the emission/reception of L+M and 1 to the emission/reception of H, P implements *AND*. However, when assigning the symbol 0 to the emission/reception of L and 1 to the emission/reception of M+H, P implements *OR*. Which automaton is relevant to computational individuation? I argue that if the content (e.g., the interpretation) of L and M is (the number) zero and the content of H is (the number) one, then P falls under the computational kind *AND*. If the content of L is zero and the content of M and H is one, however, then P falls under the computational kind *OR*. This indicates that content matters to the computational structure of P.

Or take the sensor that simultaneously implements two automata, S_1 and S_2. Suppose that certain input/output variables for S_1 correlate with the physical properties L and M, and the other input/output variables for S_1 correlate with the physical property H. Suppose, further, that some input/output variables for S_2 correlate with L and the other input/output variables for S_2 correlate with M and H. I contend that S_1 would be preferred over S_2 if it turns out (say) that outputting L+M carries one content (such as apples) and outputting H carries another (such as oranges). I am therefore suggesting that the contents correlated with the physical properties L, M, and H at least partly determine which automaton is relevant for individuative purposes. A computational taxonomy would select the automaton, such as S_1, whose implementing physical states correlate with the content of those states.

Lastly, assume that our sensor—which simultaneously implements S_1 and S_2—is removed and embedded in a different environment, where the content of the physical properties L, M, and H fit not with S_1, but with S_2. In this environment, the content of L and H remains the same, but the content of M is now the same as the content of H (e.g., oranges instead of apples). In this scenario, I believe, we would say that the automaton that is relevant for computational individuation is no longer S_1, but S_2. It is quite reasonable to say that in the new environment, a computational taxonomy would take into account S_2, and not S_1. In other words, it would take into account the automaton that fits with the current content. This example shows that a change in the content of M can alter the computational structure of the system from S_1 to S_2. Thus, content affects computational identity.

One could argue that in all the examples just given, I have *assumed*, and not actually demonstrated, that the systems have content. If so (according to the complaint), then the argument for Premise 4 assumes that content affects computational identity—which is precisely what that argument is meant to show. My reply is that both the semanticist and the non-semanticist maintain that computing systems very often carry content. The debate is rather over whether this content affects computational identity (see Chapter 7). The master argument shows that the content possessed by systems has this effect. What is true is that I do not deal directly with contentless computing systems. So one might still say that the argument does not defy NSNNS (neither semantic nor non-semantic view—see Chapter 7). The argument is consistent with the view that content enters computational individuation when the system carries content (which is what the argument allegedly shows), but that content does not enter computational individuation when the system does not carry content.

In replying to this, I would make two points. First, as noted in Chapter 7, my overall defense of the semantic view consists of two arguments that somewhat complement each other. The standard argument, which supports the C-semantic view, is that computing systems always operate on semantic properties. The master argument, which supports the E-semantic view, is that these semantic properties affect the individuation of computational types. Second, the argument for Premise 4 does not only show how content can resolve the indeterminacy of computation, but also eliminates non-semantic properties as candidates for removing computational indeterminacy. I eliminated the intrinsic implementational properties earlier; in Section 8.4, I eliminate certain extrinsic non-semantic properties. Thus, it is up to the advocate of NSNNS to show what non-semantic properties resolve the indeterminacy when no content is involved.

Thus far, I have associated computational identity with the content of input and output representations (or the input-output semantic task). Often, this is enough. But there are cases where we have no choice but to appeal to the content of internal states. For example: a given physical system simultaneously implements two automata—but the inputs and outputs of both automata are correlated with the same input and output, L and H. In this case, the content of the input and output states cannot distinguish between the implemented automata. Instead, we must examine the content of *internal states*. For example, we can assume that a certain physical state, p, implements a state, P, of one automaton, but two substates of p, p_1 and p_2, implement the states P_1 and P_2 of the other automaton. In that case, we would favor the first automaton over the second if only one content is correlated with p. We would favor the second automaton over the first if two different contents were correlated with p: one content with p_1 and another with p_2.

I do not claim that every change in content alters computational identity (as noted in Chapter 7). Assume, for example, that, in our visual system, outputting L and outputting M have *green* content, and outputting H has *red* content. Changing the content of *these* physical types (i.e., of L, M, and H) to *apple* and *orange* content, respectively, would not alter the identity of the computing system: a computational taxonomy would still consider the same S_1 automaton. Still, I would insist that the content of the physical states/properties (e.g., voltages) *always* matters to computational identity, because content is always relevant to the way we group these states/properties together into computational types. The sameness and difference of the contents of L, M, and H (in a given context) is a crucial factor in individuating computational types. If L and M have one content and H another content, we will get one computational type. If L has one content and M and H another, we will get another computational type. Thus, altering the contents of L, M, and H, as in the case just given, would not result in a new grouping of computational types, as L and M would still have one content and H another. However, attributing *apple* content to the output of L and *orange* content to the output of M and H would alter computational identity. In this case, L would have one content and M and H another. A computational taxonomy would now take into account the S_2 automaton. Either way, content always drives the formation of physical states/properties into computational types.

My view lies somewhere between that of Burge (1986, 2010) and that of Egan (1995, 2010). Like Burge, I think that computational identity is content-sensitive and can vary across contexts; unlike him, I do not assume that every change in content makes a computational difference. Like Egan, I think that changing the content of the states of the same computational vehicle does not affect computational identity—but unlike her, I do not take this to show that computational identity is content- (or context-) independent. A change in content can alter the computational vehicle, and with it computational identity.[10] This is once again consistent with the semantic view: the fact that different content sometimes leads to different computational vehicles and sometimes to the same computational vehicles is in consonance with the semantic view, which asserts that content *always* determines the computational vehicle of the system. The content of physical states/properties (e.g., voltages) always determines how we group these states/properties into a computational vehicle.

---

[10] Egan (2014) notes that the computational vehicle carries the same *mathematical content* across context. Elsewhere, I also talk about sameness and differences of specific contents in terms of *formal content* (Shagrir 2001). My claim, however, is that the formal content (sameness and differences of contents) can vary across contents, and this will alter computational vehicle and identity.

To recapitulate: If (1) a physical system may implement more than one automaton, and (2) a computational taxonomy may take into account different automata in different contexts, then (3) there must be another constraint that determines which automaton is relevant to the computational identity of the system. And this constraint, I have argued, involves the content of the system's states (4). Thus, semantic features indeed impact the computational identity of physical systems.

Let us turn now to two objections to this argument. The first objection targets Premise 2. The mistake, it says, is in the assumption that a computational taxonomy takes into account one of the implemented automata—or even any automata at all. In fact, the objection goes, the computational structure of the system is identified with a more *basic* (and non-semantic) structure. The second objection targets Premise 4. It says that we need not appeal to semantic properties for the purposes of computational individuation. Extrinsic yet non-semantic features would do the job just as well. In the following two sections I will address these objections in detail.

8.3 Objection 1: Computational Individuation Is More Basic

Some scholars have argued that a physical system has a more basic (non-semantic) computational structure than that presented in the argument. According to this view, the mistake in the argument is in assuming that we must choose between the implemented automata—for example, between *AND* and *OR*. However, we do not need to make this choice, because the computational structure of the system is not identified by any of these functions/automata. This does not mean that implemented automata and logical functions are not interesting; they might be useful for all sorts of applications in computer science and engineering. But their individuation "is over and above computational individuation . . . Computational individuation is more basic, and non-semantic" (Coelho Mollo 2018: 3492).

What is this "more basic" structure? Several proposals are given in the literature. One is a *maximal automaton*: an automaton is maximal for a given system in the case that it is not implied by another, more complex automaton implemented by the system. For example, the maximal automaton implemented by the physical system **P** is the one associated with tri-stable gates (Table 8.1), and can be described by Table 8.8.

Now, it is a mathematical fact about automata that when a system implements some (maximal) automaton, it *ipso facto* simultaneously implements simpler automata. For example, by implementing the automaton with the tri-stable gates,

Table 8.8 The "maximal" automaton implemented by physical gate P. The labels A, B, and C stand for different equivalence classes of inputs and outputs. The maximal automaton implies *AND* under the assignments of 1 to A and 0s to B and C. It implies *OR* under the assignments of 1s to A and B and 0 to C.

Input 1	Input 2	Output
A	A	A
A	B	B
A	C	B
B	A	B
B	B	B
B	C	B
C	A	B
C	B	B
C	C	C

P also simultaneously implements (under some relabeling) the simpler automata *AND* and *OR*.[11]

Joe Dewhurst (2018a) has put forward another proposal. Taking the mechanistic viewpoint, he argues that computational identity is defined not by the logical functions (*AND*, *OR*, etc.), but by the *computing mechanism*. Take our physical system P. According to Dewhurst, the computational identity of the system is set out in Table 8.1, which provides a description of the computing mechanism—the components of the system, their functions, and their interactions. This description tells us that the system consists of two processor types and three digit types, and also tells us how these components interact with one another. This is everything we need to know for the purposes of computational individuation.

Dimitri Coelho Mollo (2018) offers a third suggestion. Like Dewhurst, he, too, adopts the mechanistic view of computation. But he correctly observes that Dewhurst's proposal, which links computational identity with specific implementational (structural) physical properties such as voltages, is

---

[11] Chalmers (1996) and Scheutz (2001) introduce and discuss such maximal automata in the context of simultaneous implementation.

untenable: the individuation conditions are too fine-grained. Consequently, systems whose implementational physical properties are different cannot be computationally equivalent.[12] Instead, Coelho Mollo (2018) suggests that we identify the computing mechanism with the *functional profile* of the system P. This profile attributes different implementational properties to the same equivalence classes, thereby allowing for some multiple realization.[13] The functional profile of P is given in Table 8.8, where the labels A, B, and C simply stand for three equivalent classes of implementational properties.[14]

Lastly, one might argue that the more "basic" structure is the set $\{S_1, S_2, \ldots\}$ of all implemented automata.[15]

Reply: Let us first clarify what is at stake here. Two suggestions have been put on the table in the face of simultaneous implementation. One is that computational individuation might take into account one of the implemented automata. The other is that computational individuation always takes into account some basic structure. This structure might be a maximal automaton, the mechanistic structure of the system, its functional profile, or the entire span of implemented automata. The debate is not so much about semantic versus non-semantic individuation as it is about less basic individuation (e.g., a non-maximal automaton) versus more basic individuation (e.g., a maximal automaton). Indeed, Piccinini—whose views are discussed in Section 8.4—thinks that computational individuation might take into account a non-basic structure (such as OR), but that the individuation is still non-semantic. Thus, my reply to the objection targets the basicness aspect rather than the non-semantic aspect of the second suggestion. I argue that the proposals grouped under the second, basic-structure suggestion all suffer from the same drawback: they do not do justice to, and even put at risk, the notion of computational equivalence, and, hence, of computational individuation.

Let us return to the sensor example, which simultaneously implements S_1 and S_2. Assume that you have another sensor that is somewhat physically different from mine: it implements S_1, but not S_2. We can assume that it implements S_1 in two different physical properties—L* and H*. Assume also that the contents of the two sensors align (as described previously) with S_1. In my sensor, one kind of content (such as oranges) is aligned with M+H, and the other kind of content is aligned with L (such as apples). In your sensor, one kind of content (such as

---

[12] See also Fresco and Miłkowski (2021).
[13] Like Dewhurst, Coelho Mollo (2018) thinks that the implementational details are part of computational explanations—but he distinguishes between functional and implementational levels.
[14] See also Schiller (2018) and Fresco and Miłkowski (2021), who also propose functional solutions along these lines. However, they consider only simpler cases of indeterminacy (e.g., Sprevak 2010).
[15] This has been suggested by Miłkowski (2013).

oranges) is aligned with H* and another one with L* (such as apples). There is little doubt, in my view, that a computational taxonomy would count the two sensors as computationally equivalent, because they both implement the same automaton, S_1, in its sensing task. Note that the individuation is not merely in terms of representational commonalities and differences. As previously noted, computational individuation does not require that the two sensors have precisely the same contents. Assume that your sensor, while implementing S_1, has other color-contents; we might even consider the semantic tasks to be different. We would still say that the two sensors are computationally equivalent, since they implement the same automaton, S_1, in their tasks.

Another way to make the point is this: Assume that we manufacture a set of sensors that fulfill their sensing goal by implementing S_1. I think that we would happily deem them to be computationally equivalent. Even if we were to discover one day that the manufactured devices were somewhat different and that some of them simultaneously implemented S_2, others S_3, and so on, the equivalence verdict would not change. Given that they all continued to sense by implementing S_1, we would still deem them to be computationally equivalent. The fact that some sensors implemented other automata would be ignored for the purposes of computational individuation. I think that this practice reflects the way we individuate computing systems in general. If so, then computational individuation operates at the level of the implemented automata, at least in some cases, and not at a more basic level.

Thus far, I have argued that computational individuation might take into account one of the implemented automata rather than a more basic structure. I now want to argue that the proposals grouped under the second (basic-structure) suggestion jeopardize the notion of computational equivalence—and, hence, that of computational individuation—because they imply that different physical systems might always turn out to be computationally distinct. All it takes is one cell (say) flipping at one more value. For example, if our flip detector turns out to differentiate between inputs of 0–1 volt and inputs of 1.5–3 volts, then we are dealing with a system with a different "basic" computational structure—which is computationally different from the system we started with. And, of course, there is no reason to limit computation to digital cases. In analog cases, we can carve up the values in many more ways, as we are no longer limited to the digital threshold values. If that is the case—and assuming that one takes seriously the notion of analog computation—we might find that different individual systems always belong to different computational types.

Coelho Mollo (2018) and Dewhurst (2018a) both discuss the possibility that different physical systems are inherently computationally distinct. Coelho Mollo denies that this possibility presents a problem for computational individuation:

In consequence, devices that differ in the number of stable states (e.g., two vs. three), as in Shagrir's (2001) version of the argument from the multiplicity of computations, are never computationally equivalent. . . . This, I take, is as it should be: given their different functional profiles, those two devices will differ in their capacity to carry out logical and mathematical functions—having a richer functional structure makes the tri-stable device considerably more versatile. (2018: 3494–3495 n. 20)

Importantly, Coelho Mollo thinks that the systems can be equivalent under some other, non-computational, scheme, as does Dewhurst. In particular, they both agree that the two systems might share *semantically individuated* logical functions (and other automata). Coelho Mollo writes that the "individuation by logical function . . . may well rely on wide functions or semantic properties" (2018: 3492). Dewhurst says that "both Shagrir and Sprevak are correct when they point out that the logical status of a gate is indeterminate prior to the attribution or identification of its representational content" (2018a: 107). But both Coelho Mollo and Dewhurst insist that this individuation scheme is not computational.

There is no dispute here that systems that differ in the number of stable states (e.g., two versus three) are not equivalent in some respects: they differ in the totality of automata they implement. I also agree that this difference is reflected "in their capacity to carry out logical and mathematical functions." In my view, however, this does not obligate us to say that the system must be computationally different. I shall explain my position through the notion of *computational explanation*.

Take the tri-stable device that has the capacity to carry out *AND* and *OR*. Assume, as before, that the device performs some sensing task, and that it exercises *AND* to perform this task. It is reasonable to say, I believe, that the fact that it exercises *AND* at least partly explains this sensing task. The explanation itself might be blind to the fact that the device is tri-stable; we would provide the same explanation for a bi-stable device that exercised *AND* to perform the sensing task.

Note that the explanation itself is a formal one, and does not mention specific content (such as apples and oranges). Rather, the explanation shows how performing the *AND* function supports the sensing task. Also, note that I do not insist that the sensing task is individuated semantically. As far as I am concerned at this point, the task can also be individuated non-semantically (for example, as per Piccinini's proposal discussed in Section 8.4). The issue here is whether we treat this *AND* explanation as a computational one, even though there is a more basic, functional, tri-stable structure.

One option is to treat the formal *AND* explanation as a computational explanation of the sensing task. This is the option that I favor—and in this regard, I believe that I am in the same camp as non-semanticists such as Egan and Piccinini, who treat such formal explanations as computational. But my guess is that Coelho Mollo would reject this approach. In his view, this *AND* explanation has nothing to do with computational individuation, which is more basic. Thus, unless he concedes that computational explanation is not related to computational individuation, Coelho Mollo would not treat the *AND* explanation as computational.

The other option, then, is to deny that the *AND* explanation is computational—on the grounds that computational explanations invoke the underlying functional profile rather than the logical functions. However, although certainly coherent, this view runs counter to computational explanations in the sciences. We maintain, for example, that computing the zero-crossing of second-derivative Laplacians explains the fact that the system performs edge detection (Marr 1982); that computing integration explains the fact that the system produces signals of eye position (Robinson 1989); and that implementing *AND-XOR* in Block's machine explains the fact that the system performs addition. (Again, all these explanations are formal ones, and must be distinguished from explanations that refer to the specific content of the states.) These are all considered good computational explanations whose adequacy is independent of whether the systems have more basic functional profiles. Denying that these explanations are computational might ultimately leave us with a non-semantic notion of computation—but one that is very limited in scope.

Dewhurst has a somewhat different take on the issue of computational equivalence:

> Taken to its logical extreme, this argument might imply that no two systems are computationally equivalent. In practice, the physical structure of two computing mechanisms is always going to be distinct, and it is unclear whether we can draw any non-arbitrary boundary between the structures that are relevant or irrelevant to computational individuation. This is a serious issue, and at this point I am unsure how a proponent of the mechanistic account ought best to respond. (2018a: 110)

He offers two lines of response to the challenge of computational equivalence. One is to bite the bullet, at the risk of *reductio ad absurdum*. I will assume that this is not a good option (why not adopt the semantic view instead?). A second line of response is to point out that this problem of equivalence is one that confronts the mechanistic account in general, and is not limited to computational mechanistic accounts—for no two physical systems are equivalent in all respects. In practice, however, when we set out to account for a certain phenomenon, we appeal to experts who can identify the system's *relevant* properties. They can tell which properties (e.g., temperature) are relevant for the individuation of these systems—and, hence, are relevant for determining whether or not they are equivalent with respect to this explanandum phenomenon. The same should be true for computing systems: while it might be the case that no two sensors are physically equivalent, we are still free to ask the experts which properties are relevant when accounting for these sensors. Confining ourselves to those relevant properties, we might find that our sensors are computationally equivalent—namely, that they have precisely the same relevant computational properties.

I agree that, when we explain a certain phenomenon, we take into account the properties that are relevant (e.g., causally) to the explanandum phenomenon. It is therefore highly plausible that two physical systems that are not physically equivalent with respect to the totality of their physical properties are still equivalent in terms of some subclass of their physical properties. A bar of metal and a bar of wood are different in terms of the totality of their physical properties, as they are made of different materials. Nonetheless, the theorist can treat them as physically equivalent if they have the same temperature and their temperature is the factor relevant to the explanandum phenomenon. But how do these observations extend to the analogous computational case? What are the relevant computational properties that are analogous to the temperature of the bars? While our physicist can select from the reservoir of physical properties those that are relevant to the explanandum phenomenon, we need to know the analogous reservoir of computational properties.

To illustrate, take two systems (such as sensors) with different basic computational structures. What would make them computationally equivalent? What is the reservoir of their potentially relevant computational properties? One option (advocated by Dewhurst and Coelho Mollo) is that the reservoir includes only one such property—namely, their basic (mechanistic/functional) structure. According to this option, the two systems (sensors) are not computationally equivalent, pure and simple: if they do not share the same basic structure, then they do not share computational properties at all. And if they do not share computational properties, they also do not share relevant computational properties. In other words, the appeal to relevance has no impact on computational equivalence: the two systems would be computationally distinct, irrespective of the explanandum phenomenon.

Another option is that the reservoir of computational properties includes more than one—and perhaps many—non-basic computational properties. This option seems to accord with the first (maximality) and last (entire-span) basic-structure proposals. Those proposals are consistent with the claim that the reservoir of the computational properties includes other non-maximal automata and logical functions in addition to the basic structure. Under this option, the two systems can be counted as computationally equivalent, since they implement the same non-maximal automaton (or logical function). Yet this is precisely my claim! To clarify, the objection was that computational individuation always takes into account a more basic structure. The second option implies, however, that computational individuation actually might not take into account the more basic structure. A computational individuation can take into account only the factor relevant to the explanandum phenomenon, and this factor might be one of the automata implemented by the system. But this claim—that a computational individuation can take into account only one of the automata implemented by the system—is precisely the second premise of the master argument. We might yet disagree on whether or not the factors affecting the selection of the automaton are semantic (although I point out that Dewhurst agrees that "the logical status of a gate is indeterminate prior to the attribution or identification of its representational content"). But this is a different issue; indeed, it is the subject of Objection 2, which we shall turn to momentarily.

In summary, Dewhurst's appeal to relevance does not resolve the problem of computational equivalence. Either it has no effect on computational equivalence (as in Dewhurst's and Coelho Mollo's proposals), or it supports my claim that computational individuation might take into account only one of the implemented automata—namely, the one that is relevant to the explanandum phenomenon (as in the maximality and entire-span proposals).

8.4 Objection 2: Externalism Without Content

Piccinini (2008a; 2015: 40–44) agrees that there is a further constraint that determines which automaton is the computational structure of a system in a given context. He also agrees that this constraint takes into account features that are external to the computing systems. He denies, however, that these features must be semantic. In other words, he adopts the master argument to establish externalism about computation—namely, the view that computational individuation essentially takes into account features that are external to the system (see also Horowitz 2007 and Shea 2013). But he refuses to accept the further step that these external features are semantic, arguing instead that

> provided that the interaction between a mechanism and its context plays a role in individuating its functional (including computational) properties, a (non-semantic) functional individuation of computational states is sufficient to determine which task is being performed by a system, and therefore which computation is explanatory in a context. (2015: 43)

More specifically, Piccinini argues that the master argument rests on the premise that tasks (i.e., the explanandum input-output function) are individuated semantically. But, according to him, the semantic individuation of tasks in these cases does not entail the semantic individuation of computational states. Tasks are *also* individuated functionally (non-semantically), and it is this functional task description that is relevant to the computational individuation of the systems. Piccinini says that the proponents of the semantic account give no reason to prefer a semantic individuation of tasks to his wide functional individuation. In fact, he says, we can single out a functional individuation of a task that determines computational structure. Hence, we need not appeal to the semantic task for the individuation of computational structure.

As previously noted, Piccinini concedes that we must take into account the functional task in which computation is embedded, and that this task may depend on the interaction between the computing mechanism and its context—for example, on "which external events cause certain internal events" (2008a: 220). Nonetheless, he argues that the environment need not be very wide. In artificial computing systems, the relevant functional properties might be input devices (such as a keyboard) and output devices (such as a display). In the cognitive case, it might be sensory receptors and muscle fibers. This might be enough to determine whether a computation is being performed—and, if so, which one. On this understanding, computational taxonomy selects the relevant automaton on the basis of how the implementing mechanism interacts with events that are not part of the computing mechanism. These external features define the functional task, and it is the functional task, rather than the semantic one, that is essential for computational individuation.

Reply: Piccinini has a point, of course. The argument for Premise 4 demonstrates how content can resolve indeterminacy—that is, how it can decide which implemented automaton is relevant to computational individuation. But if Piccinini shows that some wider functional factors (such as the factors that define the functional task) suffice to define computational structure, then the argument for Premise 4 fails. In that case, we can always individuate computational states without appealing to the content of the states.

My response has two parts. I will first argue that the functional task—or at least the one proposed by Piccinini—is insufficient to resolve the indeterminacy issue. I will then suggest more tentatively that computational taxonomy will often favor the semantic task over the functional task, even if the functional task removes the indeterminacies.

Let us consider more closely the functional account proposed by Piccinini. The idea is to look at the interactions of the inputs and outputs in the context of the surrounding external mechanism. Let us assume that our *AND/OR* toy device projects to arm movement. Assume also that the behavior of our system,

Table 8.9 Our toy example automaton interacts with arm movement. Outputting H results in movement; outputting L+M results in no movement.

Input 1	Input 2	Output	Motor Output
H	H	H	Movement
H	M	M	No movement
H	L	M	No movement
M	H	M	No movement
M	M	M	No movement
M	L	M	No movement
L	H	M	No movement
L	M	M	No movement
L	L	L	No movement

conjoined with the arm movement, is as given by Table 8.9: outputting L or M produces no movement, whereas outputting H produces movement. Thus, taking into account the interaction with movements gives us a reason to pick out the *AND* automaton—where one output is implemented in H, and another output in L and M. If that is correct, then we have a non-semantic way to individuate computation.

In response, let me first remark that if this solution works, it has little to do with the "interactions of the inputs and outputs with the surrounding external mechanism." The external, peripheral mechanism that connects the computing modules and the arm makes no difference to computational identity: what matters is the arm movement alone. This is readily apparent if we leave the outputs of the computing module and the arm movements unchanged, and only alter the mechanism that links the module's outputs and the arm movements. This alteration of mechanisms will not change the computational identity of the module. The same goes for sensory processes. Assume, for example, that we replace the mechanisms that transduce the light waves hitting the retina with other transducing mechanisms that yield the same light intensities. It is clear that the computational identity of the sensor remains the same, as long as what is being transduced is the same. Different external mechanisms would make a computational difference only if what they transduce is different.[16]

---

[16] This point is discussed at length in Harbecke and Shagrir (2019).

224   THE NATURE OF PHYSICAL COMPUTATION

The more pertinent reply, however, is that the computational individuation still depends on how you individuate the arm movement. Assume that further examination shows that the no-movement actually includes some movement (call it *medium movement*). Thus, overall, the H outputs are plugged to (physical) *large-movement*, the M outputs to (physical) *medium-movement*, and the L outputs to *no-movement* (Table 8.10).

In that case, we can identify *movement* either with large-movement, or with medium-movement-plus-large-movement. As Table 8.10 shows, if we choose the former option, we end up with the *AND* automaton; if we choose the latter, we end up with the *OR* automaton. How do we decide which functional kinds are relevant to singling out a computational structure? Appealing to *movement* is no longer helpful, since we do not want to identify movement with specific physical, or even geometrical, properties, for reasons of multiple realization. In other contexts, we can have the same functional task, even if larger movement is associated with different physical properties. Nor can we correlate the movement with the implemented automata or their outputs, as each implemented automaton is correlated with a different individuation of movement. Thus, in this case (and obviously in others as well), the appeal to external, non-semantic short-arm factors does not help in choosing between the implemented *AND* and *OR* automata.

One might suggest relating to even more external factors, such as the outside environment. We could say, for example, that large-movement results in reaching apples, whereas medium-movement does not. How should we treat

Table 8.10 Our toy example automaton interacts with arm movement. In this scenario, outputting H results in large movement, outputting M results in medium movement, and outputting L results in no movement.

Input 1	Input 2	Output	Motor Output
H	H	H	Large movement
H	M	M	Medium movement
H	L	M	Medium movement
M	H	M	Medium movement
M	M	M	Medium movement
M	L	M	Medium movement
L	H	M	Medium movement
L	M	M	Medium movement
L	L	L	No movement

this proposal? Importantly, I am not saying that there cannot be a long-arm wide functional task—one that extends all the way into the environment—that resolves all cases of indeterminacy. But before adopting this or any other account, we should be convinced that the account does indeed remove these indeterminacies. This is not a trivial task: the phenomenon of simultaneous implementation is a special case of a broader phenomenon—namely, that we can group the same physical states into different formal types in different ways. The construction is merely extended to include the wider functional facts. We performed this exercise when extending the functional typing to external outputs within the embedding system (movements), and we could further extend it to functional typing that includes environmental factors. It might turn out, for example, that medium-movement results in reaching small apples.[17] A functional account should show us how the wider context helps to avoid indeterminacies without referring to the content of the system's states—that is, without assuming that the contents of the states are (for example) apples and oranges (which would make it a semantic account). As far as I know, this has not yet been shown.

I have argued that the functional account fails to single out the (correct) implemented automaton, at least in some cases. In this situation, I currently see no reason to abandon the semantic account—which provides a simple and elegant solution to the issue of indeterminacy—in favor of a non-semantic one. I could stop here, but I would like to go further. I want to argue that, at least in some cases, we would favor the semantic proposal over the functional proposal *even if the functional proposal could always single out one implemented automaton*. I admit that the argument for this is brief and tentative, but I believe that it nevertheless has some merit.

The argument goes as follows: Piccinini presents a picture of semantic and functional tasks that compete over the impact on computational individuation, arguing that computational individuation acts in accordance with the functional task. How should we understand the relationship between these semantic and functional tasks? One possibility is that the two tasks are coextensive: the semantic and functional individuations always single out the same automaton. Under this understanding, we can always replace the semantic task with the more basic, functional task. This means that the functional, non-semantic task *naturalizes* the semantic task. But, as noted in Chapter 7, this naturalization claim can hardly challenge the semantic view. The semantic view is consistent with, but not committed to, the view that all computational contents can be naturalized. The debate between semantic and non-semantic views is over whether or not the identity conditions of computation must involve content. If the identity

---

[17] See also Hemmo and Shenker (2019), who argue that this externalist-functionalist proposal leads to an infinite regress.

conditions always involve content (naturalized or otherwise), then the semantic view wins. Otherwise, it loses. Thus, if the argument is that the semantic properties that essentially affect computational individuation can be further analyzed in non-semantic, functional terms, then, as far as I can tell, the semantic view of computation has the upper hand.[18]

A second, more reasonable possibility is that the two tasks are not coextensive—that the semantic and the functional tasks single out, at least in some cases, different automata. Under this understanding, Piccinini's claim is that a computational taxonomy would prefer the functional over the semantic individuation. I would like to challenge this claim. Take the tri-stable device that simultaneously implements *AND* and *OR*. Assume that outputting M+H encodes oranges, and outputting L encodes apples. In this case, the semantic content (such as the orange-content and apple-content) implies that the relevant computational structure is *OR*. As it also turns out, however, the functional task implies that the relevant computational structure is *AND* (Table 8.11). The outputs of our detector project to certain motor devices such as arm movements. Outputs of H produce movement, and outputs of L+M produce no movement.[19]

Which automaton counts for the computational identity? I think that we would agree that if we wish to explain the semantic task—that is, how the system categorizes the stimuli into apples and oranges—we would use the *OR* structure, not the *AND* structure. The *OR* helps to explain how the system categorizes certain stimuli as apples and other stimuli as oranges—whereas the *AND* is irrelevant to the explanation, as its states are not matched with the vehicles of apple-content and orange-content. Note that I do not deny that the *AND* structure might explain the wide non-semantic functional task. Nor do I deny that, in a different context with different content, the computational structure might be the *AND* rather than the *OR*. I also agree that we can individuate the *OR* mechanism in non-semantic terms. However, the point of the latter example is just this: if the explanandum is the semantic task and the content is as described in the example, then the automaton relevant to the explanation is the *OR*, not the *AND*.

Non-semanticists therefore face a dilemma. Taking the first horn, they can admit that the *OR* explanation is a *computational* explanation of the semantic task. But in that case, the semantic task, rather than the functional one, determines that the explanatorily relevant automaton is the *OR* automaton, not

---

[18] See also Dewhurst (2016), who notes that such a naturalistic move would be self-defeating for the non-semantic view, as it would remove "one of the primary motivations for giving a non-representational account of computation in the first place" (p. 797).

[19] If one is so inclined, one might further assume that movement results in reaching the object, and that no movement means not reaching it; one may also assume that our system has never produced

AN ARGUMENT FOR THE SEMANTIC VIEW   227

Table 8.11 A discrepancy between semantic and functional tasks. The semantic task is correlated with groupings of L and M+H. The functional task is correlated with groupings of L+M and H.

Input 1	Input 2	Output	Semantic Output	Motor Output
H	H	H	Orange content	Movement
H	M	M	Orange content	No movement
H	L	M	Orange content	No movement
M	H	M	Orange content	No movement
M	M	M	Orange content	No movement
M	L	M	Orange content	No movement
L	H	M	Orange content	No movement
L	M	M	Orange content	No movement
L	L	L	Apple content	No movement

the *AND*. Assuming that this determination also tells us something about computational individuation, we can conclude that content plays an essential role in computational individuation—namely, that it plays a role in determining that the computational structure of the system is *OR*.

Taking the second horn, non-semanticists can deny that the *OR* explanation—though a formal explanation—is a *computational* one. They can say, for instance, that computational explanations do not explain semantic tasks, but rather are formal explanations of non-semantic tasks. But, as previously noted, this view is not consistent with the explanatory powers that scientists attribute to computational theories. Scientists do maintain that computational-formal theories in cognitive neuroscience and computer vision explain semantic tasks such as edge detection, shape-from-shading, structure-from-motion, and so forth.

Piccinini appears to side with the former approach: he agrees that the explanandum of computational explanations is—at least in some cases—a semantic task. But he denies that this premise leads to the semantic individuation of computational states, arguing instead that computational individuation is affected by the functional task. However, my argument shows that, on the contrary, in cases of discrepancy a computational taxonomy would favor the semantic individuative scheme.

M values, so the conflict between the semantic task and the functional task has never been apparent. Our system would still have produced M values had it encountered certain oranges, and this output would have resulted in no movement.

The overall argument can be summed up as follows: We can understand Piccinini's claim as asserting that the semantic task is naturalized by the functional task. This claim, however, is consistent with the assertion that the semantic task affects computational individuation. We can also understand Piccinini's claim as asserting that the semantic and functional tasks compete in the sense that they lead, at least sometimes, to different individuations of the system. Assuming, with Piccinini, that computational explanations account for semantic tasks, my reply is that the computational individuation would be affected by the semantic task. Either way, the semantic task does affect computational individuation, at least in some cases.

8.5 Summary

This chapter has put forward an argument in favor of the semantic view that is based on the phenomenon known as *simultaneous implementation* (described in Section 8.1). Specifically, I suggested that simultaneous implementation implies (with further premises) that the events, states, and so on of a system are individuated into computational types—essentially, by semantic factors (Section 8.2). In the latter half of the chapter, I discussed and responded to objections to the argument (Sections 8.3 and 8.4).

9
Computing as Modeling

In this chapter, I will argue that modeling is highly central to computing. This will be done as follows: I will first explicate the notion of *modeling* (Section 9.1). I will then provide a modeling characterization of *physical computation* (Section 9.2). Next, I will discuss others who have associated computing and modeling (Section 9.3), and then highlight one methodological role of modeling, which is discovering the computed function (Section 9.4). Finally, I will discuss the central role of modeling in computational explanations (Section 9.5). The conclusion is that modeling is an essential element of physical computation, at least in current computational approaches in cognitive neuroscience.

9.1 What Is Modeling?

There is a wealth of literature about models and their role in the sciences.[1] I take here a *model* to be a representational system that preserves patterns of relations in the target (represented) system. By preserving patterns of relations, I mean that there is an isomorphism—or more realistically, something less than that— between the representing system and the target system. Another way to put this is to say that the model and the target domains are *structurally similar*, in the sense that they have the same formal or mathematical structure (Swoyer 1991). While many people would agree that some degree of structural similarity is necessary for modeling, they would also argue that the demand for a full-fledged isomorphism is excessive, at least when talking about tangible models. They therefore confine the requirement to homomorphism, partial isomorphism, or even weaker similarity or mapping relations.[2] Even these weaker morphism relations should be taken with a grain of salt. Given that we often talk about domains—both of models and their targets—that are physical and biological

---

[1] See, e.g., Weisberg (2013) and Frigg and Hartmann (2020).
[2] Less-than-isomorphism characterizations are apparent in partial isomorphism (e.g., French and Ladyman 1999), homomorphism (e.g., Bartels 2006), and similarity (e.g., Giere 2004).

systems, these similarity relations involve at least some degree of approximation and idealization.

A good example of a model is a family tree (Figure 9.1). In this tree, the lines, arrows, and double arrows preserve certain familial relations, such as a sibling relationship, parenthood, or marriage. This does not mean, of course, that the relations of the models are the same as the familial relations. Hopefully, being married and being related by a double arrow are dissimilar in many respects. The similarity is at a higher, structural level, of a mathematical or formal kind. In our example, being married and being related by a double arrow are both symmetrical relations.

It is crucial to distinguish between the notions of *computational model* and *simulation* on the one hand and *computing-as-modeling* on the other (see also discussion in Chapter 1). The former notions refer to the use of computers to model and simulate the behavior and processes of physical, biological, social, and other systems. They do not, however assume that modeling is an essential feature of computing; the claim is that computers can be, and are, used for modeling and simulating other phenomena. In contrast, the notion of *computing-as-modeling* is indifferent as to the actual use of computers for the purposes of modeling or simulating other phenomena; rather, the claim is that modeling is an essential element of the characterization of computing. Although the two notions are clearly related, my focus here is not on computer models and simulations, but on computing-as-modeling.

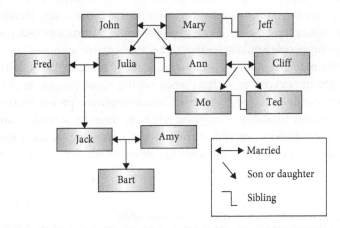

Figure 9.1 A family-tree model for determining familial links (From Ramsey, William M. 2007. *Representation Reconsidered*. Cambridge: Cambridge University Press. Reproduced with permission of Cambridge University Press through PLSclear).

9.1.1 Input-Output Mirroring

I propose that computing does not require a vast amount of modeling. It is satisfied with modeling of the input-output type, which is a minimal degree of morphism. Input-output modeling consists of two components: *input-output mirroring* of a given target, which is merely a morphism relation, and *representing*, which is that the inputs and outputs represent some entities in the target. In this section, I characterize the mirroring aspect.

Assuming that computing is a process of the physical system that transforms (physical) input variables into output variables, the mirroring condition is as follows:

Input-output mirroring: The input-output function, g, preserves a certain relationship, $\underline{R}$, in a target domain: There is a mapping from the physical process to the target domain that maps g to $\underline{R}$, x to $\underline{x}$, y to $\underline{y}$, ..., such that $g(x) = y$ iff $<\underline{x},\underline{y}> \in \underline{R}$.

Throughout this chapter, I shall use plain italicized symbols (such as x and y) to signify certain properties of the mirroring system, and underlined italicized symbols (such as $\underline{x}$ and $\underline{y}$) to signify properties of the target domain. This condition amounts to saying that there is a similarity at the more abstract (e.g., formal) level. Some might say that at the more abstract level, g and $\underline{R}$ are similar formal relations—for example, in that both are mathematical integrations. The target domain is to be understood in a very broad sense: it can be part of the immediate environment of the system, but it can also be a more distant domain, as well as future, past, or even imaginative and counterfactual scenarios. It can also be some peripheral or internal part of the mirroring system.

This input-output mirroring requirement should be qualified. First, the mirroring system can be embedded in (can be a module of) some other system. In some instances, we talk about subsystems whose inputs are received and/or outputs are projected to other parts of the system. The inputs and outputs are very often (magnitude) values of certain properties, such as voltages. Second, there are some computing systems whose activity is not described in terms of input-output processes. In these instances, the modeling relation might be characterized in terms of some other type of relation. Broadly speaking, however, it is not controversial to argue that the behavior of many computing systems is couched in terms of input-output processes—and that is my focus in this section. Lastly, we will see that, in many cases, the morphism relation is richer than input-output mirroring, in the sense that internal relations in the computing system also preserve relations in the target domain. But in other instances, computing relies on the more minimal, input-output mirroring alone.

9.1.2 Input-Output Modeling

Input-output *modeling* is input-output mirroring plus the requirement that the input and output variables, x and y, *represent* the features $\underline{x}$ and $\underline{y}$ in the target domain. As said, I take it for the purposes of this work that *modeling = mirroring + representing*. Also, I'm not getting into the debate on the relationships between mirroring and representing. Some authors argue that morphism, or a sufficient amount of it, constitutes representation, perhaps in tandem with some other conditions. Thus, they use terms such as *input-output representation* (Ramsey 2007) and *structural representation* or *S-representation* (Swoyer 1991; Ramsey 2007) to describe entities that satisfy the mirroring relations.[3] A well-known argument against the sufficiency of isomorphism to representation is that a system that is isomorphic to one target domain is immediately isomorphic to many other target domains without representing or modeling them. Another argument is that isomorphism is a symmetric relation, whereas *representing* (and *modeling*) is not.[4]

I am not in the business of analyzing the relationship between morphism and representation. In particular, I do not argue that morphism is necessary and/or sufficient for representation. I will therefore refrain from using the terms *structural representation* and *input-output representation*, which are often dogged by the philosophical baggage that morphism is necessary and/or sufficient for representation. I define *modeling* as *mirroring plus representing*, which I think is fairly uncontroversial, and I keep the notions of *representation* and *mirroring* distinct, without any commitment to the degree of overlap between them. As I noted earlier, I believe that it is reasonable to take a pluralistic stance with respect to representation in computing systems (Chapter 7), but even this assumption does not play an important role in the definition of *modeling*.

9.1.3 The Neural Integrator in the Oculomotor System

Consider an example from computational neuroscience, where the neural network is described as computing mathematical integration. The oculomotor system controls eye movements. There are several types of eye movement. *Gaze stabilization movements* stabilize the visual world on the retina when the head/

---

[3] See, e.g., Gallistel and King: "Representations are functioning homomorphisms. They require structure-preserving mappings (homomorphisms) from states of the world (the represented system) to symbols in the brain (the representing system). These mappings preserve aspects of the formal structure of the world" (2009: x).

[4] See, e.g., Suárez (2010).

body is moving: the *vestibulo-ocular reflex* (VOR) keeps the visual world stable on the retina while the head is moving, and the *optokinetic reflex* stabilizes the visual world when the head is stationary (e.g., when one is looking out of the window of a moving train). *Gaze-aligning movements* include voluntary and reflexive saccades and smooth pursuit movements that allow one to track a moving target (Glimcher 1999; Leigh and Zee 2015). Our focus is a subnetwork of the oculomotor system called the *neural integrator*. It receives neural signals that encode velocity as inputs, and transforms them into signals that encode position. The neural integrator converts eye-velocity inputs into eye-position outputs, thereby enabling the oculomotor system to move the eyes to the right position (Robinson 1989; Seung 1998; Eliasmith and Anderson 2003; Leigh and Zee 2015).

Take vestibular movements, where the eyes are moved at the same speed as, and in the opposite direction of, the head movements. A wealth of experimental evidence from the 1960s onward indicates that the vestibulo-ocular system determines the new eye position based on the inertial velocity information transduced through the canals behind our ears (the semicircular canals). In cats, monkeys, and goldfish, the network that computes *horizontal* eye movements appears to be localized in two brainstem nuclei: the *nucleus prepositus hypoglossi* (NPH) and the *medial vestibular nucleus* (MVN).[5] Robinson and others infer that this velocity-to-position function is performed by an integrator network (I discuss the logic behind this inference in Section 9.4). Thus, Robinson writes:

> That there is indeed a second integrator is without doubt, since single unit studies in the vestibular and abducens nuclei show that the firing of units in the vestibular nuclei are in fact proportional to head velocity (over the bandwidth mentioned) and single units in the abducens nuclei increase their rate of firing in a manner proportional to eye position during the slow phase of nystagmus for which the lateral rectus is an agonist. (1968: 1041)

Robinson (1989; Cannon and Robinson 1987) also hypothesizes that the same neural integrator is used for vestibular, optokinetic, saccadic, and pursuit movements (Figure 9.2).[6] The inputs arrive from different fibers that code vestibular, optokinetic, saccadic, and pursuit velocity. The integrator system produces eye-position codes by computing mathematical integration over these eye-velocity encoded inputs. The eye-velocity codes, $\dot{E}$, are projected directly to the motor neurons that must produce velocity commands to move the eyes at the right speed. But the eye-velocity codes, $\dot{E}$, are also projected to the neural

---

[5] See Robinson (1968, 1989) and Leigh and Zee (2015).
[6] See also Goldman et al. (2002).

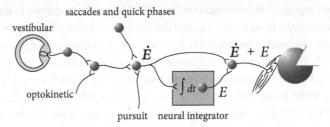

Figure 9.2 The common neural integrator. The neural integrator receives as inputs eye-velocity encoded signals, $\dot{E}$, and produces eye-position encoded outputs, E. The velocity codes, $\dot{E}$, combine the vestibular, optokinetic, saccadic, and pursuit velocities. These codes are projected directly to the motoneurons that have to produce velocity commands, but also to the neural integrator, which produces position codes projected to the motoneurons for position commands (From Cannon, Stephen C., and David A. Robinson. 1987. "Loss of the Neural Integrator of the Oculomotor System from Brain Stem Lesions in Monkey." *Journal of Neurophysiology* 57: pp. 1383–1409. Republished with permission of American Physiological Society. Permission conveyed through Copyright Clearance Center, Inc.).

integrator that produces position codes, E. The latter eye-position codes are further projected to the motor neurons for position commands.

Crucially, mathematical integration characterizes operations performed at two *very different locations*. One is in the neural system—namely, the neural integrator, which computes integration on the neural inputs to generate neural commands (which is why the system is known as an *integrator*). Another, very different location, however, is in the target domain—in our case, the eyes. The relation between position and velocity of the eye can be described in terms of integration as well. The distance between the previous and current positions of the eye is determined by integrating over its velocity with respect to time. So what we have here is input-output modeling: the input-output function of the representing sensory-motor neural system (the integrator) mirrors, or preserves, a certain relation in the target domain—namely, the distance between two successive eye positions. By computing integration, the neural function mirrors, reflects, or preserves the integration relation between eye velocity and eye positions.[7]

We can describe this morphism (mirroring) relation between the representing neural system and the represented target domain (the eyes and their properties) through the analogy of Cummins's London-Tower Bridge picture (Figure 9.3).

---

[7] To keep things simpler, I shall use the terms *distance* and *position* interchangeably. New (horizontal) position is evaluated based on the distance from the previous position.

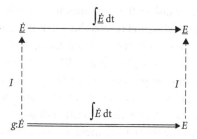

Figure 9.3 The oculomotor integrator as an input-output model. The lower span describes a causal process in the neural system (i.e., in the neural integrator) that transforms input values $\dot{E}$ to output values E (the physical function g). The upper span describes the target domain—namely, the eyes. The term $\underline{\dot{E}}$ describes the velocity of the eye, whereas the term $\underline{E}$ describes the (horizontal) distance from the previous eye position to the new eye position. The mapping, I, interprets the input signals, $\dot{E}$, as representing velocity values $\underline{\dot{E}}$, and the output signals, E, as representing position values, $\underline{E}$. Both domains share a formal structure, of mathematical integration.

The bottom span describes a causal process in the neural system (i.e., in the neural integrator), which transforms input values, $\dot{E}$, that code eye velocity, $\underline{\dot{E}}$, into output values, E, that code eye position, $\underline{E}$.[8] This input-output relation can be described mathematically as integration—mathematically speaking, the values E are the result of mathematical integration over $\dot{E}$ with respect to time. The upper span describes a certain relation in the target domain—namely, the eyes. The new position (which is the distance from the previous position), $\underline{E}$, can also be described mathematically as the result of the integration over the velocity, $\underline{\dot{E}}$, with respect to time. Thus, the mapping relation, I, which maps the input values, $\dot{E}$, to the encoded velocity values, $\underline{\dot{E}}$, and the output values, E, to the encoded distance values, $\underline{E}$, is a morphism relation.

9.1.4 The Neural Integrator as an Internal Model

As previously noted, in some cases the computing system entertains a much stronger morphism to the target system. In particular, some internal relations also preserve certain relations in the target system. The oculomotor integrator

---

[8] Note that in Figure 9.2, the term E stands for both the representing (output) neural activity and the represented eye position. Similarly, the term $\dot{E}$ stands for both the representing (input) neural activity and the represented eye velocity. This presentation is customary in neuroscience, and underscores the *modeling* assumption, as it is apparent that the integration relationship holds true both in representing and represented domains.

is a case in point. Another task of the oculomotor system is to keep the eyes still between movements. Experimental results show that normal humans are able to hold their eyes stationary at arbitrary positions for up to dozens of seconds at a time, even in complete darkness (Becker and Klein 1973; Hess et al. 1985). The brain can track the current eye position even after the stimulus has gone; in this respect, it employs a *short-term memory* of eye positions. When this memory is damaged, there is a constant drift of the eyes to a null point.

It is hypothesized that the short-term memory is located in the same network that computes eye positions—namely, the neural integrator (Seung 1998). When velocity signals are received, the network produces position signals by computing integration. But when the input signals are gone, the memory network keeps producing the same position signals until new velocity signals are received. How does the neural network implement the memory? Experimental findings show that when the eyes are still, the pattern of neural activity is constant in time, and that for every eye position, the pattern of activity is different and persistent. These findings have encouraged modelers to describe the memory system as a *multi-stable (attractor) recurrent network* (see Chapter 4 for a detailed discussion of attractor networks).

The use of attractor neural networks to implement memory is widespread (Amit 1989). The dynamics of these networks is often described in terms of an energy landscape whose minima are stable states. To implement an eye position, we can think of each state as encoding a different eye position. However, the typical multi-stable networks do not seem appropriate for memory of the eye position—because the attractors are discrete, while the encoding of the eye position in the neural activity requires a continuous, analog-graded code. Theoreticians have therefore suggested that the memory of eye position is implemented in a recurrent network with *continuous line attractor* dynamics. A new stimulus disturbs the state of the memory network away from the line of fixed points, and the network gradually relaxes on a new point along the attractor line—this point encodes the current eye position (Figure 9.4).[9]

Although the mathematical details are quite complex, the crucial features of the network are easy to explain. First, the (attractor) network has no designated input and output units: all cells are interconnected to all other units, and each cell receives external inputs from outside (velocity signals). Second, a single cell does not represent an eye position; only a collective, "total" state of the network is a candidate to be such a representation. Respectively, each point in the state-space portrait signifies not the activity of a single cell, but the activity of a collective ("total") state. At each moment, the memory network is at one of the points in

---

[9] See Canon and Robinson (1985) and Seung (1996, 1998). For a general framework, see Eliasmith and Anderson (2003: 250ff.).

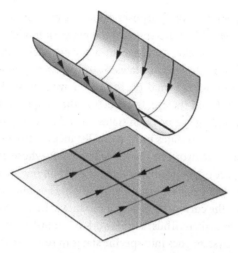

Figure 9.4 A recurrent network with a continuous line attractor. In this state-space portrait ("energy landscape"), every possible trajectory of the network (an arrow) converges to a minimum point, and these fixed points (minima) lie along a line called a line attractor (From Seung, H. Sebastian. 1996. "How the Brain Keeps the Eyes Still." *Proceedings of the National Academy of Sciences USA 93*: pp. 13339–13344. Copyright (1996) National Academy of Sciences, U.S.A.).

the landscape portrait, but it aspires toward the line attractor; the points along the line attractor are the collective states that encode eye positions.

Third, there are no synaptic changes ("learning")—at least, not in the simplified case: the weights are fixed in advance. The important idea is that the weight matrix W_{ij} has only positive feedbacks that are tuned to have a single unity eigenvalue (this produces an energy function with no curvature, as in Figure 9.4). The rest of the eigenvalues have real parts that are less than unity; this condition ensures stability—namely, that the bottom trough of the energy function is perfectly level (Seung 1996). In real biological systems, however, where these idealizations do not hold, the issues of robustness and stability become acute. There are suggestions to handle these with learning—that is, synaptic plasticity—but the effectiveness and biological reality of these suggestions are questionable (Seung 1996, 1998). However, we must also remember that the biological system itself is not perfect, and the memory of eye position is gradually corrupted over time. In human subjects, during gaze-holding in the dark there is generally a slow, systematic drift, usually less than one degree per second in normal subjects (Becker and Klein 1973; Hess et al. 1985).

The important point for our purposes is that the memory network functions as an *internal model*.[10] The state-space of the network models the space of eye positions. Each state, Si, along the line attractor encodes a different eye position, and the distance between two states, Si and Sj, corresponds to the distance between two eye positions, $\underline{Ei}$ and $\underline{Ej}$. Thus, we have a morphism between the representing network and the eye: the function that maps the stable states, the Si's, to the corresponding eye-position states, the $\underline{Ei}$'s, is type-preserving, in that the distances between two states mirror the distances between eye positions.[11] The state-space of the network can be viewed as a *map* whose line attractor corresponds to the space of eye positions. By moving from one state, Si, to another, Sj, one can reflect a transition from one eye position, $\underline{Ei}$, to another, $\underline{Ej}$. The motor neurons "read out" the current state in order to move the eyes to a new eye position and keep the eyes there. Thus, the network is not just an abstract byproduct of the fact that the system goes into specific states in response to specific inputs, but rather consists of concrete inner states that the oculomotor system can go "look up" for the purposes of problem-solving.

9.2 The Modeling Notion of Computation

We are now in a position to proceed to a characterization of physical computation. The first step is to link modeling with implementing.

9.2.1 Modeling and Implementing

As we will see, I am not the first one to associate computing with modeling. The distinctive feature of my account is that the relata of implementing and modeling can be quite different. Implementation is a relation between the physical computing system and a formalism. Modeling, by contrast, can be a relation between the physical computing system and other target systems, such as the physical world. This is the case with our oculomotor integrator. The integrator implements a particular abstract formalism (network), whose input-output

---

[10] Thus, Seung writes:
> According to modern computational theories, biological motor control is performed by an internal model.... A wealth of experimental data indicates that the internal model used for maintaining eye position is the integrator, which has been localized to specific brainstem nuclei. The nature of the internal model is also known; it appears to be a recurrent network with a continuous attractor. (1998: 1253–1254)

[11] More formally, one can see the morphism in the formula $I(\text{S-distance}(Si,Sj)) = \text{E-distance}(I(Si),I(Sj)) = \text{E-distance}(\underline{Ei},\underline{Ej})$—where I is the mirroring function, S-distance is the distance in state-space, and E-distance is the horizontal distance between angular eye positions.

(mathematical) function is integration, while it models the eyes ("target"); in particular, its input-output function models the velocity-position relation.

There are other similarities and differences between modeling and implementing. The most obvious similarity is that both modeling and implementing involve morphic mapping relations. As a kind of mirroring relation, modeling is a morphic relation from a subdomain (such as an input-output mapping function) of a physical system, P, to a given target, T. Implementing is a (homo) morphic relation between a physical system, P, and an abstract structure (such as an automaton), S.[12]

One significant difference is that modeling is a sort of representation, whereas implementing is not. As we saw earlier, Cummins and others apparently regard implementation ("instantiating") as sufficient for the representation of an abstract structure. I do not. I do not deny that the implementing relation can also be a representing relation, at least in some cases, but I do not assume that this is always the case.[13]

As said at the outset of this section, the important difference between modeling and implementing, at least for our purposes, is this: implementation, at least in the context of computation, is a relation between abstract and physical domains, whereas modeling (and mirroring) need not be. Of course, modeling can link together physical and abstract domains: a physical system P can, and sometimes does, model a formal structure S. But modeling can also link together two physical (or non-abstract) domains. Often a physical system P implements S while representing and modeling a different target T, which is the case with our oculomotor integrator.

9.2.2 The Definition of Computing

Computing consists of implementing and modeling, while modeling, in turn, consists of mirroring and representing. In addition, computing occurs if and only if the modeling and implementing relations are linked in a certain way—that is, if the shared formal structure in the input-output modeling and the mapping relation of implemented formal structure are one and the same. The oculomotor integrator computes because its shared formal structure with the

---

[12] Can we say that a physical system, P, implements a formal structure S iff P mirrors S? Mirroring entails implementing under the assumption that mirroring satisfies other constraints of implementation, such as causal, model, dispositional, and perhaps other constraints (Chapter 5). Whether implementing entails mirroring depends on certain ontological commitments about the relationships between the physical and abstract domains.

[13] See Dresner (2010) for further discussion of these issues.

eyes (mathematical integration) is the same as the mathematical function that it implements, which is also integration. To put it more precisely:

A physical system P is a computing system just in case:
   (i) **Input-output mirroring.** The input-output function, g, of a given process in P preserves a certain relation, $\underline{R}$, in a target domain T: there is a mapping from P to T that maps g to $\underline{R}$, x to $\underline{x}$, y to $\underline{y}$, ..., such that $g(x) = y$ iff $\langle \underline{x}, \underline{y} \rangle \in \underline{R}$. This means that g and $\underline{R}$ share some formal relation f.
   (ii) **Implementing.** This process of P, whose input-output function is g, implements some formalism S whose input-output (abstract) function is f.
   (iii) **Representing.** The input variables x of P represent the entities $\underline{x}$ of T, and the output variables y of P represent the entities $\underline{y}$ of T.

Our integrator (P) maps certain neural inputs to neural outputs (the function g), and this mapping input-output process mirrors the relation between movements and positions (the relation $\underline{R}$) of the eye (T). They both share the mathematical relation of integration (the function f). Our integrator also implements a certain formal structure S (such as an abstract network) whose input-output function is integration f. Finally, the input and output signals of the integrator represent the eye's velocities and positions. Thus, the integrator is a computing system. In many cases, the shared formal structure with the target is richer and includes more elements from the implemented formal structure. Assuming that the oculomotor integrator is an internal model of the eye that functions as memory, both the integrator and the eye share an abstract state-space that is implemented by the integrator. In many other cases, however, the physical system P implements some formalism S, but this formalism (e.g., algorithm) is *not* an internal model of the target (T); S is not shared by the physical system P and the target T. The only requirement is that the input-output function (relation), f, of S, is shared by P and T.

As said earlier, I am not the first to associate computing with modeling and implementing. But most who do associate them tend to characterize both implementation (or instantiation) and modeling (and representation) as relations between physical and abstract domains.[14] I do not. I do think that both relations have a shared formal structure. I also think that there are cases in which the two relations, implementing and modeling, coincide. This occurs when the physical system not only implements the formalism, but also models (and so represents) the formalism. In Cummins's example (Figures 4.1 and 4.2), the

---

[14] Cummins (1989) is an early example of this; Horsman (Horsman et al. 2014; Horsman, Kendon, and Stepney 2017) is a more recent one.

physical addition machine both implements (instantiates) the function *plus*, and also models *plus* in that the arguments and values of the physical machine represent the arguments and values of *plus*. As noted earlier, however, there are cases in which the computing system implements one domain (formalism) but represents another target. Our oculomotor integrator implements mathematical integration. However, it does not represent numbers; it represents properties of the eyes.

The physical process that is doing the g-mapping is often called the *computing process*. Interestingly, the physical input-output function, g, is seldom described as the *computed function* (Cummins is a rare exception): many take that to be the implemented mathematical function f (e.g., mathematical integration). Others describe the computed function in representational terms. The integrator, for example, is sometimes described as computing position-codes from velocity-codes. Less often, it is said that it computes the velocity-position relation ($\underline{R}$). Lastly, while implementing a certain formalism S is essential for computing, the kind of formalism represented by S matters only to the taxonomy of computational types. This does not mean that the taxonomy takes into account every implemented automaton. As argued in Chapter 8, the taxonomy takes into account the automaton that matches what is being represented.

9.2.3 Is Computing Modeling?

Is computing a type of input-output modeling? How well does the modeling definition of computation fare with the classification criteria? Let us start with *the-right-things-compute* part. The oculomotor integrator is both a computing system and an input-output model of the eye. But this can hardly support the more general claim that computing is modeling. My aim in what follows is to provide some support for the computing-is-modeling claim. In Section 9.3, I will discuss the work of other researchers who have associated and even characterized computing in terms of modeling. In the course of this discussion, we will review further examples of computing as input-output modeling. I will then turn to the methodological role of input-output modeling in computational theories in cognitive neuroscience (Section 9.4). This role is of interest because it shows how deeply the idea that computing must be accompanied by input-output modeling is entrenched—even when it is harder to pinpoint the modeling relation. Finally, I turn to discuss the explanatory role of modeling (Section 9.5), which is of interest because it locates the distinctive role (or so I argue) of computational explanations. When all these are taken together, I think that we have good evidence to link together computing and input-output modeling, at least in the context of cognitive neuroscience.

What about non-computing systems—namely, those satisfying *the-wrong-things-don't-compute* desideratum? The *modeling* definition of computing deems most physical systems to be non-computing. The *representation* condition rules out stomachs, hurricanes, rocks, planets, chairs, and many other physical systems. This does not mean that these systems cannot compute: if we assign contents to their states in ways that fulfill the other conditions, they *might* compute—but until we do so, they will not (see Chapter 7).

The combination of the *mirroring* and *implementation* conditions rules out some non-computing representational processes. Consider an addition table for kids; assume that there is a physical mechanism that connects the squares labeled by 3 and 4 to the square labeled by 7, and so on. One can insist that this process really mirrors the *plus* function, at least under a very liberal notion of mirroring.[15] But even if there is such mirroring, it is very unlikely that the physical mechanism implements the *plus* function. The implementing mechanism is just the same even when you relabel the squares. Another way to see this is the following: The *plus* function is symmetric. Thus, our mechanism also connects the other squares labeled by 4 and 3 to a square labeled by 7. But these latter three squares in the lookup table are not the same as the first three squares (also labeled by 3, 4, and 7). We could thus relabel the second 7-square to 8 without changing the mechanism. This shows that the operations of the mechanism need not be symmetric, in which case it does not implement *plus*. The same goes for many other lookup tables: the implemented formalism does not match the function that is (allegedly) mirrored by the system.

Or consider the process of screening a movie of the old-fashioned film-reel type. The machine shows one frame (representation) after another—yet screening is not computing. Even when we consider screening as implementing a formalism, the formal function that is implemented by screening is not the same as the outer (formal) relation between the represented scenes: the same screening process takes place with very different sequences of representations. You could change the order of the frames (and, thus, the outer relations between the represented scenes), and the inner screening process (and, therefore, the implemented function) would be just the same. The same goes for shredding machines, which receive paper inputs with sentences and words; stamping machines, which stamp representations; and other non-computing systems. The inner input-output function of shredding, stamping, and so on does not appear to mirror the outer relations in the target (represented) domain. Once again, the *modeling* definition of computing does not rule out the possibility that,

---

[15] One way to deal with liberality is to impose on the mapping relation the same constraints that are imposed on the implementation (mapping) relation (Chapter 5). Instead of doing this, I require that the implemented function and the shared function (of mirroring) is the same.

under certain circumstances, these systems would compute—but under current circumstances, they do not.

9.3 Others Who Have Linked Computing to Modeling

I am by no means the first to link computing with modeling. Cummins (1989) presents a similar notion of *input-output modeling* with his famous London-Tower Bridge diagram (as discussed in Chapter 4). However, he uses this notion to advance a notion of content ("interpretational semantics") that fits in with computational, mainly classical theories of cognition. The crucial feature that defines computation, according to him, lies elsewhere—in the notion of *step-satisfaction*. Ramsey (2007) also links computing with modeling: he associates the notion of representation found in classical theories (in minds and machines) with an internal model, and calls this *structural representation* or *S-representation*. Like Cummins, Ramsey aims to account for a strong enough notion of representation—specifically, mental representation—rather than computation. Both Cummins and Ramsey argue that modeling is an essential element in the notion of mental or cognitive representation.[16]

Fodor (1994) and others (Haugeland 1981b; Pylyshyn 1984) emphasize that a digital computer (or a Turing machine) has the ability to support processes that are truth-preserving. This means that in these systems, we can implement inference ("syntactic") rules that mirror semantic relations such as logical validity. Taking the inputs and outputs to be symbolic expressions (say, the input is a set of sentences K, and the output is a sentence p), the "inner" input-output function, which is the inferential relation of K to p, mirrors the semantic relations. In other words, $K \vdash p$ iff $K \vDash p$:

> Well, as Turing famously pointed out, if you have a device whose operations are transformations of symbols, and whose state changes are driven by the syntactic properties of the symbols that it transforms, it is possible to arrange things so that, in a pretty striking variety of cases, the device reliably transforms true input symbols into output symbols that are also true. (Fodor 1994: 9)

Like Cummins and Ramsey, Fodor focuses on classical machines, and, respectively, classical theories of cognition (Fodor and Pylyshyn 1988). Fodor also stresses an important role of modeling: computing makes it possible to transform

---

[16] See also Shepard and Chipman (1970); Palmer (1978); Edelman (1998, 2008); and Gallistel and King (2009).

some representations into others in ways that preserve semantic relations such as truth. Although the system is not sensitive to the content of the expressions, it does transform true symbolic expressions into other true symbolic expressions. I shall discuss this type of inference in Section 9.5.

The linkage between computing and modeling is not confined to digital and classical machines. It is central to other computational paradigms as well, in which the representations are not symbolic expressions that have truth-values (preserving truth might be said to be merely a special case of the morphism relationship). Many scholars associate *analog* computation and representation with modeling. They note that the term *analog* signifies that the computation relation is *analogous* to some target domain.[17] Thus in his book *Analog Computing*, Ulmann says that a problem is solved on an analog computer "by changing its structure in a suitable way to generate a *model*, a so-called *analog* of the problem. This analog is then used to *analyze* or *simulate* the problem to be solved" (2013: 2). Maley defines analog representation "to be one in which some quantity varies with the quantity being represented in a strictly monotonic manner, where this variation can be either discrete or continuous" (2018: 86). Elsewhere, he defines analog representation in terms of mirroring: "The basic idea is that analog representations are structurally isomorphic (or, in some cases, homomorphic) to what they represent" (Maley 2020); he associates this definition with the monotonicity condition. Analog computation, according to Maley, is "the mechanistic manipulation of analog representations" (2020). Maley does not claim that every computation is analog, but rather aims to demonstrate that the brain is an analog computer.

A paradigmatic case of an analog computer is a differential analyzer designed to solve differential equations by using wheel-and-disc mechanisms that perform integration. One example of such an equation is $md^2X/d^2t + cdX/dt + k = F$, which describes the forced response of a single degree-of-freedom spring/mass/damper system—where X is the displacement of mass m, supported by a spring of stiffness k and a viscous damper of rate c, and F is an external force applied to the mass. Another example is the tide-predicting machine designed by Lord Kelvin and constructed in 1873. The machine determines the height of the tides by integrating ten principal constituents. These constituents are made by means of toothed wheels that simulate the motion of the sun, moon, earth, and other factors that govern the tides. Ulmann (2013) and Papayannopoulos (2020b) offer impressive surveys of analog computation. Maley (2020) provides examples of

---

[17] Others associate the term *analog* with continuous values (as discussed in Section 4.2); see Maley (2011) and Papayannopoulos (2020b) for a more recent discussion and comparison of these two approaches.

analog representations—both in artifacts and in the brain—showing that the representations can be discrete or continuous.

Many other scholars have noted that the notion of *modeling*—including that of an internal model—is central to neural networks (e.g., Eliasmith and Anderson 2003; Ryder 2004; Churchland 2007; O'Brien and Opie 2009; Shagrir 2012c). The oculomotor integrator is one example; another well-known example is the cognitive maps in the hippocampus of rats, humans, and other mammals and animals. These maps, which consist of *place cells*, are used for navigation and spatial processing (O'Keefe and Nadel 1978). In his discussion of several artificial neural networks, Paul Churchland notes that there is a similarity between high-dimensional relations (such as geometrical congruence) in the state-space of the representing network and high-dimensional relations in the represented domain "in the world." Thus, when using road maps as an example of his point, he writes that "it is these interpoint distance relations that are collectively isomorphic to a family of real world similarity relations among a set of features within the domain being represented" (2007: 107). The linkage between computing and modeling is also found in Bayesian approaches, in predictive coding, in control theories, and in other theoretical frameworks.[18] Although it is questionable whether all these approaches use exactly the same notion of a model, they all take the computing system to preserve relations in, and represent, the target domain.

Thus far, I have cited philosophers and scientists who associate computing with modeling. These scholars, however, do not characterize computing in terms of modeling. In the rest of this section, I will discuss two characterizations of computing that share some affinities with the idea that computing is a type of modeling; both characterizations are made in the context of computational approaches in the brain and cognitive sciences.

9.3.1 Grush on Neural Computation

Analyzing Churchland, Koch, and Sejnowski's (1990) characterization of computation, Grush (2001) distinguishes between two components of a computing physical system (at least in the context of computational neuroscience). One component is the implementation of an abstract function or algorithm by the physical system. Grush calls this an *algorithm-semantic* (or *a-semantic*) interpretation. The second component is information-processing in the sense that the

---

[18] Thus, Griffiths, Kemp, and Tenenbaum (2008) say that the big computational question underlying the Bayesian approach is *"How does the mind build rich, abstract, veridical models of the world given only the sparse and noisy data that we observe through our senses?"* (p. 59). Clark (2015) notes the central role of generative models in the hypothesis that the brain is a prediction machine. Grush (2004) highlights the role of models in control theory.

states of the physical system carry information about objects or states of affairs in the environment. Grush calls this the *environmental-semantic* (or *e-semantic*) interpretation. He next argues that a notion of computation should include both *a-semantic* and *e-semantic* interpretations. Grush exemplifies his notion of computation through case studies. Let us look at one of them—the famous Zipser and Andersen (1988) model.

Changing reference or coordinate frames is central to many visual-motor tasks. Andersen, Essick, and Siegel (1985) argue that the *posterior parietal cortex* (PPC) of macaque monkeys is home to the information-processing task of relocating a target in body-centered or head-centered coordinates. Experimental results indicate that the PPC includes three types of cells: (1) cells that respond to eye position only (15% of the sampled cells); (2) cells that are not sensitive to eye orientation but have an activity field in retinotopic coordinates (21%); (3) cells that combine information from retinotopic coordinates with information about eye orientation (57%).

Zipser and Andersen (1988) hypothesize that the PPC combines retinotopic and extra-retinal (eye-orientation) signals in order to compute target location in head-centered coordinates, and have trained a neural network to simulate this computation. They use a three-layer network in which the two sets of input units model the behavior of the first two groups of cells, (1) and (2). The input layer projects onto a layer of hidden units, which aims to model the activity of the third group of cells, (3). The output units encode the target's position in head-centered coordinates; cells with this property were not found in the PPC. Zipser and Andersen's impressive result is that the activity of the *hidden units*, after the training period, is very similar to the response properties of the third-group cells that combine information about eye orientation and the target's retinotopic location. It transpires that these units function as *planar gain fields*, in the sense that the Gaussian retinal receptive field is modulated (linearly) by the orientation of the eye. Given this result, Zipser and Andersen hypothesized that there are head-centered target-location cells somewhere in the brain that are the correspondents of the network model's output units.

Grush refers to the computations by the third-group PPC cells as follows: "We can suppose that the function computed by an idealized posterior parietal neuron is something like $f = (e - e_p)\sigma(r - r_i)$" (p. 161). This is the *a-semantic* interpretation. It refers to the mathematical relation between the two groups of "input" PPC cells. The activity of the "output" PPC cells (group (3)) is a multiplication of the activity of the groups (1) and (2). Grush also notes, however, that this mathematical equation applies to complex relations in the environment between the things being represented; this is his *e-semantic* interpretation. The "stimulus distance from preferred direction relative to the head" (p. 161), which is represented by the output, is a multiplication of the properties encoded by the inputs—namely, the difference between actual and preferred eye orientation

($e - e_p$) and (the Gaussian of) the distance of the retinal location of stimulation from the receptive field ($\sigma(r - r_i)$). This shared structure shows that there is a morphism relation between the nervous system and the world. In fact, what we have here is a case of input-output modeling: the input-output function (of multiplication) preserves a pattern of relation—between eye orientation and stimulus retinotopic location—that can also be described in terms of multiplication. Thus, Grush's characterization of computing has affinities to modeling characterization presented here.[19]

9.3.2 Marr on Computational-Level Theories

In *Vision*, Marr (1982) famously proposes a three-level approach to the study of visual processes and to the study of cognition more generally. The "most abstract" is the computational level (CL), which "is the level of *what* the device does and *why*" (p. 22). The role of the *what* aspect is to specify what is computed; the job of the *why* aspect is to demonstrate the appropriateness and adequacy of what is being computed to the information-processing task (pp. 24–25). The algorithmic level characterizes the system of representations being used—for example, decimal versus binary—and the algorithm for the transformation from input to output. The implementation level specifies how the representations and algorithm are physically realized. Marr's levels are not levels of organization, where the entities at higher levels are composed of lower-level entities; rather, he refers to his levels as *levels of analysis*, whereby each such level provides a further understanding of the visual phenomenon.

Our focus, naturally, is the top, computational level. Marr, however, never provides a systematic and detailed account of his notion of CL. He moves on to advance a set of computational theories of specific visual tasks that have had a tremendous impact on vision research. The explication of a computational-level theory was left to philosophers, who in turn provided a number of very different interpretations.[20] A more recent interpretation emphasizes the role of the environment in Marr's notion of *computational analysis* (Shagrir 2010; Bechtel and Shagrir 2015; Shagrir and Bechtel 2017). In Shagrir and Bechtel's interpretation, the *what* element characterizes the computed (typically input-output) function in precise mathematical terms; the *why* element demonstrates that this function mirrors a relationship in the visual field, between the represented entities. I shall return in Section 9.5.1 to discuss the explanatory aims of CL. At this point, I wish to provide two examples of the modeling approach of CL.

---

[19] There are some differences as well: I do not present the first ("implementation") relation as a semantic one, and I focus only on the input-output function, not on the entire algorithm.
[20] See Shagrir and Bechtel (2017) for a detailed discussion of some of these interpretations.

When discussing the example of a cash register (1982: 22–24), Marr says that *what* is being computed by the device is *addition*. We arrive at this characterization when noticing that the machine maps digits to digits, and that this mapping satisfies the rules of commutativity, associativity, zero, and inverse. Marr then turns to demonstrate *why* computing addition is appropriate for the information-processing task by showing that the "external" relationship between the final bill and the purchased items in this particular case is *also* that of addition. In this example, the rules ("constraints") of purchasing at this store define addition. These are the rules of *zero* ("if you buy nothing, it should cost you nothing; and buying nothing and something should cost you the same as buying just the something" [p. 22]); *commutativity* ("the order in which goods are presented to the cashier should not affect the total" [p. 23]); *associativity* ("arranging the goods into two piles and paying for each pile separately should not affect the total amount you pay" [p. 23]), and *inverses* ("if you buy an item and then return it for a refund, your total expenditure should be zero" [p. 23]). These rules, according to Marr, define *addition* uniquely. Thus, what we have here is a case of input-output modeling (although Marr himself never refers to it as such). The inner function of the cash register (*addition*) mirrors the outer relation between the represented items (namely, the prices of the purchased items). Importantly, the system of representations being used—binary, decimal, Roman, or even continuous—makes no difference to the modeling relation. According to Marr, characterizing the system of representations and the algorithm that transforms them is part of the algorithmic level.

The other example is edge detection. Marr's computational theory of edge detection states that V1 cells detect edges by computing the zero-crossings of second-derivative Laplacian operators. The latter operators are applied by the ganglions and LGNs to the retinal image and are described quantitatively by the formula $\nabla^2 G * I$—where I is the image, $*$ is a convolution operator, and $\nabla^2 G$ is a filtering operator: G is a Gaussian that blurs the image, and ∇^2 is the Laplacian ($\partial^2/\partial x^2 + \partial^2/\partial y^2$). The zero-crossings signify extreme points ("sharp changes") in the arrays of intensity pixels (retinal images). This is the *what* element of the theory. The *why* element shows that detection of the zero-crossings of the second-derivative operators mirrors sharp changes in light reflection in the visual field (which often occur along physical edges, such as object boundaries). The latter changes can be described in terms of extreme points of first derivatives or zero-crossings of second derivatives of the *reflection* function. Thus, what we have here is input-output modeling. The early visual processes and certain relations in the represented visual field—such as sharp changes in light intensities that typically occur along object boundaries—share the mathematical relation of differentiation. This is another case of input-output modeling.

9.3.3 Summary

We have seen that several scholars and scientists have associated computing with modeling. Some have even characterized computing in terms of modeling (even if only implicitly). In some of the examples, such as the oculomotor integrator, the modeling is more apparent, whereas in other cases, such as the PPC network (as described by Zipser and Andersen), it takes some effort to make the modeling relationship explicit.

9.4 The Methodological Role of Modeling

In this section, I discuss the methodological role that input-output modeling plays in computational theories: it helps to reveal the mathematical input-output function that the system computes. I do not claim that modeling always plays this role. Indeed, it usually will not, if we design the system to compute this function—in which case, we already know what the system computes (which is not to say that we will not try to verify that the system functions properly). This methodology is often invoked in the context of natural systems, when we do not know in advance what is being computed. I shall provide examples from cognitive neuroscience, which show just how entrenched the modeling assumption is, at least in this field of study.

In many cases, environmental cues are used to infer the computed function. Input-output modeling plays a key role in this inference. Consider our oculomotor system. Scientists discovered that the inputs to the system are velocity signals. They also hypothesized that these signals are translated to position signals, which are crucial to move the eyes to new positions. Assuming that the velocity-position relation is that of integration, they inferred that there is a subsystem that performs this transformation by computing integration. They therefore called this system the *neural integrator*.

We can put the inference, somewhat crudely, as follows:

- Electrophysiological experiments show that certain cells (*input* cells) encode eye velocity. Other cells (*output* cells) encode eye position.
- The eye's velocity-position relation in the *target domain* is (in the abstract) that of mathematical integration.

Therefore: The input-output function computed by the neural system is integration.

But it may be noted that the conclusion does not follow from the premises: why infer that the *inner* function is that of integration from the premise that the *outer* function is that of integration? The inference becomes valid *if* we also assume that the (inner) input-output function mirrors the velocity-position relation. When making the additional (third) premise, the argument looks as follows:

- Electrophysiological experiments show that input cells encode eye velocity and output cells encode eye position.
- The eye's velocity-position relation in the *target domain* is that of mathematical integration.
- The computed input-output function mirrors the eye's velocity-position relation.

Therefore: The (mathematical) input-output function computed by the neural system is integration.

The advantage of this methodology is that we can learn about the inner function of the nervous system—which is often hidden and hard to decipher—from the outer function, which is often readily apparent. This is not the end of the scientific investigation, of course. Further studies are conducted to confirm the conclusion and to locate the integrator in the nervous system. More studies aim to characterize how the system performs integration—namely, the mechanisms that conduct the input-output transformation. The important message, however, is that the input-output modeling assumption is entrenched in cognitive neuroscience. Theoreticians such as Robinson, Seung, and many others are deeply convinced that there must be *an integrator* within the oculomotor system that mirrors the velocity-position relation (see the quotation from Robinson earlier). They take it to be obvious that if the outer relation between the represented entities is that of integration, the nervous system must also mirror this relation somewhere by computing integration.

Another striking example is path integration. Homing is the ability of animals and humans to return to their departure point. Animals use external cues—environmental stimuli and events—to navigate back home.[21] But experimental results show that homing occurs even when all the external cues are removed. Cues about initial reference and self-motion suffice to calculate the animal's relative spatial location—a phenomenon known as *path integration*.[22] The input of the calculation is angular velocity signals, which are provided by the

---

[21] This ability is possessed by various animals. One well-known example is the desert ant (*Cataglyphis fortis*), which returns home after venturing out hundreds of yards.
[22] See Mittelstaedt and Mittelstaedt (1982); Collett and Collett (2000); Etienne and Jeffery (2004); Conklin and Eliasmith (2005); McNaughton et al. (2006); and Gallistel and King (2009).

vestibular system or other systems. The computation of path integration mirrors the velocity-position relation of the animal's locomotion—and thereby enables the system to keep track of the animal's relative position.

In this case, there might not be a specific neural subsystem that computes integration. Nevertheless, scientists take it for granted that integration must occur within the navigational system, even if it is spread over different parts of the system.[23] This assumption is very explicit, for example, in the following paragraph from a review paper by Etienne and Jeffery (2004):

> How is information about angular motion processed? Recently it has been found that cells in the dorsal tegmentum code for angular velocity (Sharp et al., 2001; Bassett and Taube, 2001), information they receive from the semicircular canals via the vestibular nuclei. The picture that seems to be emerging is that information about angular acceleration in the horizontal plane is collected and converted to an angular velocity signal by the semicircular canals, then passed on to the dorsal tegmentum and integrated again on its way through the mammillary nuclei and thalamus (Bassett and Taube, 2001). This provides an angular distance measure that updates the head direction signal appropriately. (p. 183)

Etienne and Jeffery are describing here a process with two integration steps. The inputs to the vestibular system are signals of angular acceleration; these are converted to angular velocity signals (first integral). The latter signals are then converted again into the angular distance measure (second integral). What the authors describe here is a double-mirroring process. In the first step, the nervous system converts input signals that encode acceleration to output signals that encode velocity by computing mathematical integration. This input-output function mirrors the acceleration-velocity relation, which is a relation of mathematical integration. In the second step, the nervous system converts input signals that encode velocity (these are the outputs of the first step in the process) to output signals that encode position by computing mathematical integration. This input-output function mirrors the velocity-position relation, which is of mathematical integration as well. Taken together, the overall input-output function of the double integral mirrors the acceleration-position relation. This function consists of a sequence of two input-output integration functions: the first mirrors the acceleration-velocity relation, while the second mirrors the velocity-position relation.

What is more interesting for our purposes is that Etienne and Jeffery take it as self-evident that the computation of input signals that encode acceleration to

---

[23] More recently, it has been suggested that path integration in rats is computed by the grid cells situated in the dorsolateral medial entorhinal cortex (dMEC) (Hafting et al. 2005).

output signals that encode velocity is a computation of integration. They assume, in other words, that the relevant computation mirrors the acceleration-velocity relation and therefore must be integration. They make the same assumption about the second integral: they take it as obvious that the computation of input signals that encode velocity to output signals that encode position is one of integration. They infer that the nervous system must compute a double integral. This inference—from outer relation to inner function—is valid *if* we also assume that the nervous system is an input-output model of the animal's movement. The assumption, more precisely, is that the overall input-output function of the double integral mirrors the acceleration-position relation, and that this function consists of a sequence of two input-output integration functions—the first mirroring the acceleration-velocity relation and the second mirroring the velocity-position relation. Without this assumption, Etienne and Jeffery cannot reach their conclusion that the system computes a double integral. Again, this assumption of input-output modeling is not made explicitly. It is an implicit assumption about our brain-world relations that underpins their scientific investigation.

A third example is Marr's computational theories. Marr and his students appeal to physical external factors ("physical constraints") to discover the mathematical function being computed. These *physical constraints* are physical facts and features in the physical *environment* of the perceiving individual (1982: 22–23) that limit the range of functions that the system could compute to perform a given visual task successfully. Shimon Ullman puts this point succinctly in his manuscript on visual motion: "In formulating the computational theory, a major portion concerns the discovery of the implicit assumptions utilized by the visual system. Briefly, these are valid assumptions about the environment that are incorporated into the computation" (1979: 3–4).[24] Returning to the example of edge detection, the discovery that early visual processes compute differentiation (whether of the first or second degree) is made through the observation that in our perceived environment, sharp changes in light reflectance occur along physical edges, such as the boundaries of objects. This contextual feature puts substantial constraints on the mathematical function being computed—namely, that it must have to do with some form of differentiation. The implicit assumption, again, is that by computing differentiation, the visual system mirrors the relevant relationship in the visual field.[25]

---

[24] See also Hildreth and Ullman, who write that a computational theory includes "an analysis of how properties of the physical world constrain how problems in vision are solved" (1989: 582).

[25] The methodological role of the physical constraints is related to a top-down methodology, which is often associated with Marr's framework. The idea behind this methodology is that scientific investigation proceeds from the top down—from the computational level down to the algorithmic and implementation ones. A key plank of this approach is that it would be practically impossible to extract the computed mathematical function by abstracting from neural mechanisms. Rather, the way to go is to extract what the system computes from relevant cues in the physical world that

We see, therefore, that the methodology of discovering the input-output function from outer relations in the target system is fairly common in computational cognitive neuroscience. But it is certainly not the only way to discover the computed function. When scientists do not know or are unsure about the outer relation, they cannot infer about the inner function, and therefore they use different methodologies to discover it. Thus, in Zipser and Andersen (1988), we do not see a progression from the outer function to the inner function; rather, the fact that the input-output function in the nervous system is that of multiplication is discovered through the training of the (artificial) neural network that simulates the (real) neural computation. That the inner relation models the outer relation is featured only later, in the analysis of the neural network. The point, however, is that the frequent use of this methodology (as in the first three cases) indicates that these scientists assume that computing goes hand in hand with input-output modeling.

9.5 Computational Explanations

Many scholars have noticed that a main function of models is *surrogative reasoning*. This means that we use models to reason about the target domain; our inferences about the target are made by looking at the model, not at the target. Take the family tree (Figure 9.1): we can infer, for example, whether or not John is the grandparent of Mo by examining the model alone. This is possible precisely because the relations in the model preserve, or mirror, relations in the target domain.[26] The flipside of this is that modeling helps to explain how a computing system attains a certain information-processing task—in other words, how it moves from certain input representations to the "right" output representations. Take our neural integrator: one might ask why the inner algorithmic and neural mechanisms of the integrator transform representations of eye velocity into representations of eye positions. The answer, I maintain, is that the integrator performs an inference that is similar to surrogative reasoning: the inner mechanisms support an input-output function that models the velocity-position relation. Because of this modeling relation, the inner mechanisms, when starting with representations of eye velocity, must end up with representations of eye position.

constrain the computed function. The *modeling* assumption therefore plays a central role in this top-down approach.

[26] See Swoyer (1991) for a general discussion about the relationship between modeling and surrogative reasoning. See Grush (2004) for a discussion about modeling and surrogative reasoning in the brain.

Note that inner mechanisms alone do not provide such an explanation. Inner mechanisms can certainly tell us *how* the function is being computed. Specifying the algorithm tells us how the input values are mapped to output values, and specifying the underlying neural structures tells us how the neural mechanism enables this computation. But our question is not about the inner mechanisms that give rise to the input-output function, but rather about the system-world relations involved. The question is about the relations between the inner input-output function and the information-processing task that is defined, at least partly, by the target system—such as eyes. The question is why the network computes integration, and not (say) factorization or exponentiation, in order to move the eye to the new (desired) position. Saying that the computed function leads to representations of positions only reiterates the *why* question. After all, computing integration in a very different environment would not lead to representations of positions. If you remove the neural integrator—with the same algorithmic and neural mechanisms—to a very different environment, one with other relations between velocity and position, then performing the same input-output function (integration) might no longer provide codes of eye position.

Input-output modeling answers the question of *why* these inner mechanisms are appropriate for the information-processing task. The inner mechanisms support an input-output function that preserves the velocity-position relation—namely, the (integration) relation between eye movement and eye position in the target domain. When you compute integration over eye-velocity encoded inputs, you mirror the integration relation between velocity and position; hence, you generate representations of a new eye position as output. Mechanisms that support factorization, exponentiation, or other functions would not result in moving the eyes to the right place—precisely because they do not preserve relations in the target domain that are relevant to eye movements.

Woodward (2003) famously proposes that causal information is explanatory by virtue of allowing answers to *what-if-things-had-been-different* questions. Others have recently suggested that such *what-if-things-had-been-different* questions are also valuable in non-causal contexts (Chirimuuta 2014; Rusanen and Lappi 2016; Elber-Dorozko 2018). Input-output modeling answers relevant *what-if-things-had-been-different* questions. We can see, for example, that if we intervene in input-output modeling, then the system will no longer produce codes of eye position. We can intervene in input-output modeling either by changing the inner input-output function, or by changing the velocity-position relation. In neither case does the system produce codes of eye position any longer:

- If the system had not computed integration, but rather exponentiation, the system would not have produced codes of eye position.
- If the world had changed so that the eye's velocity-position relation were not integration, but exponentiation, the system would not have produced codes of eye position when computing integration.[27]

I am certainly not the first one to assign this sort of explanatory role to modeling in computing systems. As we have seen, Cummins, Fodor, Churchland, Ramsey, and others have done just that. My further claim is that input-output modeling is the distinctive feature of computational explanations. I do not deny that some specification of the mediating mechanisms is part of computational explanations. Whether this specification takes the form of functional, mechanistic, dynamical, or other analyses is a matter of debate. My view is that a computational explanation is hospitable to any of these analyses. But I also maintain that none of these analyses is distinctive to computational explanations. In previous chapters, I argued that we can find each of these analyses in non-computational explanations as well. What makes computational explanations distinctive is that the specification of mechanism is augmented with input-output modeling—which explains why these mechanisms are appropriate to the explanandum information-processing task.

Consider again the oculomotor integrator. A computational theory aims to explain how the system produces position signals from velocity signals (the explanandum). A computational explanation might look like this:

- The system computes the mathematical input-output function of integration.
- The system computes integration by implementing a certain formalism/algorithm (e.g., Seung's network).
- Computing integration mirrors the velocity-position relation.

My account does not differ from other accounts with respect to the first two components. A computational explanation of information-processing task specifies, in formal (e.g., mathematical) terms, the computed function and the mediating mechanism. My point is that this specification is also found in other formal, but non-computational, explanations of physical systems. The distinctive component of computational explanations is the last one. Its role is to explain why computing integration is relevant to the information-processing task.

[27] One could argue that the claim that the velocity-position relation is that of integration is physically, or even mathematically, necessary. My point here is conceptual, but at any rate, there are other examples from visual theories in which the mirrored relations are contingent (see Shagrir 2018).

While I will not provide a full account of computational explanations here, I will attempt to state the reasoning behind the explanatory role of the mirroring component more precisely. This component aims at the following phenomenon: The computation starts with an input of neural values $\dot{E}$ that encode some distal features $\underline{\dot{E}}$ (that is, eye velocities). It performs a certain input-output mapping, g, whose output is other neural values E that encode another distal feature $\underline{E}$ (i.e., eye positions). The explanandum question is why this mapping, which starts from neural values that encode eye velocity, terminates in neural values that encode eye position. To put it succinctly:

(1) $I(\dot{E}) = \underline{\dot{E}}$ (the neural input activity $\dot{E}$ encodes $\underline{\dot{E}}$).
(2) $g(\dot{E}) = E$ (g maps input neural values $\dot{E}$ to output neural values E).

Conclusion: $I(E) = \underline{E}$.

When put this way, the question is about the inference from (1) and (2) to the conclusion. The answer is in no way trivial: if we change the environment, then the same mapping g (and the very same mediating mechanisms), which starts from the same velocity-coded neural input values $\dot{E}$, will still terminate with the same neural output values E—but E might no longer encode eye position, or anything else. Why, then, does mapping g, which starts from neural input values $\dot{E}$ (which encode eye velocity), end up with neural codes of eye position?

The reasoning requires input-output modeling. The third premise states that g preserves (mirrors) some relation $\underline{R}$ in the target system (e.g., that both g and $\underline{R}$ are, in the abstract, integration relations), and that this mirroring is also a representing relation. The fourth premise states that the mirrored relation, $\underline{R}$, relates velocities and positions:

(3) $\underline{R}(I(\dot{E}), I(g(\dot{E})))$ (g models some $\underline{R}$).[28]
(4) $\underline{R}(\underline{\dot{E}},\underline{E})$ ($\underline{R}$ is the velocity-position relation).

From (1)–(4), we can reach the conclusion.
From (1) and (3) follows (5):

(5) $\underline{R}(\underline{\dot{E}}, I(g(\dot{E})))$.

From (5) and (2) follows (6):

(6) $\underline{R}(\underline{\dot{E}}, I(E))$.

---

[28] (P3) is implied by the conjunction of the mirroring condition, $g(x) = y$ iff $<\underline{x},\underline{y}> \in \underline{R}$; and the representing condition, that $I(x) = \underline{x}$ and $I(y) = \underline{y}$, and that the inputs, x, are $\dot{E}$ values (which is implicit in (P2)).

From Premise 6 and Premise 4, it follows that:

Conclusion: $I(E) = \underline{E}$.

In the rest of this section, I compare and contrast this account of computational explanations with related accounts of computational explanations.

9.5.1 Marr's Computational-Level Explanations

My account is inspired by Marr's notion of *computational-level explanation*—or at least by how we interpret it (Shagrir 2010; Bechtel and Shagrir 2015; Shagrir and Bechtel 2017). Notably, Marr (1982: 22) refers to CL as a "level of explanation." He says: "The key observation is that neurophysiology and psychophysics have as their business to *describe* the behavior of cells or of subjects but not to *explain* such behavior" (1982: 15). He continues:

> There must exist an additional level of understanding at which the character of the information-processing tasks carried out during perception are analyzed and understood in a way that is independent of the particular mechanisms and structures that implement them in our heads. This was what was missing—the analysis of the problem as an information processing task. (p. 19)

And he concludes: "It is the top level, the level of computational theory, which is critically important from an information-processing point of view" (p. 27).

As said previously, the CL consists of two aspects, the *what* and the *why*. The *what* aspect specifies the mathematical function that is being computed. In the case of the cash register, it is *addition*. But Marr goes on to state that this characterization is only one half of the computational explanation: "The other half of this level of explanation has to do with the question of *why* the cash register performs addition and not, for instance, multiplication when combining the prices of the purchased items to arrive at a final bill" (p. 22). After all, we can certainly think of stores where the cashier executes multiplication and not addition. Establishing that the relation between the purchased items and the final bill is that of addition, Marr draws the conclusion that the input-output addition mapping in the cash register is appropriate for the task in this particular store. This explanation appeals to the fact that this "internal" mapping (of addition), defined over digits, corresponds to an "external" relation between the represented items (in the abstract), that is, between the prices of purchased items and the final bill.

Or take edge detection. The *what* aspect specifies that early visual processes compute the zero-crossings of $\nabla^2 G * I$. This computation leads to the detection of "visual edges" that are extracted from sharp changes in the retinal images. But this is only part of the explanation. We still want to know why this computation and not another—factorization or exponentiation, for example—leads to the representations of "physical edges," for example, object boundaries. This concern is emphasized by Marr and Hildreth, who say that "the concept of an 'edge' has a partly visual and partly physical meaning. One of our main purposes . . . is to make explicit this dual dependence" (1980: 211). The role of the *why* aspect is to address this question. It shows that the detection of visual edges mirrors a pertinent relation in the visual field. This mirroring (morphism) is exemplified by the (alleged) fact that the visual system and the visual field have a shared mathematical description (or structure). On the one hand, the visual system computes the zero-crossings of second-derivative operations (over the retinal pixels) to detect edges. On the other hand, the reflection function in the visual field changes sharply along physical edges such as object boundaries. These changes can be described in terms of extreme points of first-derivatives or zero-crossings of second derivatives. Thus, even if he does not state this explicitly, Marr's CL explanation is rooted in input-output modeling. The *what* aspect specifies the mathematical input-output function. The *why* aspect shows that this mathematical relation also holds between the represented inputs and outputs.

One can argue that specifying mechanisms, especially at the algorithmic level, is an integral part of computational explanations. I agree. Marr was a bit hasty in contrasting computational and algorithmic explanations (and, some would argue, also with implementational ones). However, it should also be noted that Marr does not offer CL as an alternative to algorithmic and implementational explanations, but rather as a *complementary* explanation. More importantly, by calling the top level *computational*, Marr highlights what is unique and distinctive in computational explanations. Mechanistic descriptions, both algorithmic and/or implementational, can also be found in non-computational explanations of physical systems. The distinct character of computational explanations is in modeling the environment, and this character is captured at the computational level (CL).

9.5.2 Egan's Function-Theoretic Explanations

Frances Egan (2017) argues that computational theories put forward *function-theoretic (FT) explanations*. The aim of these theories is to explain a particular information-processing capacity. They achieve this by providing a characterization of the mathematical function being computed. The computational core (i.e., individuation conditions) of computational theories, according to Egan, is formal and non-semantic. Egan grounds her notion of FT explanation in her interpretation of Marr's notion of CL explanations. She associates Marr's "computational level" with "the specification of *the function computed*" (Egan 1991: 196–197; see also 1995: 185)—namely, the input-output mathematical function computed by the system. Thus, for example, she notes:

> Marr's (1982) theory of early vision explains edge detection by positing the computation of the Laplacian of a Gaussian of the retinal array. The mechanism takes as input intensity values at points in the image and calculates the rate of intensity change over the image. (Egan 2017: 145)

From a computational point of view, this mathematical characterization is an exhaustive description of the retina's activity. Egan cites Marr, who says:

> Take the retina. I have argued that from a computational point of view, it signals $\nabla^2 G * I$ (the X channels) and its time derivative $\partial/\partial t(\nabla^2 G * I)$ (the Y channels). From a computational point of view, this is a precise specification of what the retina does. (1982: 337)

Egan admits that a full-fledged cognitive explanation requires the attachment of the FT characterization to the environment (such as a visual field) in which the system operates. She also observes that "one way to connect the abstract FT characterization to the target cognitive capacity is to attribute representational contents that are appropriate to the relevant cognitive domain" (2017: 147). But she argues that the latter attribution—of representational content—is not an integral part of the computational theory. In an earlier paper, she says:

> *Qua* computational device, it does not matter that input values represent *light intensities* and output values the rate of change of *light intensity*. The computational theory characterizes the visual filter as a member of a well understood class of mathematical devices that have nothing essentially to do with the transduction of light. (2010: 255)

In other words, we invoke representational content only *after* the computational-level theory has accomplished its task of specifying the mathematical function. The cognitive, intentional characterization is what Egan terms a *gloss* on the mathematical characterization provided by the computational theory. This intentional characterization "forms a bridge between the abstract, mathematical characterization that constitutes the explanatory core of the theory and the intentionally characterized pre-theoretic explananda that define the theory's cognitive domain" (2010: 256–257). But beyond this gloss, the representational content is immaterial to computational explanation and individuation.[29]

I agree with Egan on many points. I agree that the aim of computational theories is to explain information-processing capacities. I also agree that the explanatory core of computational theories is formal, and that an important part of this theory is the mathematical characterization of the computed (input-output) function. This aspect, in my view, coincides with the *what* of Marr's CL explanations. Egan is also correct in asserting that "the intentional characterization"—that retinal photoreceptors encode light intensities and that sets of V1 cells ("visual edges") encode physical edges—is one of the pre-theoretic explananda. It is often determined long before we invoke computational theory—for example, by electrophysiological experiments (e.g., Hubel and Wiesel 1962). I also agree with Egan (2017) that the formal theory need not map to "mechanism" in the sense required by mechanistic explanations. Some ("algorithmic") characterization of the mediating mechanism might be an integral part of computational explanation. But this characterization can take different forms—such as functional, dynamic, and sometimes mechanistic.

Where I disagree with Egan is on one crucial point: I think that the conjunction of the formal theory with the intentional characterization (and even with some characterization of the mediating mechanism) does not yet fully explain the information-processing task. We see that computing differentiation (zero-crossings of second-derivative Laplacians) leads to the activity of cells that encode physical edges (such as object boundaries). The fact that V1 cells detect edges is indicated by electrophysiological experiments. But what we do not see is *why* that is the case. We do not understand why computing differentiation does not lead to representation of, say, colors. And we do not understand why the system computes differentiation—and not, say, factorization or exponentiation—in order to generate representations of edges. Marr himself highlights these points. When Marr says, "From a computational point of view, this is a precise specification of what the retina does," he refers to *what* the retina does—not the *why*. After characterizing *what* early visual processes do, Marr says that "the term *edge* has a partly physical meaning—it makes us think of a real physical boundary, for example" (p. 68). He adds:

> All we have discussed so far are the zero values of a set of roughly band-pass second-derivative filters. We have no right to call these edges, or, if we do have a right, then we must say so and why. (p. 68)

[29] Contrary to Egan, many have argued that Marr's computational theories involve content—such as Burge (1986); Kitcher (1988); Segal (1989, 1991); Sterelny (1990); Davies (1991); Morton (1993); Shapiro (1993, 1997); Peacocke (1994); Silverberg (2006); and Sprevak (2010). They disagree with one another as to whether this content is "wide" or "narrow."

In short, then, we still must address the interrelations between the formal characterization of the (computed) function and the "intentional characterization" of the inputs and outputs.

How might we answer these queries? As suggested previously, the CL theories answer these *why* questions by pointing out that the input-output function also mirrors the relation in the target, between the entities represented by the inputs and outputs. When we see that the mathematical characterization of the external relation is also in terms of differentiation, we understand why differentiation—and not factorization (etc.)—leads to the detection of edges rather than colors (etc.). This additional part—the formal characterization of the morphism between the visual system and the visual field—is a crucial aspect of computational explanations. This mirroring relation constitutes the *why* aspect in Marr's CL explanations; more importantly, it is the distinctive aspect of computational theories that distinguishes them from other, non-computational mathematical characterizations of physical systems.

9.5.3 Chirimuuta's Optimality Explanations

In recent papers, Mazvita Chirimuuta (2014, 2018) has introduced the notion of *I*-minimal models in the context of computational explanations. These computational models are minimal in the sense that "they typically abstract away from many biophysical details of the neural system" (2014: 128). My focus here is on the *I*-aspect of *I*-minimal, which alludes to *interpretive models* (Dayan and Abbott 2001). These models are used alongside phenomenal (descriptive) and mechanistic ones and aim to explain why nervous systems operate as they do (see Section 6.3.4).

Although both Chirimuuta (as per Dayan and Abbott) and Marr agree that computational theories aim to answer questions such as why nervous systems operate the way they do, their answers go in different directions. Marr—at least in our interpretation—answers *why* questions in terms of input-output modeling. Chirimuuta answers *why* questions in terms of efficient coding principles. Her main example of such an optimality explanation is the normalization equation that models the cross-orientation suppression of simple cell response in the primary visual cortex and in other systems.[30] According to Chirimuuta, the computational explanation of this suppression behavior is anchored in the fact that this behavior is more efficient (optimal) in that it enables the network to transmit more information (see Section 6.3.4).

---

[30] Chirimuuta says that "the use of the term 'normalization' in neuroscience retains much of its original mathematical-engineering sense. It indicates a mathematical operation—a computation—not a biological mechanism" (2014: 142).

How does Chirimuuta's account, articulated in terms of efficient coding principles, square with my account of computation, which is in terms of modeling? My tentative answer is that the accounts are different because the *why* questions are somewhat different. But the questions are not unrelated. I am more concerned with questions such as: Why is a certain function f appropriate (or not) for a certain task? Why is f appropriate and not g or h? By contrast, Chirimuuta is concerned with a further question: Take all the functions $f_1, f_2, \ldots$ that are appropriate for the task. Why choose f_i rather than the other fs?[31] To see the difference, let us return to the theory of edge detection. One question we can ask, as Marr does, is why this computation is appropriate for detecting edges. The answer, I have suggested, is provided by the concept of modeling. In particular, this input-output function preserves sharp changes in reflectance and illumination in the visual field that happen to occur along physical edges (such as object boundaries) and that can be described in terms of differentiation. Other functions that do not preserve the pertinent relationship—such as factorization and exponentiation—are obviously not appropriate for edge detection.

However, there are other functions that might also be appropriate for the task. As Marr noticed, the visual system might detect edges by computing the extreme points of first-derivative operators, the second-order directional derivatives, or other appropriate functions. Thus, there is a further question: Why compute the zero-crossings of second-derivative Laplacian operators rather than other derivative (directional) operators that would also be appropriate? I think that Chirimuuta is concerned with this further question. Assuming that the task is responding to oriented lines ("edges"), her question is: Why compute the normalization equation (cross-orientation suppression) rather than, say, a simple linear response to the receptive-field properties?

The answer to that question often has to do with the efficiency of computation. Given that there is a limit to the amount of information processing possible in the brain, the expected simple-linear-response function might not be consistent with the brain's actual limitations. In that case, we appeal to efficient-coding principles and other canons of information theory. Indeed, Marr discusses this point of efficiency in some detail in his theory of edge detection (1982: 56ff.), where he writes that "the great advantage of using it [Laplacian operator] is economy of computation" (p. 56). The computation of the directional derivative operators is costly, whereas the use of Laplacian operators is efficient and satisfactory.[32]

My tentative proposal, then, is that computational theories might be concerned with a family of *why* questions about the operations of the system. The

---

[31] A similar question arises in relation to the various algorithms supporting the same function: *why* one algorithm is used rather than another.

[32] See van Rooij et al. (2019) for a more general discussion of intractability and cognition.

more basic questions are about the appropriateness of these operations to the task, and these are answered in terms of modeling. Other questions address the advantage of certain appropriate (i.e., modeling) operations over other appropriate operations, and these questions are answered in terms of optimality. There might be other kinds of questions as well, but they all depend, in my view, on the basic idea that computing is modeling.

9.6 Summary

I started the chapter with a characterization of *input-output modeling* (Section 9.1). A process is said to input-output model a given target when its input-output function and some relation in the target have a shared formal structure. This characterization led to a modeling definition of physical computation (Section 9.2). According to the definition I provided, a system computes if it implements a formalism whose input-output function is shared with a certain relation in the target (represented) domain. The next step was to show that modeling is often associated with computing (Section 9.3), that it plays a major methodological role in discovering what function is being computed (Section 9.4), and that it enhances a distinctive account of computational explanation (Section 9.5). This may not be enough to show that this modeling notion is consistent with every notion of computation that we have today, but it does demonstrate that the modeling notion of computation is forceful and pervasive—particularly in computational approaches in cognitive neuroscience.

Conclusion

According to the proposed account, a physical system computes just in case:

- The system implements a formalism whose input-output function is f.
- The system's input-output function mirrors some relation in a target domain.
- The mirroring input-output function and the mirrored (target) relation share the formal function (relation) f.
- The system's inputs and outputs represent the entities of the mirrored relation in the target domain.

How does this characterization square with the desiderata set out in Chapter 1? The main desideratum of the account is to correctly classify physical systems into *computing* and *non-computing* systems. In the category *the-right-things-compute*, the account deems artifact systems such as smartphones, laptops, and robots—as well as natural cognitive and nervous systems—to be computing systems. In the category *the-wrong-things-don't-compute*, the account deems stomachs, hurricanes, rocks, and many other non-representational systems to be non-computing systems. It also deems as non-computing representational systems such as screening and stamping, whose implemented input-output function f does not match the formal (shared) function underlying the relevant mirroring relation (if there is one at all) between the system and the represented target. This does not mean that these systems cannot possibly compute: if we were to assign content to the states of the stomach, and the other conditions are met, then the stomach could be regarded as computing. But as long as these systems do not satisfy these requirements, they do not compute. This result is consistent with the view of *very limited pancomputationalism*.

A key takeaway of the book is that the features that meet the classification desiderata are not the same as those that meet the taxonomy desideratum (which lists the features relevant to the classification of types of computation). Functional or architectural profiles do not distinguish computing from non-computing systems (or so I argue, mainly in Chapter 4), but they do distinguish one type of computation from another. Semantic properties, on the other hand, matter both to the identification of computation and to the identification of computational types. Semantic properties determine which of the implemented formalisms constitute the system's computational vehicle (or so I argue in Chapter 8).

The account meets a milder objectivity desideratum (PO1 & PO2). In fact, all but one of the conditions of computation are entirely objective and non-semantic. Both the implementation and mirroring conditions are defined in terms of morphism, plus a few additional (e.g., causal and counterfactual) constraints. Scientists discover which formalisms are being implemented; they do not assign them. The semantic properties of some computations might not be objective, however. It is reasonable to maintain that the contents of the states of smartphones and laptops are mind-dependent, in the sense that they are assigned by the designers or users. The semantic properties of other computing systems might be entirely objective (mind-independent). If the contents of cognitive and/or brain states are objective, then all the computational properties of these systems are objective as well. Thus, the account is consistent with the claims that the computational properties of some computing systems are entirely objective (PO1) and that some computational properties of all computing systems are entirely objective (PO2).

Finally, to the explanatory role of computation (the *utility* desideratum). Arguably, the explananda of computational theories are semantic, so-called information-processing tasks. One part of the explanation is the mathematical function that underpins the input-output semantic task (the *what* aspect in Marr's computational-level explanations). Many would argue that another part of the explanation is the process that mediates the inputs and outputs, described in abstract terms (which corresponds to Marr's *algorithmic level*). These two components, however, are not exclusive to computational explanations—we find them in many mathematical explanations of physical systems. The distinctive element of the computational explanation lies in demonstrating why the computed mathematical function is appropriate to the explanandum information-processing task (the *why* aspect in Marr's computational-level explanations). This, I have argued, is provided when we show that the mathematical input-output function preserves (mirrors) the relation between the entities represented by the inputs and outputs.

The proposed account meets Smith's scope criteria, as it provides the conditions for real-world examples of computing (the *empirical criterion*) and it acknowledges related concepts such as *implementation, algorithm,* and the semantic properties of computing systems (the *conceptual criterion*). However, an important conclusion of the book is that an account of physical computation need not—and in fact should not—be anchored in computability theory, automata theory, proof theory, and so on, all of which address certain kinds of computation and do not aim at characterizing computation in the physical world. Lastly, the proposed account aims to make sense of the claim that the mind/brain computes (the *cognitive criterion*), and highlights the methodological and explanatory roles of computation in current cognitive neuroscience (especially in Chapter 9). It does not aim, however, to reduce content to computation, which is the agenda of some philosophical theories of mind. In fact, the main claim of the book is that computation is defined by its semantic properties.

I will conclude by noting that I have left aside many important issues—such as accounting for miscomputation; more detailed discussions of other forms of ("natural" and "unconventional") computation; issues of usability and other epistemic constraints on computation; assessing the importance of computational complexity to an account of physical computation; and discussing the ethical implications of recent AI techniques. These topics certainly warrant careful consideration. In this book, however, my aim was more specific: to advance a semantic account that meets the basic desiderata of a theory of physical computation.

Acknowledgments

I started to think about physical computation after reading Itamar Pitowsky's seminal article on the subject (1990). Itamar was my M.A. advisor at the Hebrew University, and later on became a colleague and a close friend. Over the course of our friendship, I was fortunate to have many significant discussions with him, and we eventually published a joint paper on physical hypercomputation (Shagrir and Pitowsky 2003). Itamar had a brilliant mind and an engaging personality. He passed away prematurely in 2010. His friends and colleagues miss him greatly.

Jack Copeland and I have collaborated on many projects and papers in the past fifteen years. Our joint work on Gandy machines, accelerating machines, and computability theses has culminated in Chapter 3 of this book. But Jack's contribution goes far beyond this. During my visits to Christchurch and Jack's visits to Jerusalem, I had the opportunity to analyze, discuss, and argue with him about almost every topic that appears in the book. Gualtiero Piccinini and I have debated physical computation for almost twenty years now. Although we disagree about some of the fundamentals, we have had many enjoyable conversations over the years. Gualtiero's comments and criticism of my work continuously helped to shape and reshape my views. I am also grateful to Eli Dresner, Frances Egan, Nir Fresco, Jens Harbecke, Carl Posy, Nick Shea, Wilfried Sieg, and Mark Sprevak for intensive discussions on computation over the years.

My work has benefited from several projects and research groups. During my doctoral studies at the University of California at San Diego, I aimed to demonstrate that the notion of computation prevalent in computational approaches in the brain and cognitive sciences is different from the way computation is understood in central philosophical theories (computational theory of mind, computational functionalism, and so on). I was lucky to be part of a vibrant community of philosophers and scientists who grappled extensively with issues of computation and the brain. I'm thankful to my teachers at UCSD: my supervisor Patricia Churchland, Gila Sher, Steve Yalowitz, Paul Churchland, Philip Kitcher, Patricia Kitcher, Sandy Mitchel, Adrian Cussins, Francis Crick, David Zipser, and others. No less important was the interaction with my fellow graduate students, among them Georg Schwarz, Valerie Hardcastle, Rick Grush, Jonathan Gunderson, Bruce Glymour, Adina Roskies, Aare Laasko, Brian Keeley, Gillian Barker, Steve Quartz, Kyle Stanford, and Joe Ramsey.

In the spring of 2011, I was a member of the Computation and the Brain group at the Israel Institute for Advanced Studies (IIAS) at the Hebrew University of Jerusalem, together with Eli Dresner, Arnon Levy, Hilla Jacobson, Bill Bechtel, Adele Abrahamsen, Frances Egan, Bob Matthews, and Gualtiero Piccinini. We had many lively discussions on the relationship between computational explanations and models and other mechanistic explanations in the cognitive and brain sciences. This has been a wonderfully fruitful experience that resulted in two joint papers with Bill (Bechtel and Shagrir 2015; Shagrir and Bechtel 2017) and one with Gualtiero (Piccinini and Shagrir 2014). A follow-up endeavor was our German-Israeli research group on Causation and Computation in Cognitive Neuroscience (2015–2018). The group's members were Jens Harbecke, Vera Hoffmann-Kolss, Jan Philipp Köster, Carlos Zednik, and my students Lotem Elber-Dorozko, Ori Hacohen, and Shahar Hechtlinger. My work on joint papers with Jens (Harbecke and Shagrir 2019) and Lotem (Elber-Dorozko and Shagrir 2019) played a role in my critique of the mechanistic account (Chapter 6).

In the fall of 2015, I was a member of the Computability: Historical, Logical, and Philosophical Foundations group at IIAS, together with Jack Copeland, Eli Dresner, Nir Fresco, Carl Posy, Diane Proudfoot, Stewart Shapiro, and Moshe Vardi. We had extremely fruitful discussions on the relations between effective computation, which was the focus of the work of the founders of computability, and more recent computational paradigms; these discussions are reflected in Copeland, Dresner, Proudfoot, and Shagrir (2017). We also intensively debated the relations between the notions of effective computation and physical computation, the scope of triviality results, and the nature of the implementation relation. I wrote the first drafts of Chapters 5 and 8 during this time.

I was privileged to spend a sabbatical in 2004 as a fellow at the Center of Philosophy of Science at the University of Pittsburgh; to visit the University of Canterbury, New Zealand, three times, in 2008, 2012, and 2016 (in the latter two years as an Erskine Fellow); and to spend the past few summers at Clare Hall College, Cambridge, where I wrote large parts of the book.

Special thanks go to Eli Dresner, Frances Egan, Arnon Levy, and Gualtiero Piccinini, who read and commented on the entire manuscript. I'm also grateful to those who were generous enough to participate in the reading group on my book in the summer of 2020: Dimitri Coelho Mollo, Joe Dewhurst, Gordana Dodig-Crnkovic, Lotem Elber-Dorozko, Aya Evron, Nir Fresco, Ariel Furstenberg, Ori Hacohen, Jens Harbecke, Meir Hemmo, Arnon Levy, Marcin Miłkowski, Jonathan Najenson, Philippos Papayannopoulos, Carl Posy, Ofra Rechter, Lavi Rosenthal, Paul Schweizer, Orly Shenker, Adam Singer, and Gal Vishne. The discussion and weekly comments mailed to me by the participants were crucial for the final revisions.

In addition to those mentioned above, I benefited from intellectual interaction with many other scholars and scientists who contributed in one way or another to this work: Scott Aaronson, Darren Abramson, Dorit Aharonov, Ehud Ahissar, Merav Ahissar, Ken Aizawa, Colin Allen, Neal Anderson, Arnon Avron, Yemima Ben Menahem, Michael Ben-Or, Udi Boker, Sacha Bourgeois-Gironde, Selmer Bringsjord, Mark Burgin, Meir Buzaglo, Cris Calude, Doug Campbell, Rosa Cao, David Chalmers, Carol Cleland, Matteo Colombo, Leo Cory, Carl Craver, Ron Chrisley, Nahum Dershowitz, Yuval Dolev, Chris Eliasmith, Jonathan Bowen, Martin Davis, Currie Figdor, Thomas Forster, Carl Gillett, Yosef Grodzinsky, Yuri Gurevich, Amit Hagar, Amir Horowitz, Paul Humphreys, Andreas Hüttemann, Michael John-Turp, David Kaplan, Colin Klein, Beate Krickel, Saul Kripke, Otto Lappi, Yakir Levin, Yonatan Loewenstein, Holger Lyre, Peter Machamer, Corey Maley, Ruth Manor, Jonathan Mills, Vincent Müller, Tom Polger, Hilary Putnam, Paula Quinon, Michael Rabin, Michael Rescorla, Michael Roubach, Anna-Mari Rusanen, Gil Sagi, Richard Samuels, Andrea Scarantino, Matthias Scheutz, Susan Schneider, Larry Shapiro, Hava Siegelmann, Aaron Sloman, Brian Smith, Haim Sompolinsky, Etye Steinberg, Mark Steiner, Tali Tishby, Tony Travis, Ray Turner, Shimon Ullman, Michael Weisberg, Jan Woleński, and Jonathan Yaari.

I could not have completed the manuscript without the constant encouragement of the editors of the Oxford Studies in Philosophy of Science series. I'm grateful to Paul Humphreys, Peter Ohlin, and Kyle Stanford for their patience while I filled a number of time-consuming administrative positions during the past decade.

Many thanks to Sara Tropper and Jonathan Orr-Stav, who edited the manuscript; to Marc Sherman, who prepared the index; to Noa Weiss, who compiled the bibliography; to Maya Lahat Kerman and Gili Meisler, who assisted with the final touches; to Ori Kerman who generously offered his expertise, and advised me with the cover design; and to Zehava Cohen, who created the original figures and reproduced previously published figures.

This work was supported by research grants from the Israel Science Foundation (830/18) and the German-Israeli Foundation for Scientific Research and Development. This book was published with the support of the Israel Science Foundation.

Parts of this work draw on previous publications. Chapter 2 draws on (1) Shagrir, Oron. 2006. "Gödel on Turing on Computability." In *Church's Thesis After 70 Years*, edited by Adam Olszewski, Jan Wolenski, and Robert Janusz, pp. 393–419. Heusenstamm: Ontos Verlag; and (2) Copeland, B. Jack, and Oron Shagrir. 2013. "Turing Versus Gödel on Computability and the Mind." In *Computability: Turing, Gödel, Church, and Beyond*, edited by B. Jack Copeland, Carl Posy, and Oron Shagrir, pp. 1–33. Cambridge, MA: MIT Press.

Chapter 3 is based on my joint work with Jack Copeland on machine computation: (1) Copeland, B. Jack, and Oron Shagrir. 2007. "Physical Computation: How General Are Gandy's Principles for Mechanisms?" *Minds and Machines 17*: pp. 217–231; (2) Copeland, B. Jack, and Oron Shagrir. 2011. "Do Accelerating Turing Machines Compute the Uncomputable?" *Minds and Machines 21*: pp. 221–239; (3) Copeland, B. Jack, and Oron Shagrir. 2019. "The Church-Turing Thesis: Logical Limit or Breachable Barrier?" *Communications of the ACM 62*: pp. 66–74.

Chapter 6 draws on (1) Shagrir, Oron. 2017. "Review of *Physical Computation: A Mechanistic Account* by Gualtiero Piccinini." *Philosophy of Science 84*: pp. 604–612; and (2) Elber-Dorozko, Lotem, and Oron Shagrir. 2019. "Integrating Computation into the Mechanistic Hierarchy in the Cognitive and Neural Sciences." *Synthese*, May 13, 2019.

Chapters 7 and 8 draw on Shagrir, Oron. 2020. "In Defense of the Semantic View of Computation." *Synthese 197*: pp. 4083–4108.

Chapter 9 draws on (1) Shagrir, Oron. 2012. "Structural Representations and the Brain." *The British Journal for the Philosophy of Science 63*: pp. 519–545; (2) Shagrir, Oron, and William Bechtel. 2017. "Marr's Computational Level and Delineating Phenomena." In *Explanation and Integration in Mind and Brain Science*, edited by D. M. Kaplan, pp. 190–214. New York: Oxford University Press; (3) Shagrir, Oron. 2018. "The Brain as an Input–Output Model of the World." *Minds and Machines 28*: pp. 53–75.

I dedicate this book to my beloved family: my wife, Iris, and my sons, Tomer and Eyal. I cannot thank them enough for their love, support, and understanding throughout the years.

Bibliography

Aaronson, Scott. 2013. "Why Philosophers Should Care About Computational Complexity." In *Computability: Turing, Gödel, Church, and Beyond*, edited by B. J. Copeland, C. Posy, and O. Shagrir, pp. 261–327. Cambridge, MA: MIT Press.

Adams, Rod. 2011. *An Early History of Recursive Functions and Computability from Gödel to Turing*. Boston: Docent Press.

Aharonov, Dorit, and Umesh V. Vazirani. 2013. "Is Quantum Mechanics Falsifiable? A Computational Perspective on the Foundations of Quantum Mechanics." In *Computability: Turing, Gödel, Church, and Beyond*, edited by B. J. Copeland, C. Posy, and O. Shagrir, pp. 329–349. Cambridge, MA: MIT Press.

Aizawa, Kenneth, and Carl Gillett. 2009. "The (Multiple) Realization of Psychological and Other Properties in the Sciences." *Mind & Language 24*: pp. 181–208.

Amit, Daniel J. 1989. *Modeling Brain Function: The World of Attractor Neural Networks*. Cambridge: Cambridge University Press.

Amit, Daniel J., and Stefano Fusi. 1994. "Learning in Neural Networks with Material Synapses." *Neural Computation 6*: pp. 957–982.

Amit, Daniel J., Hanoch Gutfreund, and Haim Sompolinsky. 1985. "Spin-Glass Models of Neural Networks." *Physical Review A 32*: pp. 1007–1018.

Andersen, Holly. 2014a. "A Field Guide to Mechanisms: Part I." *Philosophy Compass 9*: pp. 274–283.

Andersen, Holly. 2014b. "A Field Guide to Mechanisms: Part II." *Philosophy Compass 9*: pp. 284–293.

Andersen, Richard A., Greg K. Essick, and Ralph M. Siegel. 1985. "Encoding of Spatial Location by Posterior Parietal Neurons." *Science 230*: pp. 456–458.

Anderson, James A., Andras Pellionisz, and Edward Rosenfeld (eds.). 1990. *Neurocomputing 2: Directions for Research*. Cambridge, MA: MIT Press.

Anderson, James A., and Edward Rosenfeld (eds.). 1988. *Neurocomputing: Foundations of Research*. Cambridge, MA: MIT Press.

Andréka, Hajnal, Judit X. Madarász, István Németi, Péter Németi, and Gergely Székely. 2018. "Relativistic Computation." In *Physical Perspectives on Computation, Computational Perspectives on Physics*, edited by M. E. Cuffaro and S. C. Fletcher, pp. 195–218. Cambridge: Cambridge University Press.

Andréka, Hajnal, István Németi, and Péter Németi. 2009. "General Relativistic Hypercomputing and Foundation of Mathematics." *Natural Computing 8*: pp. 499–516.

Arora, Sanjeev, and Boaz Barak. 2009. *Computational Complexity: A Modern Approach*. Cambridge: Cambridge University Press.

Astrachan, Owen L. 2000. *A Computer Science Tapestry: Exploring Programming and Computer Science with C++*. New York: McGraw-Hill.

Avigad, Jeremy, and Vasco Brattka. 2014. "Computability and Analysis: The Legacy of Alan Turing." In *Turing's Legacy: Developments from Turing's Ideas in Logic*, edited by R. Downey, pp. 1–47. Cambridge: Cambridge University Press.

Barrett, David. 2014. "Functional Analysis and Mechanistic Explanation." *Synthese 191*: pp. 2695–2714.

Barrett, Jeffrey A., and Wayne Aitken. 2010. "A Note on the Physical Possibility of Transfinite Computation." *The British Journal for the Philosophy of Science* 61: pp. 867–874.

Bartels, Andreas. 2006. "Defending the Structural Concept of Representation." *THEORIA. Revista de Teoría, Historia y Fundamentos de la Ciencia* 21: pp. 7–19.

Bassett, Joshua P., and Jeffrey S. Taube. 2001. "Neural Correlates for Angular Head Velocity in the Rat Dorsal Tegmental Nucleus." *Journal of Neuroscience* 21: pp. 5740–5751.

Bechtel, William, and Adele A. Abrahamsen. 2002. *Connectionism and the Mind: Parallel Processing, Dynamics, and Evolution in Networks*. 2nd ed. Oxford: Blackwell.

Bechtel, William, and Robert C. Richardson. 1993. *Discovering Complexity: Decomposition and Localization as Strategies in Scientific Research*. Princeton: Princeton University Press.

Bechtel, William, and Oron Shagrir. 2015. "The Non-Redundant Contributions of Marr's Three Levels of Analysis for Explaining Information-Processing Mechanisms." *Topics in Cognitive Science* 7: pp. 312–322.

Becker, Wolfgang, and Horst-Manfred Klein. 1973. "Accuracy of Saccadic Eye Movements and Maintenance of Eccentric Eye Positions in the Dark." *Vision Research* 13: pp. 1021–1034.

Beggs, Edwin J., and John. V. Tucker. 2006. "Embedding Infinitely Parallel Computation in Newtonian Kinematics." *Applied Mathematics and Computation* 178: pp. 25–43.

Bernstein, Ethan, and Umesh Vazirani. 1997. "Quantum Complexity Theory." *SIAM Journal on Computing* 26: pp. 1411–1473.

Bishop, John Mark. 2009. "A Cognitive Computation Fallacy? Cognition, Computations and Panpsychism." *Cognitive Computation* 1: pp. 221–233.

Blackmon, James. 2013. "Searle's Wall." *Erkenntnis* 78: pp. 109–117.

Blake, Ralph M. 1926. "The Paradox of Temporal Process." *The Journal of Philosophy* 23: pp. 645–654.

Blass, Andreas, Nachum Dershowitz, and Yuri Gurevich. 2009. "When Are Two Algorithms the Same?" *Bulletin of Symbolic Logic* 15: pp. 145–168.

Blass, Andreas, and Yuri Gurevich. 2003. "Abstract State Machines Capture Parallel Algorithms." *ACM Trans. on Computational Logic* 4: pp. 578–651.

Blass, Andreas, and Yuri Gurevich. 2006. "Ordinary Interactive Small-Step Algorithms I." *ACM Trans. on Computational Logic* 7: pp. 363–419.

Blass, Andreas, and Yuri Gurevich. 2007a. "Ordinary Interactive Small-Step Algorithms II." *ACM Trans. on Computational Logic* 8: article 15.

Blass, Andreas, and Yuri Gurevich. 2007b. "Ordinary Interactive Small-Step Algorithms III." *ACM Trans. Computational Logic* 8: article 16.

Block, Ned. 1978. "Troubles with Functionalism." In *Perception and Cognition*, edited by W. Savage, pp. 9–26. Minnesota: University of Minnesota Press.

Block, Ned. 1986. "Advertisement for a Semantics for Psychology." *Midwest Studies in Philosophy* 10: pp. 615–678.

Block, Ned. 1990. "Can the Mind Change the World?" In *Meaning and Method: Essays in Honor of Hilary Putnam*, edited by G. Boolos, pp. 137–170. Cambridge: Cambridge University Press.

Blum, Lenore, Felipe Cucker, Michael Shub, and Steve Smale. 1998. *Complexity and Real Computation*. Berlin: Springer.

Boden, Margaret A. 2006. *Mind as Machine: A History of Cognitive Science*, vols. 1 *and* 2, New York: Oxford University Press.

Boghossian, Paul A. 1989. "The Rule-Following Considerations." *Mind 98*: pp. 507–549.
Boker, Udi, and Nahum Dershowitz. 2009. "The Influence of Domain Interpretations on Computational Models." *Journal of Applied Mathematics and Computation 215*: pp. 1323–1339.
Boker, Udi, and Nachum Dershowitz. Forthcoming. "What Is the Church-Turing Thesis?" In *Axiomatic Thinking I*, edited by F. Ferreira, R. Kahle, and G. Sommaruga. Cham: Springer.
Bontly, Thomas. 1998. "Individualism and the Nature of Syntactic States." *The British Journal for the Philosophy of Science 49*: pp. 557–574.
Boolos, George S., and Richard S. Jeffrey. 1989. *Computability and Logic*. 3rd ed. Cambridge: Cambridge University Press.
Boone, Worth, and Gualtiero Piccinini. 2016. "Mechanistic Abstraction." *Philosophy of Science 83*: pp. 686–697.
Botvinick, Matthew M., Yael Niv, and Andrew G. Barto. 2009. "Hierarchically Organized Behavior and Its Neural Foundations: A Reinforcement Learning Perspective." *Cognition 113*: pp. 262–280.
Bournez, Olivier, Manuel L. Campagnolo, Daniel S. Graça, and Emmanuel Hainry. 2006. "The General Purpose Analog Computer and Computable Analysis Are Two Equivalent Paradigms of Analog Computation." In *International Conference on Theory and Applications of Models of Computation*, edited by J. Y. Cai, S. B. Cooper, A. Li, pp. 631–643. Berlin, Heidelberg: Springer.
Brabazon, Anthony, Mark O'Neill, and Seán McGarraghy. 2015. *Natural Computing Algorithms*. Berlin: Springer.
Braverman, Mark, and Stephen Cook. 2006. "Computing over the Reals: Foundations for Scientific Computing." *Notices of the AMS 53*: pp. 318–329.
Bringsjord, Selmer, Owen Kellett, Andrew Shilliday, Joshua Taylor, Bram van Heuveln, Yingrui Yang, Jeffrey Baumes, and Kyle Ross. 2006. "A New Gödelian Argument for Hypercomputing Minds Based on the Busy Beaver Problem." *Applied Mathematics and Computation 176*: pp. 516–530.
Bringsjord, Selmer, and Michael John Zenzen. 2003. *Superminds: People Harness Hypercomputation, and More*. Dordrecht: Kluwer.
Brown, Curtis. 2012. "Combinatorial-State Automata and Models of Computation." *Journal of Cognitive Science 13*: p. 51–73.
Buckner, Cameron, and James Garson. 2019. "Connectionism." *The Stanford Encyclopedia of Philosophy*, edited by E. N. Zalta. https://plato.stanford.edu/archives/fall2019/entries/connectionism/.
Burge, Tyler. 1986. Individualism and Psychology. *The Philosophical Review 95*: pp. 3–45.
Burge, Tyler. 2010. *Origins of Objectivity*. New York: Oxford University Press.
Button, Tim. 2009. "SAD Computers and Two Versions of the Church-Turing Thesis." *The British Journal for the Philosophy of Science 60*: pp. 765–792.
Buzaglo, Meir. 2002. *The Logic of Concept Expansion*. Cambridge: Cambridge University Press.
Calude, Cristian S., Michael J. Dinneen, Monica Dumitrescu, and Karl Svozil. 2010. "Experimental Evidence of Quantum Randomness Incomputability." *Physical Review A 82*: pp. 022102-1–022102-8.
Calude, Cristian S., and Boris Pavlov. 2002. "Coins, Quantum Measurements, and Turing's Barrier." *Quantum Information Processing 1*: pp. 107–127.

Calude, Cristian S., and Ludwig Staiger. 2010. "A Note on Accelerated Turing Machines." *Mathematical Structures in Computer Science 20*: pp. 1011–1017.

Calude, Cristian S. and Karl Svozil. 2008. "Quantum Randomness and Value Indefiniteness." *Advanced Science Letters 1*: pp. 165–168.

Campbell, Douglas Ian, and Yi Yang. 2019. "Does the Solar System Compute the Laws of Motion?" *Synthese*, May 31, 2019.

Cannon, Stephen C., and David A. Robinson. 1985. "An Improved Neural-Network Model for the Neural Integrator of the Oculomotor System: More Realistic Neuron Behavior." *Biological Cybernetics 53*: pp. 93–108.

Cannon, Stephen C., and David A. Robinson. 1987. "Loss of the Neural Integrator of the Oculomotor System from Brain Stem Lesions in Monkey." *Journal of Neurophysiology 57*: pp. 1383–1409.

Cao, Rosa. 2018. "Computational Explanations and Neural Coding." In *The Routledge Handbook of the Computational Mind*, edited by M. Sprevak and M. Colombo, pp. 283–296. London: Routledge.

Carandini, Matteo, and David J. Heeger. 1994. "Summation and Division by Neurons in Primate Visual Cortex." *Science 264*: pp. 1333–1336.

Carandini, Matteo, and David. J. Heeger. 2012. "Normalization as a Canonical Neural Computation." *Nature Reviews Neuroscience 13*: pp. 51–62.

Care, Charles. 2010. *Technology for Modelling: Electrical Analogies, Engineering Practice, and the Development of Analogue Computing*. Berlin: Springer.

Chaitin, Gregory J. 1977. "Algorithmic Information Theory." *IBM Journal of Research and Development 21*: pp. 350–359.

Chalmers, David J. 1994. "On Implementing a Computation." *Minds and Machines 4*: pp. 391–402.

Chalmers, David J. 1996. "Does a Rock Implement Every Finite-State Automaton?" *Synthese 108*: pp. 309–333.

Chalmers, David J. 2004. "Epistemic Two-Dimensional Semantics." *Philosophical Studies 118*: pp. 153–226.

Chalmers, David J. 2011. "A Computational Foundation for the Study of Cognition." *Journal of Cognitive Science 12*: pp. 323–357.

Chalmers, David. 2012. "The Varieties of Computation: A Reply." *Journal of Cognitive Science 13*: pp. 211–248.

Chater, Nick, Joshua B. Tenenbaum, and Alan Yuille. 2006. "Probabilistic Models of Cognition: Conceptual Foundations." *Trends in Cognitive Sciences 10*: pp. 287–291.

Chemero, Anthony. 2009. *Radical Embodied Cognitive Science*. Cambridge, MA: MIT Press.

Chirimuuta, Mazviita. 2014. "Minimal Models and Canonical Neural Computations: The Distinctness of Computational Explanation in Neuroscience." *Synthese 191*: pp. 127–153.

Chirimuuta, Mazviita. 2018. "Explanation in Computational Neuroscience: Causal and Non-Causal." *The British Journal for the Philosophy of Science 69*: pp. 849–880.

Chomsky, Noam. 1965. *Aspects of the Theory of Syntax*. Cambridge, MA: MIT Press.

Chomsky, Noam. 1980. *Rules and Representations*. New York: Columbia University Press.

Chrisley, Ronald L. 1994. "Why Everything Doesn't Realize Every Computation." *Minds and Machines 4*: pp. 403–420.

Church, Alonzo. 1933. "A Set of Postulates for the Foundation of Logic II." *Annals of Mathematics 34*: pp. 839–864.

Church, Alonzo. 1936a. "An Unsolvable Problem of Elementary Number Theory." *American Journal of Mathematics 58*: pp. 345–363. Reprinted in *The Undecidable: Basic Papers on Undecidable Propositions, Unsolvable Problems and Computable Functions*, edited by M. Davis (1965), pp. 88–107. Page references are to Davis.

Church, Alonzo. 1936b. "A Note on the Entscheidungsproblem." *Journal of Symbolic Logic 1*: pp. 40–41.

Church, Alonzo. 1937a. "Review of Turing (1936)." *Journal of Symbolic Logic 2*: pp. 42–43.

Church, Alonzo. 1937b. "Review of Post (1936)." *Journal of Symbolic Logic 2*: p. 43.

Church, Alonzo. 1941. *The Calculi of Lambda-Conversion*. Princeton: Princeton University Press.

Churchland, Patricia S. 1986. *Neurophilosophy: Toward a Unified Science of the Mind-Brain*. Cambridge, MA: MIT Press.

Churchland, Patricia S., Christof Koch, and Terrence J. Sejnowski. 1990. "What Is Computational Neuroscience?" In *Computational Neuroscience*, edited by E. L. Schwartz, pp. 46–55. Cambridge, MA: MIT Press.

Churchland, Patricia S., and Terrence J. Sejnowski. 1992. *The Computational Brain*. Cambridge, MA: MIT Press.

Churchland, Paul M. 1981. "Eliminative Materialism and Propositional Attitudes." *The Journal of Philosophy 78*: pp. 67–90.

Churchland, Paul M. 1989. *A Neurocomputational Perspective: The Nature of Mind and the Structure of Science*. Cambridge, MA: MIT Press.

Churchland, Paul M. 2007. *Neurophilosophy at Work*. Cambridge: Cambridge University Press.

Clark, Andy. 2013. "Whatever Next? Predictive Brains, Situated Agents, and the Future of Cognitive Science." *Behavioral and Brain Sciences 36*: pp. 181–253.

Clark, Andy. 2015. *Surfing Uncertainty: Prediction, Action, and the Embodied Mind*. New York: Oxford University Press.

Cleland, Carol E. 1993. "Is the Church-Turing Thesis True?" *Minds and Machines 3*: pp. 283–312.

Cleland, Carol E. 2002. "On Effective Procedures." *Minds and Machines 12*: pp. 159–179.

Cobham, Alan. 1964. "The Intrinsic Computational Difficulty of Functions." In *Logic, Methodology and Philosophy of Science, Proceedings of the 1964 International Congress*, edited by Y. Bar-Hillel, pp. 24–30. Amsterdam: North-Holland.

Coelho Mollo, Dimitri. 2018. "Functional Individuation, Mechanistic Implementation: The Proper Way of Seeing the Mechanistic View of Concrete Computation." *Synthese 195*: pp. 3477–3497.

Coelho Mollo, Dimitri. 2019. "Are There Teleological Functions to Compute?" *Philosophy of Science 86*: pp. 431–452.

Coelho Mollo, Dimitri. Forthcoming. "Against Computational Perspectivalism." *The British Journal for the Philosophy of Science*.

Cohen, Rina S., and Arie Y. Gold. 1978. "ω-Computations on Turing Machines." *Theoretical Computer Science 6*: pp. 1–23.

Collett, Matthew, and Thomas S. Collett. 2000. "How do Insects Use Path Integration for Their Navigation?" *Biological Cybernetics 83*: pp. 245–259.

Colombo, Matteo. 2021. "(Mis) Computation in Computational Psychiatry." In *Neural Mechanisms: New Challenges in the Philosophy of Neuroscience* (Studies in Brain and Mind, vol 17), edited by F. Calzavarini and M. Viola, pp. 427–448. Cham: Springer.

Conklin, John, and Chris Eliasmith. 2005. "A Controlled Attractor Network Model of Path Integration in the Rat." *Journal of Computational Neuroscience 18*: pp. 183–203.
Copeland, B. Jack. 1996. "What Is Computation?" *Synthese 108*: pp. 335–359.
Copeland, B. Jack. 1997. "The Broad Conception of Computation." *American Behavioral Scientist 40*: pp. 690–716.
Copeland, B. Jack. 1998. "Even Turing Machines Can Compute Uncomputable Functions." In *Unconventional Models of Computation*, edited by C. S. Calude, J. Casti, and M. J. Dinneen, pp. 150–164. Berlin: Springer.
Copeland, B. Jack. 2000. "Narrow Versus Wide Mechanism: Including a Re-Examination of Turing's Views on the Mind-Machine Issue." *Journal of Philosophy 97*: pp. 1–32.
Copeland, B. Jack. 2002a. "Accelerating Turing Machines." *Minds and Machines 12*: pp. 281–301.
Copeland, B. Jack. 2002b. "Hypercomputation." *Minds and Machines 12*: pp. 461–502.
Copeland, B. Jack. 2003. "Computation." In *The Blackwell Guide to the Philosophy of Computing and Information*, edited by L. Floridi, pp. 3–17. Oxford: Blackwell.
Copeland, B. Jack. 2004a. *The Essential Turing: Seminal Writings in Computing, Logic, Philosophy, Artificial Intelligence, and Artificial Life Plus the Secrets of Enigma*. Oxford, New York: Oxford University Press.
Copeland, B. Jack. 2004b. "Computable Numbers: A Guide." In *The Essential Turing: Seminal Writings in Computing, Logic, Philosophy, Artificial Intelligence, and Artificial Life, Plus the Secrets of Enigma*, edited by B. Jack Copeland, pp. 5–57. New York: Oxford University Press.
Copeland, B. Jack. 2004c. "Hypercomputation: Philosophical Issues." *Theoretical Computer Science 317*: pp. 251–267.
Copeland, B. Jack. 2006. "Turing's Thesis." In *Church's Thesis After 70 Years*, edited by A. Olszewski, J. Wolenski, and R., pp. 147–174. Heusenstamm: Ontos Verlag.
Copeland, B. Jack. 2012. *Turing: Pioneer of the Information Age*. New York: Oxford University Press.
Copeland, B. Jack. 2015. "The Church-Turing Thesis." In *The Stanford Encyclopedia of Philosophy*, edited by E. N. Zalta. https://plato.stanford.edu/archives/sum2015/entries/church-turing/.
Copeland, B. Jack, Eli Dresner, Diane Proudfoot, and Oron Shagrir. 2016. "Time to Reinspect the Foundations?" *Communications of the ACM 59*: pp. 34–36.
Copeland, B. Jack, and Diane Proudfoot. 2010. "Deviant Encodings and Turing's Analysis of Computability." *Studies in History and Philosophy of Science 41*: 247–252.
Copeland, B. Jack, and Oron Shagrir. 2007. "Physical Computation: How General Are Gandy's Principles for Mechanisms?" *Minds and Machines 17*: pp. 217–231.
Copeland, B. Jack, and Oron Shagrir. 2011. "Do Accelerating Turing Machines Compute the Uncomputable?" *Minds and Machines 21*: pp. 221–239.
Copeland, B. Jack, and Oron Shagrir. 2013. "Turing Versus Gödel on Computability and the Mind." In *Computability: Turing, Gödel, Church, and Beyond*, edited by B. J. Copeland, C. Posy, and O. Shagrir, pp. 1–33. Cambridge, MA: MIT Press.
Copeland, B. Jack, and Oron Shagrir. 2019. "The Church-Turing Thesis: Logical Limit or Breachable Barrier?" *Communications of the ACM 62*: pp. 66–74.
Copeland, B. Jack, Oron Shagrir, and Mark D. Sprevak. 2018. "Zuse's Thesis, Gandy's Thesis, and Penrose's Thesis." In *Physical Perspectives on Computation, Computational Perspectives on Physics*, edited by M. Cuffaro and S. Fletcher, pp. 39–59. Cambridge: Cambridge University Press.

Copeland, B. Jack, Mark D. Sprevak, and Oron Shagrir. 2017. "Is the Whole Universe a Computer?" In *The Turing Guide: Life, Work, Legacy*, edited by B. J Copeland, J. Bowen, M. Sprevak, and R. Wilson, pp. 445–462. New York: Oxford University Press.

Copeland, B. Jack, and Richard Sylvan. 1999. "Beyond the Universal Turing Machine." *Australasian Journal of Philosophy 77*: pp. 46–66.

Crane, Tim. 1990. "The Language of Thought: No Syntax Without Semantics." *Mind & Language 5*: pp. 187–212.

Crane, Tim. 2016. *The Mechanical Mind: A Philosophical Introduction to Minds, Machines and Mental Representation*. 3rd ed. London: Routledge.

Craver, Carl F. 2007. *Explaining the Brain: Mechanisms and the Mosaic Unity of Neuroscience*. New York: Oxford University Press.

Craver, Carl F. 2013. "Functions and Mechanisms: A Perspectivalist View." In *Functions*, edited by P. Huneman, pp. 133–158. Berlin: Springer.

Craver, Carl F. 2016. "The Explanatory Power of Network Models." *Philosophy of Science 83*: pp. 698–709.

Craver, Carl F., and David M. Kaplan. 2020. "Are More Details Better? On the Norms of Completeness for Mechanistic Explanations." *The British Journal for the Philosophy of Science 71*: pp. 287–319.

Cummins, Robert C. 1975. "Functional Analysis." *The Journal of Philosophy 72*: pp. 741–765.

Cummins, Robert C. 1983. *The Nature of Psychological Explanation*. Cambridge, MA: MIT Press.

Cummins, Robert C. 1989. *Meaning and Mental Representation*. Cambridge, MA: MIT Press.

Cummins, Robert C. 1996. *Representations, Targets and Attitudes*. Cambridge, MA: MIT Press.

Cummins, Robert C. 2000. "'How Does It Work?' Versus 'What Are the Laws?': Two Conceptions of Psychological Explanation." In *Explanation and Cognition*, edited by F. C. Keil and R. A. Wilson, pp. 117–145. Cambridge, MA: MIT Press.

Cummins, Robert C., and Georg Schwarz. 1991. "Connectionism, Computation, and Cognition." In *Connectionism and the Philosophy of Mind*, edited by T. Horgan and J. Tienson, pp. 60–73. Dordrecht: Kluwer.

Da Costa, Newton C. A., and Steven French. 2003. *Science and Partial Truth: A Unitary Approach to Models and Scientific Reasoning*. New York: Oxford University Press.

Daniels, Norman. 2020. "Reflective Equilibrium." *The Stanford Encyclopedia of Philosophy*, edited by E. N. Zalta. https://plato.stanford.edu/archives/sum2020/entries/reflective-equilibrium/.

Dasgupta, Dipankar. 1993. "An Overview of Artificial Immune Systems and Their Applications." In *Artificial Immune Systems and Their Applications*, edited by D. Dasgupta, pp. 3–21. Berlin: Springer.

Dasgupta, Dipankar. 1997. "Artificial Neural Networks and Artificial Immune Systems: Similarities and Differences." *1997 IEEE International Conference on Systems, Man, and Cybernetics. Computational Cybernetics and Simulation, vol. 1*. pp. 873–878.

Davies, Brian E. 2001. "Building Infinite Machines." *The British Journal for the Philosophy of Science 52*: pp. 671–682.

Davies, Martin. 1991. "Individualism and Perceptual Content." *Mind 100*: pp. 461–484.

Davis, Martin. 1958. *Computability and Unsolvability*. New York: McGraw-Hill.

Davis, Martin (ed.). 1965. *The Undecidable: Basic Papers on Undecidable Propositions, Unsolvable Problems and Computable Functions*. New York: Raven.

Davis, Martin. 1982. "Why Gödel Didn't Have Church's Thesis." *Information and Control* 54: pp. 3–24.

Davis, Martin. 2000. *The Universal Computer: The Road from Leibniz to Turing*. New York: W. W. Norton.

Dayan, Peter, and Laurence F. Abbott. 2001. *Theoretical Neuroscience: Computational and Mathematical Modeling of Neural Systems*. Cambridge, MA: MIT Press.

Dean, Walter. 2016. "Algorithms and the Mathematical Foundations of Computer Science." In *Gödel's Disjunction: The Scope and Limits of Mathematical Knowledge*, edited by L. Horsten and P. Welch, pp. 19–66. New York: Oxford University Press.

Dean, Walter. 2019. "Computational Complexity Theory and the Philosophy of Mathematics." *Philosophia Mathematica 27*: pp. 381–439.

De Mol, Liesbeth. 2012. "Generating, Solving and the Mathematics of Homo Sapiens: Emil Post's Views on Computation." In *A Computable Universe: Understanding Computation & Exploring Nature as Computation*, edited by H. Zenil, pp. 45–62. River Edge, NJ: World Scientific.

Demopoulos, William. 1987. "On Some Fundamental Distinctions of Computationalism." *Synthese 70*: pp. 79–96.

Dennett, Daniel C. 1987. *The Intentional Stance*. Cambridge, MA: MIT Press.

Dennett, Daniel C. 1991. *Consciousness Explained*. Boston: Little, Brown.

Dershowitz, Nachum, and Yuri Gurevich. 2008. "A Natural Axiomatization of Computability and Proof of Church's Thesis." *Bulletin of Symbolic Logic 14*: pp. 299–350.

Deutsch, David. 1985. "Quantum Theory, the Church-Turing Principle and the Universal Quantum Computer." *Proceedings of the Royal Society of London 400*: pp. 97–117.

Dewhurst, Joe. 2014. "Mechanistic Miscomputation: A Reply to Fresco and Primiero." *Philosophy & Technology 27*: pp. 495–498.

Dewhurst, Joe. 2016. "Review of *Physical Computation: A Mechanistic Account* by Gualtiero Piccinini." *Philosophical Psychology 29*: pp. 795–797.

Dewhurst, Joe. 2018a. "Individuation Without Representation." *The British Journal for the Philosophy of Science 69*: pp. 103–116.

Dewhurst, Joe. 2018b. "Computing Mechanisms Without Proper Functions." *Minds and Machines 28*: pp. 569–588.

Dietrich, Eric. 1990. "Computationalism." *Social Epistemology 4*: pp. 135–154.

Dodig-Crnkovic, Gordana. 2017. "Nature as a Network of Morphological Infocomputational Processes for Cognitive Agents." *The European Physical Journal Special Topics 226*: pp. 181–195.

Dodig-Crnkovic, Gordana, and Vincent C. Müller. 2011. "A Dialogue Concerning Two World Systems: Info-Computational vs. Mechanistic." In *Information and Computation: Essays on Scientific and Philosophical Understanding of Foundations of Information and Computation*, edited by G. Dodig-Crnkovic and M. Burgin, pp. 149–184. Boston: World Scientific.

Dresner, Eli. 2010. "Measurement-Theoretic Representation and Computation-Theoretic Realization." *The Journal of Philosophy 107*: pp. 275–292.

Dretske, Fred. 1981. *Knowledge and the Flow of Information*. Cambridge, MA: MIT Press.

Dretske, Fred. 1988. *Explaining Behavior: Reasons in a World of Causes*. Cambridge, MA: MIT Press.

Dreyfus, Hubert L. 1972. *What Computers Can't Do: A Critique of Artificial Reason.* Cambridge, MA: MIT Press.
Earman, John. 1986. *A Primer on Determinism.* Dordrecht: Reidel.
Earman, John., and John D. Norton. 1993. "Forever Is a Day: Supertasks in Pitowsky and Malament-Hogarth Spacetimes." *Philosophy of Science 60*: pp. 22–42.
Edelman, Shimon. 1998. "Representation Is Representation of Similarities." *Behavioral and Brain Sciences 21*: pp. 449–467.
Edelman, Shimon. 2008. *Computing the Mind: How the Mind Really Works.* New York: Oxford University Press.
Edmonds, Jack. 1965. "Path, Trees and Flowers." *Canadian Journal of Mathematics 17*: pp. 449–467.
Egan, Frances. 1991. "Must Psychology Be Individualistic?" *The Philosophical Review 100*: pp. 179–203.
Egan, Frances. 1994. "Individualism and Vision Theory." *Analysis 54*: pp. 258–264.
Egan, Frances. 1995. "Computation and Content." *The Philosophical Review 104*: pp. 181–204.
Egan, Frances. 2010. "Computational Models: A Modest Role for Content." *Studies in History and Philosophy of Science 41*: pp. 253–259.
Egan, Frances. 2012. "Metaphysics and Computational Cognitive Science: Let's Not Let the Tail Wag the Dog." *Journal of Cognitive Science 13*: pp. 39–49.
Egan, Frances. 2014. "How to Think About Mental Content." *Philosophical Studies 170*: pp. 115–135.
Egan, Frances. 2017. "Function-Theoretic Explanation and the Search for Neural Mechanisms." In *Explanation and Integration in Mind and Brain Science*, edited by D. M. Kaplan, pp. 145–163. New York: Oxford University Press.
Elber-Dorozko, Lotem. 2018. "Manipulation Is Key: On Why Non-Mechanistic Explanations in the Cognitive Sciences Also Describe Relations of Manipulation and Control." *Synthese 195*: pp. 5319–5337.
Elber-Dorozko, Lotem, and Oron Shagrir. 2019. "Integrating Computation into the Mechanistic Hierarchy in the Cognitive and Neural Sciences." *Synthese*, May 13, 2019.
Eliasmith, Chris. 2007. "Attractor Networks." *Scholarpedia 2*: p. 1380.
Eliasmith, Chris. 2013. *How to Build a Brain: A Neural Architecture for Biological Cognition.* New York: Oxford University Press.
Eliasmith, Chris, and Charles H. Anderson. 2003. *Neural Engineering: Computation, Representation and Dynamics in Neurobiological Systems.* Cambridge, MA: MIT Press.
Etesi, Gábor, and István Németi. 2002. "Non-Turing Computations via Malament-Hogarth Space-Times." *International Journal of Theoretical Physics 41*: pp. 341–370.
Etienne, Ariane S., and Kathryn J. Jeffery. 2004. "Path Integration in Mammals." *Hippocampus 14*: pp. 180–192.
Feferman, Solomon. 2013. "About and Around Computing Over the Reals." In *Computability: Turing, Gödel, Church, and Beyond*, edited by B. J. Copeland, C. Posy, and O. Shagrir, pp. 55–76. Cambridge, MA: MIT Press.
Fernau, Henning. 2010. "Minimum Dominating Set of Queens: A Trivial Programming Exercise?" *Discrete Applied Mathematics 158*: pp. 308–318.
Figdor, Carrie. 2009. "Semantic Externalism and the Mechanics of Thought." *Minds and Machines 19*: pp. 1–24.
Figdor, Carrie. 2018. *Pieces of Mind: The Proper Domain of Psychological Predicates.* New York: Oxford University Press.

Floridi, Luciano. 1999. *Philosophy and Computing: An Introduction*. London: Routledge.
Floridi, Luciano. 2011. *The Philosophy of Information*. New York: Oxford University Press.
Fodor, Jerry A. 1968. *Psychological Explanation: An Introduction to the Philosophy of Psychology*. New York: Random House.
Fodor, Jerry A. 1975. *The Language of Thought*. New York: Thomas Y. Crowell.
Fodor, Jerry A. 1980. "Methodological Solipsism Considered as a Research Strategy in Cognitive Psychology." *Behavioral and Brain Sciences 3*: pp. 63–73.
Fodor, Jerry A. 1987. *Psychosemantics: The Problem of Meaning in the Philosophy of Mind*. Cambridge, MA: MIT Press.
Fodor, Jerry A. 1990. *A Theory of Content and Other Essays*. Cambridge, MA: MIT Press.
Fodor, Jerry A. 1994. *The Elm and the Expert*. Cambridge, MA: MIT Press.
Fodor, Jerry A. 2000. *The Mind Doesn't Work That Way: The Scope and Limits of Computational Psychology*. Cambridge, MA: MIT Press.
Fodor, Jerry A., and Ned Block. 1973. "Cognitivism and the Analog/Digital Distinction." Unpublished manuscript.
Fodor, Jerry A., and Zenon W. Pylyshyn. 1988. "Connectionism and Cognitive Architecture: A Critical Analysis." *Cognition 28*: pp. 3–71.
Folina, Janet. 1998. "Church's Thesis: Prelude to a Proof." *Philosophia Mathematica 6*: pp. 302–323.
Folina, Janet. 2006. "Church's Thesis and the Variety of Mathematical Justifications." In *Church's Thesis After 70 Years*, edited by A. Olszewski, J. Wolenski, and R. Janusz, pp. 220–241. Heusenstamm: Ontos Verlag.
Fortnow, Lance, and Steve Homer. 2003. "A Short History of Computational Complexity." *Bulletin of the European Association for Theoretical Computer Science 80*: pp. 95–133.
French, Steven, and James Ladyman. 1999. "Reinflating the Semantic Approach." *International Studies in the Philosophy of Science 13*: pp. 103–121.
Fresco, Nir. 2008. "An Analysis of the Criteria for Evaluating Adequate Theories of Computation." *Minds and Machines 18*: pp. 379–401.
Fresco, Nir. 2014. *Physical Computation and Cognitive Science*. Berlin: Springer.
Fresco, Nir. 2015. "Objective Computation Versus Subjective Computation." *Erkenntnis 80*: pp. 1031–1053.
Fresco, Nir, B. Jack Copeland, and Marty J. Wolf. Forthcoming. "The Indeterminacy of Computation."*Synthese*.
Fresco, Nir, and Marcin Miłkowski. 2021. "Mechanistic Computational Individuation Without Biting the Bullet." *The British Journal for the Philosophy of Science 72*: pp. 431–438.
Fresco, Nir, and Giuseppe Primiero. 2013. "Miscomputation." *Philosophy & Technology 26*: pp. 253–272.
Fresco, Nir, and Marty J. Wolf. 2014. "The Instructional Information Processing Account of Digital Computation." *Synthese 191*: pp. 1469–1492.
Frigg, Roman, and Stephan Hartmann. 2020. "Models in Science." In *Stanford Encyclopedia of Philosophy*, edited by E. N. Zalta. https://plato.stanford.edu/.
Frigg, Roman, and Julian Reiss. 2009. "The Philosophy of Simulation: Hot New Issues or Same Old Stew?" *Synthese 169*: pp. 593–613.
Gallistel, Charles R., and John Gibbon. 2002. *The Symbolic Foundations of Conditioned Behavior*. Mahwah, NJ: Erlbaum.

Gallistel, Charles R., and Adam Philip King. 2009. *Memory and the Computational Brain: Why Cognitive Science Will Transform Neuroscience.* Hoboken, NJ: John Wiley & Sons.

Gandy, Robin O. 1980. "Church's Thesis and Principles for Mechanisms." In *The Kleene Symposium*, edited by S. C. Kleene, J. Barwise, H. J. Keisler, and K. Kunen, pp. 123–148. Amsterdam: North-Holland.

Gandy, Robin O. 1988. "The Confluence of Ideas in 1936." In *The Universal Turing Machine*, edited by R. Herken, pp. 51–111. New York: Oxford University Press.

Gandy, Robin O. 2001. Preface to "On Computable Numbers, with an Application to the Entscheidungs Problem." In *Mathematical Logic: The Collected Works of Turing*, edited by R. O. Gandy and C. E. M. Yates, pp. 9–17. Amsterdam: North-Holland.

Garzon, Max. 1995. *Models of Massive Parallelism: Analysis of Cellular Automata and Neural Networks.* Berlin: Springer.

Gherardi, Guido. 2008. "Computability and Incomputability of Differential Equations." In *Deduction, Computation, Experiment,* edited by R. Lupacchini and G. Corsi, pp. 223–242. Milan: Springer.

Gherardi, Guido. 2011. "Alan Turing and the Foundations of Computable Analysis." *Bulletin of Symbolic Logic 17*: pp. 394–430.

Giere, Ronald N. 2004. "How Models Are Used to Represent Reality." *Philosophy of Science 71*: pp. 742–752.

Glennan, Stuart. 2002. "Rethinking Mechanistic Explanation." *Philosophy of Science 69*: pp. S342–S353.

Glimcher, Paul W. 1999. "Oculomotor Control." In *The MIT Encyclopedia of the Cognitive Sciences*, edited by R. A. Wilson and F. C. Kiel, pp. 618–620. Cambridge, MA: MIT Press.

Gödel, Kurt. 1929. "On the Completeness of the Calculus of Logic." In *Collected Works I: Publications 1929–1936* (1986), edited by S. Feferman, J. W. Dawson, Jr., S. C. Kleene, G. H. Moore, R. M. Solovay, and J. van Heijenoort, pp. 61–101. New York: Oxford University Press.

Gödel, Kurt. 1931. "On Formally Undecidable Propositions of Principia Mathematica and Related Systems I." *Monatshefte für Mathematik und Physik 38*: pp. 173–198. In *Collected Works I: Publications 1929–1936* (1986), edited by S. Feferman, J. W. Dawson, Jr., S. C. Kleene, G. H. Moore, R. M. Solovay, and J. van Heijenoort, pp. 144–195. New York: Oxford University Press.

Gödel, Kurt. 1932. "Undecidable Diophantine Propositions." In *Collected Works III: Unpublished Essays and Lectures* (1995), edited by S. Feferman, J. W. Dawson, Jr., W. Goldfarb, C. Parsons, and R. M. Solovay, pp. 164–175. New York: Oxford University Press.

Gödel, Kurt. 1933. "The Present Situation in the Foundations of Mathematics." In *Collected Works III: Unpublished Essays and Lectures* (1995), edited by S. Feferman, J. W. Dawson, Jr., W. Goldfarb, C. Parsons, and R. M. Solovay, pp. 45–53. New York: Oxford University Press.

Gödel, Kurt. 1934. "On Undecidable Propositions of Formal Mathematical Systems." In *Collected Works I: Publications 1929–1936* (1986), edited by S. Feferman, J. W. Dawson, Jr., S. C. Kleene, G. H. Moore, R. M. Solovay, and Jean van Heijenoort, pp. 346–369. New York: Oxford University Press.

Gödel, Kurt. 193?. "Undecidable Diophantine Propositions." In *Collected Works III: Unpublished Essays and Lectures* (1995), edited by S. Feferman, J. W. Dawson, Jr., W. Goldfarb, C. Parsons, and R. M. Solovay, pp. 164–175. New York: Oxford University Press.

Gödel, Kurt. 1946. "Remarks Before the Princeton Bicentennial Conference on Problems in Mathematics." In *Collected Works II: Publications 1938–1974* (1990), edited by S. Feferman, J. W. Dawson, Jr., S. C. Kleene, G. H. Moore, R. M. Solovay, and J. van Heijenoort, pp. 150–153. New York: Oxford University Press.

Gödel, Kurt. 1951. "Some Basic Theorems on the Foundations of Mathematics and Their Implications." In *Collected Works III: Unpublished Essays and Lectures* (1995), edited by S. Feferman, J. W. Dawson, Jr., W. Goldfarb, C. Parsons, and R. M. Solovay, pp. 304–323. New York: Oxford University Press.

Gödel, Kurt. 1956. "Letter to John von Neumann, March 20th, 1956." In *Collected Works V: Correspondence, H-Z* (2003), edited by S. Feferman, J. W. Dawson Jr, W. Goldfarb, C. Parsons, and W. Sieg, pp. 373–377. New York: Oxford University Press.

Gödel, Kurt. 1963. "Note Added to Gödel (1931)." In *Collected Works I: Publications 1929–1936* (1986), edited by S. Feferman, J. W. Dawson, Jr., S. C. Kleene, G. H. Moore, R. M. Solovay, and J. van Heijenoort, p. 195. New York: Oxford University Press.

Gödel, Kurt. 1964. Postscriptum to Gödel (1934). In *Collected Works I: Publications 1929–1936* (1986), edited by S. Feferman, J. W. Dawson, Jr., S. C. Kleene, G. H. Moore, R. M. Solovay, and J. van Heijenoort, pp. 369–370. New York: Oxford University Press.

Gödel, Kurt. 1972. "Some Remarks on the Undecidability Results." In *Collected Works II: Publications 1938–1974* (1990), edited by S. Feferman, J. W. Dawson, Jr., S. C. Kleene, G. H. Moore, R. M. Solovay, and J. van Heijenoort, pp. 305–306. New York: Oxford University Press.

Gödel, Kurt. 1986. *Collected Works I: Publications 1929–1936*. Edited by Solomon Feferman, John W. Dawson, Jr., Stephen. C. Kleene, Gregory H. Moore, Robert M. Solovay, and Jean van Heijenoort. New York: Oxford University Press.

Gödel, Kurt. 1990. *Collected Works II: Publications 1938–1974*. Edited by Solomon Feferman, John W. Dawson, Jr., Stephen C. Kleene, Gregory H. Moore, Robert M. Solovay, and Jean van Heijenoort. New York: Oxford University Press.

Gödel, Kurt. 1995. *Collected Works III: Unpublished Essays and Lectures*. Edited by Solomon Feferman, John W. Dawson, Jr., Warren Goldfarb, Charles Parsons, and Robert M. Solovay. New York: Oxford University Press.

Gödel, Kurt. 2003. *Collected Works V: Correspondence, H-Z*. Edited by Solomon Feferman, John W. Dawson Jr, Warren Goldfarb, Charles Parsons, and Wilfried Sieg. New York: Oxford University Press.

Godfrey-Smith, Peter. 2009. "Triviality Arguments Against Functionalism." *Philosophical Studies 145*: pp. 273–295.

Goel, Vinod. 1991. "Notationality and the Information-Processing Mind." *Minds and Machines 1*: pp. 129–165.

Goff, Philip, William Seager, and Sean Allen-Hermanson, "Panpsychism." *The Stanford Encyclopedia of Philosophy*, edited by E. N. Zalta. https://plato.stanford.edu/archives/sum2020/entries/panpsychism/.

Goldman, Mark S., Albert Compte, and Xiao-Jing Wang. 2009. "Neural Integrator Models." In *Encyclopedia of Neuroscience 6*, edited by L. R. Squire, pp. 165–178. Oxford: Academic Press.

Goldman, Mark S., Chris R. S. Kaneko, Guy Major, Emre Aksay, David W. Tank, and H. Sebastian Seung. 2002. "Linear Regression of Eye Velocity on Eye Position and Head Velocity Suggests a Common Oculomotor Neural Integrator." *Journal of Neurophysiology 88*: pp. 659–665.

Goldreich, Oded. 2008. *Computational Complexity: A Conceptual Perspective.* Cambridge: Cambridge University Press.

Goodman, Nelson. 1968. *Languages of Art: An Approach to a Theory of Symbols.* Indianapolis, IN: The Bobbs-Merrill Company.

Graça, Daniel Silva, and José Félix Costa. 2003. "Analog Computers and Recursive Functions Over the Reals." *Journal of Complexity 19*: pp. 644–664.

Grice, Paul. 1957. "Meaning." *The Philosophical Review 66*: pp. 377–388.

Grier, David A. 2005. *When Computers Were Human.* Princeton: Princeton University Press.

Griffiths, Thomas L., Nick Chater, Charles Kemp, Amy Perfors, and Joshua B. Tenenbaum. 2010. "Probabilistic Models of Cognition: Exploring Representations and Inductive Biases." *Trends in Cognitive Sciences 14*: pp. 357–364.

Griffiths, Thomas L., Charles Kemp, and Joshua B. Tenenbaum. 2008. "Bayesian Models of Cognition." In *The Cambridge Handbook of Computational Psychology*, edited by R. Sun, pp. 59–100. Cambridge: Cambridge University Press.

Grush, Rick. 2001. "The Semantic Challenge to Computational Neuroscience." In *Theory and Method in the Neurosciences*, edited by P. K. Machamer, R. Grush, and P. McLaughlin, pp. 155–172. Pittsburgh: University of Pittsburgh Press.

Grush, Rick. 2004. "The Emulation Theory of Representation: Motor Control, Imagery, and Perception." *Behavioral and Brain Sciences 27*: pp. 377–442.

Grzegorczyk, Andrzej. 1955. "Computable Functional." *Fundamenta Mathematicae 42*: pp. 168–202.

Grzegorczyk, Andrzej. 1957. "On the Definitions of Computable Real Continuous Functions." *Fundamenta Mathematicae 44*: pp. 61–71.

Gurari, Eitan M. 1989. *An Introduction to the Theory of Computation.* New York: Computer Science Press.

Gurevich, Yuri. 2000. "Sequential Abstract State Machines Capture Sequential Algorithms." *ACM Transactions on Computational Logic 1*: pp. 77–111.

Gurevich, Yuri. 2012. "What Is an Algorithm?" In *SOFSEM: Theory and Practice of Computer Science, LNCS 7147*, edited by M. Bieliková, G. Friedrich, G. Gottlob, S. Katzenbeisser, and G. Turán, pp. 31–42. Berlin: Springer.

Gurevich, Yuri. 2019. "Unconstrained Church-Turing Thesis Cannot Possibly Be True." *The Bulletin of the European Association for Theoretical Computer Science 27*: pp. 46–59.

Hafting, Torkel, Marianne Fyhn, Sturla Molden, May-Britt Moser, and Edvard I. Moser. 2005. "Microstructure of a Spatial Map in the Entorhinal Cortex." *Nature 436*: pp. 801–806.

Hagar, Amit, and Alex Korolev. 2007. "Quantum Hypercomputation—Hyper or Computation?" *Philosophy of Science 74*: pp. 347–363.

Haimovici, Sabrina. 2013. "A Problem for the Mechanistic Account of Computation." *Journal of Cognitive Science 14*: pp. 151–181.

Hamkins, Joel D. 2002. "Infinite Time Turing Machines." *Minds and Machines 12*: pp. 521–539.

Hamkins, Joel D., and Andy Lewis. 2000. "Infinite Time Turing Machines." *The Journal of Symbolic Logic 65*: pp. 567–604.

Harbecke, Jens. 2020. "The Methodological Role of Mechanistic-Computational Models in Cognitive Science." *Synthese*, February 17, 2020.

Harbecke, Jens, and Oron Shagrir. 2019. "The Role of the Environment in Computational Explanations." *European Journal for Philosophy of Science 9*: article 37.

Hardcastle, Valerie Gray. 1995. "Computationalism." *Synthese 105*: pp. 303–317.
Hardcastle, Valerie Gray. 1999. "Understanding Functions: A Pragmatic Approach." In *Where Biology Meets Philosophy: Philosophical Essays*, edited by V. G. Hardcastle, pp. 27–43. Cambridge, MA: MIT Press.
Harel, David. 1992. *Algorithmics: The Spirit of Computing*. 2nd ed. Reading, MA: Addison-Wesley.
Haugeland, John. 1978. "The Nature and Plausibility of Cognitivism." *Behavioral and Brain Sciences 1*: pp. 215–226.
Haugeland, John. 1981a. "Analog and Analog." *Philosophical Topics 12*: pp. 213–225.
Haugeland, John. 1981b. "Semantic Engines: An Introduction to Mind Design." In *Mind Design*, edited by J. Haugeland, pp. 1–34. Cambridge, MA: MIT Press.
Heeger, David J. 1992. "Normalization of Cell Responses in Cat Striate Cortex." *Visual Neuroscience 9*: pp. 181–197.
Hemmo, Meir, and Orly Shenker. 2019. "The Physics of Implementing Logic: Landauer's Principle and the Multiple-Computations Theorem." *Studies in History and Philosophy of Modern Physics 68*: pp. 90–105.
Herbrand, Jacques. 1931. "On the Consistency of Arithmetic." In *Jacques Herbrand Logical Writings* (1971), edited by W. D. Goldfarb, pp. 282–298. Cambridge, MA: Harvard University Press.
Hertz, John, Anders Krogh, and Richard G. Palmer. 1991. *Introduction to the Theory of Neural Computation*. Redwood City, CA: Addison-Wesley.
Hess, Robert F., Curtis L. Baker Jr., James N. VerHoeve, Ulker Tulunay-Keesey, and Thomas D. France. 1985. "The Pattern Evoked Electroretinogram: Its Variability in Normals and Its Relationship to Amblyopia." *Investigative Ophthalmology & Visual Science 26*: pp. 1610–1623.
Hilbert, David. 1902. "Mathematical Problems: Lecture Delivered Before the International Congress of Mathematicians at Paris in 1900." *Bulletin of the American Mathematical Society 8*: pp. 437–479.
Hilbert, David. 1926. "Über das Unendliche." *Mathematische Annalen 95*: pp. 161–190. Lecture given in Münster, June 4, 1925.
Hilbert, David, and Wilhelm F. Ackermann. 1928. *Grundzuge der Theoretischen Logik*. Berlin: Springer.
Hilbert, David, and Paul Bernays. 1939. *Grundlagen der Mathematik II*. Berlin: Springer.
Hildreth, Ellen C., and Shimon Ullman. 1989. "The Computational Study of Vision." In *Foundations of Cognitive Science*, edited by M. I. Posner, pp. 581–630. Cambridge, MA: MIT Press.
Hinton, Geoffrey E., and James A. Anderson. 2014. *Parallel Models of Associative Memory: Updated Edition*. New York: Psychology Press.
Hoffmann, Geoffrey W. 2008. "Immune Network Theory." https://phas.ubc.ca/~hoffmann/ni.html.
Hogarth, Mark L. 1992. "Does General Relativity Allow an Observer to View an Eternity in a Finite Time?" *Foundations of Physics Letters 5*: pp. 173–181.
Hogarth, Mark L. 1994. "Non-Turing Computers and Non-Turing Computability." *Proceedings of the Biennial Meeting of the Philosophy of Science Association 1*: pp. 126–138.
Hogarth, Mark L. 2004. "Deciding Arithmetic Using SAD Computers." *The British Journal for the Philosophy of Science 55*: pp. 681–691.

Hopcroft, John E., and Jeffrey D. Ullman. 1979. *Introduction to Automata Theory, Languages, and Computation*. Reading, MA: Addison-Wesley.

Hopfield, John J. 1982. "Neural Networks and Physical Systems with Emergent Collective Computational Abilities." *Proceedings of the National Academy of Science, USA 79*: pp. 2554–2558.

Hopfield, John J., and David W. Tank. 1985. "'Neural' Computation of Decisions in Optimization Problems." *Biological Cybernetics 52*: pp. 141–152.

Horowitz, Amir. 2007. "Computation, External Factors, and Cognitive Explanations." *Philosophical Psychology 20*: pp. 65–80.

Horsman, Clare, Susan Stepney, Rob C. Wagner, and Viv Kendon. 2014. "When Does a Physical System Compute?" *Proceedings of the Royal Society A: Mathematical, Physical and Engineering Sciences 470*: p. 20140182.

Horsman, Dominic, Viv Kendon, and Susan Stepney. 2017. "The Natural Science of Computing." *Communications of the ACM 60*: pp. 31–34.

Horst, Steven. 2015. "The Computational Theory of Mind." *The Stanford Encyclopedia of Philosophy*, edited by E. N. Zalta. https://plato.stanford.edu/archives/sum2015/entries/computational-mind/.

Hubel, David H., and Torsten N. Wiesel. 1962. "Receptive Fields, Binocular Interaction and Functional Architecture in the Cat's Visual Cortex." *The Journal of Physiology 160*: pp. 106–154.

Humphreys, Paul. 2004. *Extending Ourselves: Computational Science, Empiricism, and Scientific Method*. New York: Oxford University Press.

Huneman, Philippe. 2010. "Topological Explanations and Robustness in Biological Sciences." *Synthese 177*: pp. 213–245.

Hutto, Daniel D., and Erik Myin. 2012. *Radicalizing Enactivism: Basic Minds Without Content*. Cambridge, MA: MIT Press.

Hutto, Daniel D., and Erik Myin. 2017. *Evolving Enactivism: Basic Minds Meet Content*. Cambridge, MA: MIT Press.

Hutto, Daniel D., Erik Myin, Anco Peeters, and Farid Zahnoun. 2018. "The Cognitive Basis of Computation: Putting Computation in its Place." In *The Routledge Handbook of the Computational Mind*, edited by M. Sprevak and M. Colombo, pp. 272–282. London: Routledge.

Illari, Phyllis McKay, and Jon Williamson. 2012. "What Is a Mechanism? Thinking About Mechanisms Across the Sciences." *European Journal for Philosophy of Science 2*: pp. 119–135.

Jerne, Niels K. 1974. "Towards a Network Theory of the Immune System." *The Annual Review of Immunology 125*: pp. 373–389.

Kaplan, David M. 2011. "Explanation and Description in Computational Neuroscience." *Synthese 183*: pp. 339–373.

Kaplan, David M. 2017. "Neural Computation, Multiple Realizability, and the Prospects for Mechanistic Explanation." In *Explanation and Integration in Mind and Brain Science*, edited by D. M. Kaplan, pp. 164–189. New York: Oxford University Press.

Kaplan, David M., and Carl F. Craver. 2011. "The Explanatory Force of Dynamical and Mathematical Models in Neuroscience: A Mechanistic Perspective." *Philosophy of Science 78*: pp. 601–627.

Kitcher, Patricia S. 1988. "Marr's Computational Theory of Vision." *Philosophy of Science 55*: pp. 1–24.

Kleene, Stephen C. 1936. "General Recursive Functions of Natural Numbers." *Mathematische Annalen 112*: pp. 727–742.
Kleene, Stephen C. 1938. "On Notation for Ordinal Numbers." *Journal of Symbolic Logic 3*: pp. 150–155.
Kleene, Stephen C. 1943. "Recursive Predicates and Quantifiers." *Transactions of the American Mathematical Society 53*: pp. 41–73.
Kleene, Stephen C. 1952. *Introduction to Metamathematics*. Amsterdam: North-Holland.
Kleene, Stephen C. 1981. "Origins of Recursive Function Theory." *Annals of the History of Computing 3*: pp. 52–67.
Klein, Colin. 2008. "Dispositional Implementation Solves the Superfluous Structure Problem." *Synthese 165*: pp. 141–153.
Klein, Colin. 2012. "Two Paradigms for Individuating Implementations." *Journal of Cognitive Science 13*: pp. 167–179.
Knuth, Donald E. 1973. *The Art of Computer Programming: Fundamental Algorithms, vol. 1*. Reading, MA: Addison-Wesley.
Koch, Christof. 1999. *Biophysics of Computation: Information Processing in Single Neurons*. New York: Oxford University Press.
Koepke, Peter. 2005. "Turing Computations on Ordinals." *Bulletin of Symbolic Logic 11*: pp. 377–397.
Kolmogorov, Andrei N. 1958. "On the Notion of Algorithm." *Uspekhi Mat. Nauk 8* (4): pp. 175–176 (in Russian). English translation in *Selected Works of A. N. Kolmogorov: Mathematics and Its Applications*, edited by A. N. Shiryayev (1993; vol 27, p. 1). Dordrecht: Springer.
Kolmogorov, Andrei N. 1965. "Three Approaches to the Quantitative Definition of 'Information.'" *Problems of Information Transmission 1*: pp. 3–11.
Kolmogorov, Andrei N., and Vladimir A. Uspensky. 1963. "On the Definition of an Algorithm." *Uspehi Mat. Nauk 13*: pp. 3–28 (in Russian). English translation in *American Mathematical Society Translations, Series II* (1963) *29*: pp. 217–245.
Komar, Arthur. 1964. "Undecidability of Macroscopically Distinguishable States in Quantum Field Theory." *Physical Review 133*: pp. B542–544.
Kreisel, Georg. 1965. "Mathematical Logic." In *Lectures on Modern Mathematics, vol. 3*, edited by T. L. Saaty, pp. 95–195. Hoboken, NJ: John Wiley & Sons.
Kreisel, Georg. 1967. "Mathematical Logic: What Has It Done for the Philosophy of Mathematics?" In *Bertrand Russell: Philosopher of the Century*, edited by R. Schoenman, pp. 201–272. Crows Nest, Australia: George Allen and Unwin.
Kreisel, Georg. 1972. "Which Number Theoretic Problems Can be Solved in Recursive Progressions on Π1/1-Paths Through O?" *The Journal of Symbolic Logic 37*: pp. 311–334.
Kripke, Saul A. 1982. *Wittgenstein on Rules and Private Language: An Elementary Exposition*. Cambridge, MA: Harvard University Press.
Kripke, Saul A. 2013. "The Church-Turing 'Thesis' as a Special Corollary of Gödel's Completeness Theorem." In *Computability: Turing, Gödel, Church, and Beyond*, edited by B. J. Copeland, C. Posy, and O. Shagrir, pp. 77–104. Cambridge, MA: MIT Press.
Krizhevsky, Alex, Ilya Sutskever, and Geoffrey E. Hinton. 2012. "Imagenet Classification with Deep Convolutional Neural Networks." *Advances in Neural Information Processing Systems 2*: pp. 1097–1105.
Lacombe, Daniel. 1955. "Extension de la Notion de Fonction Récursive aux Fonctions d'une ou Plusieurs Variables Réelles III." *Comptes Rendus Académie des Sciences Paris 241*: pp. 151–153.

Ladyman, James. 2009. "What Does it Mean to Say That a Physical System Implements a Computation?" *Theoretical Computer Science 410*: pp. 376–383.
LeCun, Yann, Yoshua Bengio, and Geoffrey E. Hinton. 2015. "Deep Learning." *Nature 521*: pp. 436–444.
Lee, Jonny. 2021. "Mechanisms, Wide Functions, and Content: Towards a Computational Pluralism." *The British Journal for the Philosophy of Science 72*: pp. 221–244.
Leigh, R. John, and David S. Zee. 2015. *The Neurology of Eye Movements*. 5th ed. New York: Oxford University Press.
Levin, Janet. 2018. "Functionalism." In *The Stanford Encyclopedia of Philosophy*, edited by E. N. Zalta. https://plato.stanford.edu/archives/fall2018/entries/functionalism/.
Levy, Arnon. 2009. "Carl F. Craver. Explaining What? Review of Explaining the Brain: Mechanisms and the Mosaic Unity of Neuroscience." *Biology & Philosophy 24*: pp. 137–145.
Levy, Arnon. 2013. "Three Kinds of New Mechanism." *Biology & Philosophy 28*: pp. 99–114.
Levy, Arnon, and William Bechtel. 2013. "Abstraction and the Organization of Mechanisms." *Philosophy of Science 80*: pp. 241–261.
Lewis, David. 1971. "Analog and Digital." *Noûs 5*: pp. 321–327.
Lewis, Harry R., and Christos H. Papadimitriou. 1981. *Elements of the Theory of Computation*. Englewood Cliffs, NJ: Prentice-Hall.
London, M., and M. Häusser. 2005. "Dendritic Computation." *Annual Review of Neuroscience 28*: pp. 503–532.
Löwe, Benedikt. 2001. "Revision Sequences and Computers with an Infinite Amount of Time." *Journal of Logic and Computation 11*: pp. 25–40.
Lycan, William G. 1981. "Form, Function, and Feel." *The Journal of Philosophy 78*: pp. 24–50.
Machamer, Peter K., Lindley Darden, and Carl F. Craver. 2000. "Thinking About Mechanisms." *Philosophy of Science 67*: pp. 1–25.
Maley, Corey J. 2011. "Analog and Digital, Continuous and Discrete." *Philosophical Studies 155*: pp. 117–131.
Maley, Corey J. 2018. "Toward Analog Neural Computation." *Minds and Machines 28*: pp. 77–91.
Maley, Corey J. 2020. "Analog Computation and Representation." *The British Journal for the Philosophy of Science*. Available at: https://www.journals.uchicago.edu/doi/pdf/10.1086/715031
Maley, Corey J., and Gualtiero Piccinini. 2017. "A Unified Mechanistic Account of Teleological Functions for Psychology and Neuroscience." In *Explanation and Integration in Mind and Brain Science*, edited by D. M. Kaplan, pp. 236–256. New York: Oxford University Press.
Manchak, John Byron. 2020. "Malament–Hogarth Machines." *The British Journal for the Philosophy of Science 71*: 1143–1153.
Manchak, John Byron, and Bryan W. Roberts. 2016. "Supertasks." In *The Stanford Encyclopedia of Philosophy*, edited by E. N. Zalta. https://plato.stanford.edu/archives/win2016/entries/spacetime-supertasks/.
Mancosu, Paolo. 1999. "Between Russell and Hilbert: Behmann on the Foundations of Mathematics." *Bulletin of Symbolic Logic 5*: pp. 303–330.
Marr, David. 1977. "Artificial Intelligence: A Personal View." *Artificial Intelligence 9*: pp. 37–48.

Marr, David C. 1982. *Vision: A Computational Investigation into the Human Representation and Processing of Visual Information*. New York: W. H. Freeman.

Marr, David, and Ellen C. Hildreth. 1980. "Theory of Edge Detection." *Proceedings of the Royal Society of London, Series B, Biological Sciences 207*: pp. 187–217.

Matthews, Robert J., and Eli Dresner. 2017. "Measurement and Computational Skepticism." *Noûs 51*: pp. 832–854.

Maudlin, Tim. 1989. "Computation and Consciousness." *The Journal of Philosophy 86*: pp. 407–432.

Mazur, Stanislaw. 1963. *Computable Analysis*. Warsaw: Rozprawy Matematyczne.

McClelland, James L., David E. Rumelhart, and Geoffrey E. Hinton. 1986. "The Appeal of Parallel Distributed Processing." In *Parallel Distributed Processing: Explorations in the Microstructure of Cognition, volume 1: Foundations*. edited by D. E. Rumelhart, J. L. McClelland, and the PDP Research Group, pp. 3–44. Cambridge, MA: MIT Press.

McClelland, James L., David E. Rumelhart, and the PDP Research Group (eds.). 1986. *Parallel Distributed Processing: Explorations in the Microstructure of Cognition. Vol. 2: Psychological and Biological Models*. Cambridge, MA: MIT Press.

McCulloch, Warren S., and Walter H. Pitts. 1943. "A Logical Calculus of the Ideas Immanent in Nervous Activity." *Bulletin of Mathematical Biophysics 5*: pp. 115–133.

McNaughton, Bruce L., Francesco P. Battaglia, Ole Jensen, Edvard I. Moser, and May-Britt Moser. 2006. "Path Integration and the Neural Basis of the 'Cognitive Map.'" *Nature Reviews Neuroscience 7*: pp. 663–678.

Melnyk, Andrew. 1996. "Searle's Abstract Argument Against Strong AI." *Synthese 108*: pp. 391–419.

Mendelson, Elliott. 1990. "Second Thoughts About Church's Thesis and Mathematical Proofs." *The Journal of Philosophy 87*: pp. 225–233.

Miłkowski, Marcin. 2011. "Beyond Formal Structure: A Mechanistic Perspective on Computation and Implementation." *Journal of Cognitive Science 12*: pp. 359–379.

Miłkowski, Marcin. 2013. *Explaining the Computational Mind*. Cambridge, MA: MIT Press.

Miłkowski, Marcin. 2017. "The False Dichotomy Between Causal Realization and Semantic Computation." *Internetowy Magazyn Filozoficzny Hybris 38*: pp. 1–21.

Millhouse, Tyler. 2019. "A Simplicity Criterion for Physical Computation." *The British Journal for the Philosophy of Science 70*: pp. 153–178.

Millikan, Ruth Garrett. 1984. *Language, Thought, and Other Biological Categories: New Foundations for Realism*. Cambridge, MA: MIT Press.

Mills, Jonathan W. 2008. "The Nature of the Extended Analog Computer." *Physica D: Nonlinear Phenomena 237*: pp. 1235–1256.

Milner, Robin. 1971. "An Algebraic Definition of Simulation Between Programs." In *Proceedings of the Second International Joint Conference on Artificial Intelligence*, edited by D. C. Cooper, pp. 481–489. London: The British Computer Society.

Minsky, Marvin L. 1967. *Computation: Finite and Infinite Machines*. Englewood Cliffs, NJ: Prentice-Hall.

Minsky, Marvin, and Seymour Papert. 1969. *Perceptrons: An Introduction to Computational Geometry*. Cambridge, MA: MIT Press.

Mittelstaedt, Horst, and Marie-Luise Mittelstaedt. 1982. "Homing by Path Integration." In *Avian Navigation*, edited by F. Papi and H. G. Wallraff, pp. 290–297. Berlin: Springer.

Morgan, Alex, and Gualtiero Piccinini. 2018. "Towards a Cognitive Neuroscience of Intentionality." *Minds and Machines 28*: pp. 119–139.

Morton, Peter A. 1993. "Supervenience and Computational Explanation in Vision Theory." *Philosophy of Science* 60: pp. 86–99.
Moschovakis, Yiannis N. 1984. "Abstract Recursion as a Foundation for the Theory of Algorithms." *Computation and Proof Theory* 1104: pp. 289–364.
Moschovakis, Yiannis N. 1998. "On Founding the Theory of Algorithms." In *Truth in Mathematics*, edited by H. G. Dales and G. Oliveri, pp. 71–104. Oxford: Clarendon Press.
Moschovakis, Yiannis N. 2001. "What Is an Algorithm?" In *Mathematics Unlimited—2001 and Beyond*, edited by B. Engquist and W. Schmid, pp. 919–936. Berlin: Springer.
Moschovakis, Yiannis N., and Vasilis Paschalis. 2008. "Elementary Algorithms and Their Implementations." In *New Computational Paradigms: Changing Conceptions of What Is Computable*, edited by S. B. Cooper, B. Löwe, and A. Sorbi, pp. 87–118. Berlin: Springer.
Müller, Vincent C., and Matej Hoffmann. 2017. "What Is Morphological Computation? On How the Body Contributes to Cognition and Control." *Artificial Life* 23: pp. 1–24.
Nagin, Paul, and John Impagliazzo. 1995. *Computer Science: A Breadth-First Approach with Pascal*. Hoboken, NJ: John Wiley & Sons.
Neander, Karen. 1991. "Functions as Selected Effects: The Conceptual Analyst's Defense." *Philosophy of Science* 58: pp. 168–184.
Neander, Karen. 2017. *A Mark of the Mental: In Defense of Informational Teleosemantics*. Cambridge, MA: MIT Press.
Németi, István, and Gyula Dávid. 2006. "Relativistic Computers and the Turing Barrier." *Journal of Applied Mathematics and Computation* 178: pp. 118–142.
Newell, Allen. 1980. "Physical Symbol Systems." *Cognitive Science* 4: pp. 135–183.
Newell, Allen, and Herbert A. Simon. 1976. "Computer Science as Empirical Inquiry: Symbols and Search." *Communications of the Association for Computing Machinery* 19: pp. 113–126.
Norton, John D. 1999. "A Quantum Mechanical Supertask." *Foundations of Physics* 29: pp. 1265–1302.
Norton, John D. 2012. "Approximation and Idealization: Why the Difference Matters." *Philosophy of Science* 79: pp. 207–232.
O'Brien, Gerard, and Jon Opie. 2006. "How do Connectionist Networks Compute?" *Cognitive Processing* 7: pp. 30–41.
O'Brien, Gerard, and Jon Opie. 2009. "The Role of Representation in Computation." *Cognitive Processing* 10: pp. 53–62.
Odifreddi, Piergiorgio. 1989. *Classical Recursion Theory: The Theory of Functions and Sets of Natural Numbers*. Amsterdam: Elsevier.
O'Keefe, John, and Nadel Lynn. 1978. *The Hippocampus as a Cognitive Map*. Oxford: Clarendon Press.
Ord, Toby. 2006. "The Many Forms of Hypercomputation." *Applied Mathematics and Computation* 178: pp. 143–153.
O'Reilly, Randall C., and Yuko Munakata. 2000. *Computational Explorations in Cognitive Neuroscience: Understanding the Mind by Simulating the Brain*. Cambridge, MA: MIT Press.
Orlandi, Nico. 2014. *The Innocent Eye: Why Vision Is Not a Cognitive Process*. New York: Oxford University Press.
Orlandi, Nico. 2018. "Perception Without Computation?" In *The Routledge Handbook of the Computational Mind*, edited by M. Sprevak and M. Colombo, pp. 410–423. London: Routledge.

Palmer, Stephen E. 1978. "Fundamental aspects of cognitive representation." In *Cognition and Categorization*, edited by E. Rosch and B. B. Lloyd, pp. 259–303. Hillsdale, NJ: Lawrence Erlbaum.

Papayannopoulos, Philippos. 2018. "Computing, Modelling, and Scientific Practice: Foundational Analyses and Limitations." Doctoral dissertation, University of Western Ontario.

Papayannopoulos, Philippos. 2020a. "Unrealistic Models for Realistic Computations: How Idealisations Help Represent Mathematical Structures and Found Scientific Computing." *Synthese*, May 4, 2020.

Papayannopoulos, Philippos. 2020b. "Computing and Modelling: Analog vs. Analogue." *Studies in History and Philosophy of Science 83*: pp. 103–120.

Peacocke, Christopher. 1994. "Content, Computation and Externalism." *Mind & Language 9*: pp. 303–335.

Penrose, Roger. 1989. *The Emperor's New Mind: Concerning Computers, Minds and the Laws of Physics*. New York: Oxford University Press.

Penrose, Roger. 1994. *Shadows of the Mind: A Search for the Missing Science of Consciousness*. New York: Oxford University Press.

Piccinini, Gualtiero. 2003a. "Alan Turing and the Mathematical Objection." *Minds and Machines 13*: pp. 23–48.

Piccinini, Gualtiero. 2003b. "Computations and Computers in the Sciences of Mind and Brain." Doctoral dissertation, University of Pittsburgh.

Piccinini, Gualtiero. 2004a. "The First Computational Theory of Mind and Brain: A Close Look at McCulloch and Pitts's 'Logical Calculus of Ideas Immanent in Nervous Activity.'" *Synthese 141*: pp. 175–215.

Piccinini, Gualtierio. 2004b. "Functionalism, Computationalism, and Mental Contents." *Canadian Journal of Philosophy 34*: pp. 375–410.

Piccinini, Gualtiero. 2007. "Computing Mechanisms." *Philosophy of Science 74*: pp. 501–526.

Piccinini, Gualtiero. 2008a. "Computation Without Representation." *Philosophical Studies 137*: pp. 205–241.

Piccinini, Gualtiero. 2008b. "Computers." *Pacific Philosophical Quarterly 89*: pp. 32–73.

Piccinini, Gualtiero. 2008c. "Some Neural Networks Compute, Others Don't." *Neural Networks 21*: pp. 311–321.

Piccinini, Gualtiero. 2009. "Computationalism in the Philosophy of Mind." *Philosophy Compass 4*: pp. 515–532.

Piccinini, Gualtiero. 2011. "The Physical Church-Turing Thesis: Modest or Bold?" *The British Journal for the Philosophy of Science 62*: pp. 733–769.

Piccinini, Gualtiero. 2015. *Physical Computation: A Mechanistic Account*. New York: Oxford University Press.

Piccinini, Gualtiero. 2017. "Computation in Physical Systems." In *The Stanford Encyclopedia of Philosophy*, edited by E. N. Zalta. http://plato.stanford.edu/entries/computation-physicalsystems/.

Piccinini, Gualtieto, and Neal G. Anderson. 2018. "Ontic Pancomputationalism." In *Physical Perspectives on Computation, Computational Perspectives on Physics*, edited by M. Cuffaro and S. Fletcher, pp. 23–38. Cambridge: Cambridge University Press.

Piccinini, Gualtiero, and Sonya Bahar. 2013. "Neural Computation and the Computational Theory of Cognition." *Cognitive Science 37*: pp. 453–488.

Piccinini, Gualtiero, and Carl Craver. 2011. "Integrating Psychology and Neuroscience: Functional Analyses as Mechanism Sketches." *Synthese 183*: pp. 283–311.

Piccinini, Gualtiero, and Andrea Scarantino. 2011. "Information Processing, Computation, and Cognition." *Journal of Biological Physics 37*: pp. 1–38.

Piccinini, Gualtiero, and Oron Shagrir. 2014. "Foundations of Computational Neuroscience." *Current Opinion in Neurobiology 25*: 25–30.

Pitowsky, Itamar. 1990. "The Physical Church Thesis and Physical Computational Complexity." *Iyyun 39*: pp. 81–99.

Pitowsky, Itamar. 1996. "Laplace's Demon Consults an Oracle: The Computational Complexity of Prediction." *Studies in History and Philosophy of Science Part B: Studies in History and Philosophy of Modern Physics 27*: pp. 161–180.

Pitowsky, Itamar. 2002. "Quantum Speed-Up of Computations." *Philosophy of Science 69*: S168–S177.

Plotkin, Gordon D. 2004. "The Origins of Structural Operational Semantics." *The Journal of Logic and Algebraic Programming 60*: pp. 3–15.

Pnueli, Amir, Michael Siegel, and Eli Singerman. 1998. "Translation Validation." In *Proceedings of the 4th International Conference on Tools and Algorithms for the Construction and Analysis of Systems* (TACAS), *LNCS 1384*, edited by S. Bernhard, pp. 151–166. Berlin: Springer.

Polger, Thomas W., and Lawrence A. Shapiro. 2016. *The Multiple Realization Book*. New York: Oxford University Press.

Post, Emil L. 1936. "Finite Combinatory Processes – Formulation I." *Journal of Symbolic Logic 1*: pp. 103–105. Reprinted in *The Undecidable: Basic Papers on Undecidable Propositions, Unsolvable Problems and Computable Functions*, edited by M. Davis (1965), pp. 288–291.

Post, Emil L. 1947. "Recursive Unsolvability of a Problem of Thue." *Journal of Symbolic Logic 12*: pp. 1–11.

Potgieter, Petrus H., and Elemér E. Rosinger. 2010. "Output Concepts for Accelerated Turing Machines." *Natural Computing 9*: pp. 853–864.

Pour-El, Marian B. 1974. "Abstract Computability and Its Relation to the General Purpose Analog Computer (Some Connections Between Logic, Differential Equations and Analog Computers)." *Transactions of the American Mathematical Society 199*: pp. 1–28.

Pour-El, Marian B., and Ian J. Richards. 1981. "The Wave Equation with Computable Initial Data Such That Its Unique Solution Is Not Computable." *Advances in Mathematics 39*: pp. 215–239.

Pour-El, Marian B., and Ian J. Richards. 1989. *Computability in Analysis and Physics*. Berlin: Springer.

Putnam, Hilary. 1967. "Psychological Predicates." In *Art, Mind, and Religion*, edited by W. H. Capitan and D. D. Merrill, pp. 37–48. Pittsburgh, PA: University of Pittsburgh Press. Reprinted as "The Nature of Mental States." in *Mind, Language and Reality, Philosophical Papers, vol. 2*, edited by H. Putnam, pp. 429–440. Cambridge: Cambridge University Press.

Putnam, Hilary. 1988. *Representation and Reality*. Cambridge, MA: MIT Press.

Putnam, Hilary. 1997. "Functionalism: Cognitive Science or Science Fiction?" In *The Future of the Cognitive Revolution*, edited by D. M. Johnson and C. E. Erneling, pp. 32–44. New York: Oxford University Press.

Putnam, Hilary. 1999. *The Threefold Cord: Mind, Body, and World*. New York: Columbia University Press.

Pylyshyn, Zenon W. 1984. *Computation and Cognition: Toward a Foundation for Cognitive Science*. Cambridge, MA: MIT Press.
Quine, Willard Van Orman. 1960. *Word and Object*. Cambridge, MA: MIT Press.
Quinon, Paula. 2021. "Can Church's Thesis Be Viewed as a Carnapian Explication?" *Synthese 198*: pp. 1047–1074.
Ramsey, William M. 2007. *Representation Reconsidered*. Cambridge: Cambridge University Press.
Ramsey, William. 2016. "Untangling Two Questions About Mental Representation." *New Ideas in Psychology 40*: pp. 3–12.
Rapaport, William J. 1999. "Implementation Is Semantic Interpretation." *The Monist 82*: pp. 109–130.
Rathkopf, Charles. 2018. "Network Representation and Complex Systems." *Synthese 195*: pp. 55–78.
Rescorla, Michael. 2007. "Church's Thesis and the Conceptual Analysis of Computability." *Notre Dame Journal of Formal Logic 48*: pp. 253–280.
Rescorla, Michael. 2012. "Are Computational Transitions Sensitive to Semantics?" *Australasian Journal of Philosophy 90*: pp. 703–721.
Rescorla, Michael. 2013. "Against Structuralist Theories of Computational Implementation." *The British Journal for the Philosophy of Science 64*: pp. 681–707.
Rescorla, Michael. 2014. "A Theory of Computational Implementation." *Synthese 191*: pp. 1277–1307.
Rescorla, Michael. 2015. "The Computational Theory of Mind." In *The Stanford Encyclopedia of Philosophy*, edited by E. N. Zalta. https://plato.stanford.edu/archives/win2015/entries/computational-mind/.
Rescorla, Michael. 2016. "Book Review: Gualtiero Piccinini//Physical Computation." *The British Journal for the Philosophy of Science, Review of Books*.
Rescorla, Michael. 2017. "Levels of Computational Explanation." In *Philosophy and Computing: Essays in Epistemology, Philosophy of Mind, Logic, and Ethics*, edited by T. M. Powers, pp. 5–28. Cham: Springer.
Robinson, David A. 1968. "The Oculomotor Control System: A Review." *Proceedings of the IEEE 56*: pp. 1032–1049.
Robinson, David A. 1989. "Integrating with Neurons." *Annual Review of Neuroscience 12*: pp. 33–45.
Rogers, Hartley Jr. 1987. *Theory of Recursive Functions and Effective Computability*. Cambridge, MA: MIT Press (second edition reprint of the 1967 edition).
Rosen, Gideon. 2020. "Abstract Objects." *The Stanford Encyclopedia of Philosophy* (Spring 2020 Edition), edited by E. N. Zalta. https://plato.stanford.edu/archives/spr2020/entries/abstract-objects/.
Roth, Martin. 2005. "Program Execution in Connectionist Networks." *Mind and Language 20*: pp. 448–467.
Rozenberg, Grzegorz, Thomas Bäck, and Joost N. Kok (eds.). 2012. *Handbook of Natural Computing*. Berlin, Heidelberg: Springer.
Rubel, Lee A. 1985. "The Brain as an Analog Computer." *Journal of Theoretical Neurobiology 4*: pp. 73–81.
Rubel, Lee A. 1993. "The Extended Analog Computer." *Advances in Applied Mathematics 14*: pp. 39–50.
Rumelhart, David E., Geoffrey E. Hinton, and Ronald J. Williams. 1986. "Learning Representations by Back-Propagating Errors." *Nature 323*: pp. 533–536.

Rumelhart, David E., and James L. McClelland. 1986. "On Learning the Past Tenses of English Verbs." In *Parallel Distributed Processing: Explorations in the Microstructure of Cognition. Vol. 2: Psychological and Biological Models.*, edited by J. L. McClelland, D. E. Rumelhart, and the PDP Research Group, pp. 216–271. Cambridge, MA: MIT Press.

Rumelhart, David E., James L. McClelland, and the PDP Research Group (eds.). 1986. *Parallel Distributed Processing: Explorations in the Microstructure of Cognition. Vol. 1: Foundations.* Cambridge, MA: MIT Press.

Rumelhart, David E., Paul Smolensky, James L. McClelland, and Geoffrey E. Hinton. 1986. "Schemata and Sequential Thought Processes in PDP Models." In *Parallel Distributed Processing: Explorations in the Microstructure of Cognition. Vol. 2: Psychological and Biological Models.*, edited by J. L. McClelland, D. E. Rumelhartand the PDP Research Group, pp. 7–57. Cambridge, MA: MIT Press.

Rusanen, Anna-Mari, and Otto Lappi. 2016. "On Computational Explanations." *Synthese 193*: pp. 3931–3949.

Russell, Bertrand A. W. 1915. *Our Knowledge of the External World as a Field for Scientific Method in Philosophy.* Chicago: Open Court.

Ryder, Dan. 2004. "SINBAD Neurosemantics: A Theory of Mental Representation." *Mind & Language 19*: pp. 211–240.

Scarpellini, Bruno. 1963. "Zwei unentscheidbare probleme der analysis." *Zeitschrift für mathematische Logik und Grundlagen der Mathematik 9*: pp. 265–289.

Scheutz, Matthias. 1999. "When Physical Systems Realize Functions." *Minds and Machines 9*: pp. 161–196.

Scheutz, Matthias. 2001. "Computational Versus Causal Complexity." *Minds and Machines 11*: pp. 534–566.

Scheutz, Matthias. 2012. "What It Is Not to Implement a Computation: A Critical Analysis of Chalmers's Notion of Implementation." *Journal of Cognitive Science 13*: pp. 75–106.

Schiller, Henry Ian. 2018. "The Swapping Constraint." *Minds and Machines 28*: pp. 605–622.

Schmidhuber, Jürgen. 2000. "Algorithmic Theories of Everything." ArXiv: quant-ph/0011122.

Schmidhuber, Jürgen. 2015. "Deep Learning in Neural Networks: An Overview." *Neural Networks 61*: pp. 85–117.

Schneider, Susan. 2011. *The Language of Thought: A New Philosophical Direction.* Cambridge, MA: MIT Press.

Schwarz, Georg. 1992. "Connectionism, Processing, Memory." *Connection Science 4*: pp. 207–226.

Schweizer, Paul. 2014. "Algorithms Implemented in Space and Time." In *Selected Papers From th 50th Anniversary Convention of the AISB*, pp. 128–135. London: The AISB.

Schweizer, Paul. 2016. "In What Sense Does the Brain Compute?" In *Computing and Philosophy: Selected Papers from IACAP 2014*, edited by V. C. Müller, pp. 63–79. New York: Springer.

Schweizer, Paul. 2019a. "Triviality Arguments Reconsidered." *Minds and Machines 29*: pp. 287–308.

Schweizer, Paul. 2019b. "Computation in Physical Systems: A Normative Mapping Account." In *On the Cognitive, Ethical, and Scientific Dimensions of Artificial Intelligence: Themes from IACAP 2016*, edited by D. Berkich and M. V. d'Alfonso, pp. 27–47. Berlin: Springer.

Scott, Dana S., and Christopher Strachey. 1971. *Toward a Mathematical Semantics for Computer Languages*, vol. 1. Oxford: Oxford University Computing Laboratory, Programming Research Group.
Searle, John R. 1980. "Minds, Brains and Programs." *Behavioral and Brain Sciences 3*: pp. 417–457.
Searle, John R. 1992. *The Rediscovery of the Mind*. Cambridge. MA: MIT Press.
Searle, John R. 1995. *The Construction of Social Reality*. New York: Free Press.
Segal, Gabriel. 1989. "Seeing What Is Not There." *The Philosophical Review 98*: pp. 189–214.
Segal, Gabriel. 1991. "Defense of a Reasonable Individualism." *Mind 100*: pp. 485–494.
Sejnowski, Terrence J., Christof Koch, and Patricia S. Churchland. 1988. "Computational Neuroscience." *Science 241*: pp. 1299–1306.
Serban, Maria. 2015. "The Scope and Limits of a Mechanistic View of Computational Explanation." *Synthese 192*: pp. 3371–3396.
Seung, H. Sebastian. 1996. "How the Brain Keeps the Eyes Still." *Proceedings of the National Academy of Sciences USA 93*: pp. 13339–13344.
Seung, H. Sebastian. 1998. "Continuous Attractors and Oculomotor Control." *Neural Networks 11*: pp. 1253–1258.
Seung, H. Sebastian, Daniel D. Lee, Ben Y. Reis, and David W. Tank. 2000. "Stability of the Memory of Eye Position in a Recurrent Network of Conductance-Based Model Neurons." *Neuron 26*: pp. 259–271.
Shadmehr, Reza, and Steven P. Wise. 2005. *The Computational Neurobiology of Reaching and Pointing: A Foundation for Motor Learning*. Cambridge, MA: MIT Press.
Shagrir, Oron. 1992. "A Neural Net with Self-Inhibiting Units for the N-Queens Problem." *International Journal of Neural Systems 3*: pp. 249–252.
Shagrir, Oron. 2001. "Content, Computation and Externalism." *Mind 110*: pp. 369–400.
Shagrir, Oron. 2002. "Effective Computation by Humans and Machines." *Minds and Machines 12*: pp. 221–240.
Shagrir, Oron. 2005. "The Rise and Fall of Computational Functionalism." In *Hilary Putnam*, edited by Y. Ben-Menahem, pp. 220–250. Cambridge: Cambridge University Press.
Shagrir, Oron. 2006a. "Gödel on Turing on Computability." In *Church's Thesis After 70 Years*, edited by A. Olszewski, J. Wolenski, and R. Janusz, pp. 393–419. Heusenstamm: Ontos Verlag.
Shagrir, Oron. 2006b. "Why We View the Brain as a Computer." *Synthese 153*: pp. 393–416.
Shagrir, Oron. 2010. "Marr on Computational-Level Theories." *Philosophy of Science 77*: pp. 477–500.
Shagrir, Oron. 2012a. "Computation, Implementation, Cognition." *Minds and Machines 22*: pp. 137–148.
Shagrir, Oron. 2012b. "Can a Brain Possess Two Minds?" *Journal of Cognitive Science 13*: pp. 145–165.
Shagrir, Oron. 2012c. "Structural Representations and the Brain." *The British Journal for the Philosophy of Science 63*: pp. 519–545.
Shagrir, Oron. 2014. "Review of *Explaining the Computational Theory of Mind*, by Marcin Miłkowski." Notre Dame Review of Philosophy.
Shagrir, Oron. 2016. "Advertisement for the Philosophy of the Computational Sciences." In *The Oxford Handbook of Philosophy of Science*, edited by P. Humphreys, pp. 15–42. New York: Oxford University Press.

Shagrir, Oron. 2017. "Review of *Physical Computation: A Mechanistic Account* by Gualtiero Piccinini." *Philosophy of Science 84*: pp. 604–612.

Shagrir, Oron. 2018. "The Brain as an Input-Output Model of the World." *Minds and Machines 28*: pp. 53–75.

Shagrir, Oron. 2020. "In Defense of the Semantic View of Computation." *Synthese 197*: pp. 4083–4108.

Shagrir, Oron, and William Bechtel. 2017. "Marr's Computational Level and Delineating Phenomena." In *Explanation and Integration in Mind and Brain Science*, edited by D. M. Kaplan, pp. 190–214. New York: Oxford University Press.

Shagrir, Oron, and Itamar Pitowsky. 2003. "Physical Hypercomputation and the Church–Turing Thesis." *Minds and Machines 13*: pp. 87–101.

Shannon, Claude E. 1948. "A Mathematical Theory of Communication." *Bell System Technical Journal 27*: pp. 379–423.

Shannon, Claude E., and John McCarthy (eds.). 1956. *Automata Studies*. Annals of Mathematics Studies 34. Princeton: Princeton University Press.

Shannon, Claude E., and Warren Weaver. 1949. *The Mathematical Theory of Communication*. Champaign, IL: University of Illinois Press.

Shapiro, Lawrence A. 1993. "Content, Kinds, and Individualism in Marr's Theory of Vision." *The Philosophical Review 102*: pp. 489–513.

Shapiro, Lawrence A. 1997. "A Clearer Vision." *Philosophy of Science 64*: pp. 131–153.

Shapiro, Lawrence A. 2000. "Multiple Realizations." *The Journal of Philosophy 97*: pp. 635–654.

Shapiro, Lawrence A. 2017. "Mechanism or Bust? Explanation in Psychology." *The British Journal for the Philosophy of Science 68*: pp. 1037–1059.

Shapiro, Stewart. 1981. "Understanding Church's Thesis." *Journal of Philosophical Logic 10*: pp. 353–365.

Shapiro, Stewart. 1982. "Acceptable Notation." *Notre Dame Journal of Formal Logic 23*: pp. 14–20.

Shapiro, Stewart. 1984. "On an 'Empiricist' Philosophy of Mathematics." *Philosophia 14*: pp. 213–223.

Shapiro, Stewart. 1993. "Understanding Church's Thesis, Again." *Acta Analytica 11*: pp. 59–77.

Shapiro, Stewart. 2013. "The Open Texture of Computability." In *Computability: Turing, Gödel, Church, and Beyond*, edited by B. J. Copeland, C. Posy, and O. Shagrir, pp. 153–181. Cambridge, MA: MIT Press.

Sharp, Patricia E., Amanda Tinkelman, and Jeiwon Cho. 2001. "Angular Velocity and Head Direction Signals Recorded from the Dorsal Tegmental Nucleus of Gudden in the Rat: Implications for Path Integration in the Head Direction Cell Circuit." *Behavioral Neuroscience 115*: pp. 571–588.

Shea, Nicholas. 2013. "Naturalising Representational Content." *Philosophy Compass 8*: pp. 496–509.

Shea, Nicholas. 2018. *Representation in Cognitive Science*. New York: Oxford University Press.

Shepard, Roger N., and Susan Chipman. 1970. "Second-Order Isomorphism of Internal Representations: Shapes of States." *Cognitive Psychology 1*: pp. 1–17.

Sher, Gila Y. 1991. *The Bounds of Logic: A Generalized Viewpoint*. Cambridge, MA: MIT Press.

Sher, Gila Y. 1996. "Did Tarski Commit 'Tarski's Fallacy'?" *The Journal of Symbolic Logic* 61: pp. 653–686.

Shor, Peter W. 1994. "Algorithms for Quantum Computation: Discrete Logarithms and Factoring." In *Proceedings of the 35th Annual Symposium on Foundations of Computer Science*, pp. 124–134. Los Alamitos, CA: IEEE Computer Society Press.

Shor, Peter W. 1997. "Polynomial-Time Algorithms for Prime Factorization and Discrete Logarithms on a Quantum Computer." *SIAM Journal on Computing 26*: pp. 1484–1509.

Sieg, Wilfried. 1994. "Mechanical Procedures and Mathematical Experience." In *Mathematics and Mind*, edited by A. George, pp. 71–117. New York: Oxford University Press.

Sieg, Wilfried. 1997. "Step by Recursive Step: Church's Analysis of Effective Calculability." *Bulletin of Symbolic Logic 3*: pp. 154–180.

Sieg, Wilfried. 2002. "Calculations by Man and Machine: Conceptual Analysis." In *Reflections on the Foundations of Mathematics: Essays in Honor of Solomon Feferman*, edited by W. Sieg, R. Sommer, and C. L. Talcott, pp. 390–409. Natick, MA: Association for Symbolic Logic.

Sieg, Wilfried. 2006. "Gödel on Computability." *Philosophia Mathematica 14*: pp. 189–207.

Sieg, Wilfried. 2008. "Church Without Dogma: Axioms for Computability." In *New Computational Paradigms*, edited by S. B. Cooper, B. Löwe, and A. Sorbi, pp. 139–152. Berlin: Springer.

Sieg, Wilfried. 2009. "On Computability." In *Handbook of the Philosophy of Mathematics*, edited by A. D. Irvine, pp. 535–630. Amsterdam: Elsevier.

Sieg, Wilfried. 2013a. *Hilbert's Programs and Beyond*. New York: Oxford University Press.

Sieg, Wilfried. 2013b. "Gödel's Philosophical Challenge to Turing." In *Computability: Turing, Gödel, Church, and Beyond*, edited by B. J. Copeland, C. Posy, and O. Shagrir, pp. 183–202. Cambridge, MA: MIT Press.

Sieg, Wilfried, and John Byrnes. 1999. "An Abstract Model for Parallel Computations: Gandy's Thesis." *The Monist 82*: pp. 150–164.

Siegelmann, Hava T. 1995. "Computation Beyond the Turing Limit." *Science 268*: pp. 545–548.

Siegelmann, Hava T. 1999. *Neural Networks and Analog Computation: Beyond the Turing Limit*. Boston: Birkhäuser.

Silverberg, Arnold. 2006. "Chomsky and Egan on Computational Theories of Vision." *Minds and Machines 16*: pp. 495–524.

Simon, Herbert A. 1962. "The Architecture of Complexity." *Proceedings of the American Philosophical Society 106*: pp. 467–482.

Smith, Brian C. 1996. *On the Origin of Objects*. Cambridge, MA: MIT Press.

Smith, Brian C. 2002. "The Foundations of Computing." In *Computationalism: New Directions*, edited by M. Scheutz, pp. 23–58. Cambridge, MA: MIT Press.

Smith, Brian C. 2010. Introduction to *Age of Significance*. Available at https://web.archive.org/web/20161019202810/http://ageofsignificance.org/aos/en/toc.html.

Smolensky, Paul. 1988. "On the Proper Treatment of Connectionism." *Behavioral and Brain Sciences 11*: pp. 1–23.

Smolensky, Paul, and Géraldine Legendre. 2006. *The Harmonic Mind: From Neural Computation to Optimality-Theoretic Grammar*, vols. 1–2. Cambridge, MA: MIT Press.

Smolensky, Paul, Géraldine Legendre, and Yoshiro Miyata. 1992. "Principles for an Integrated Connectionist/Symbolic Theory of Higher Cognition." Technical Report CU-CS-600-92. Department of Computer Science, University of Colorado at Boulder.

Soare, Robert I. 1996. Computability and Recursion. *Bulletin of Symbolic Logic* 2: pp. 284–321.
Sosic, Rok, and Jun Gu. 1990. "A Polynomial Time Algorithm for the N-Queens Problem." *SIGART Bulletin* 1: pp. 7–11.
Sosic, Rok, and Jun Gu. 1991. "3,000,000 Queens in Less Than One Minute." *SIGART Bulletin* 2: pp. 22–24.
Sprevak, Mark. 2010. "Computation, Individuation, and the Received View on Representation." *Studies in History and Philosophy of Science* 41: pp. 260–270.
Sprevak, Mark. 2012. "Three Challenges to Chalmers on Computational Implementation." *Journal of Cognitive Science* 13: pp. 107–143.
Sprevak, Mark. 2013. "Fictionalism About Neural Representations." *The Monist* 96: pp. 539–560.
Sprevak, Mark. 2018. "Triviality Arguments About Computational Implementation." In *Routledge Handbook of the Computational Mind*, edited by M. Sprevak and M. Colombo, pp. 175–191. London: Routledge.
Sterelny, Kim. 1990. *The Representational Theory of Mind: An Introduction*. Oxford: Blackwell.
Stern, Peter, and John Travis. 2006. "Of Bytes and Brains." *Science* 314: p. 75.
Stewart, Ian. 1991. "Deciding the Undecidable." *Nature* 352: pp. 664–665.
Stich, Stephen P. 1983. *From Folk Psychology to Cognitive Science: The Case Against Belief*. Cambridge, MA: MIT Press.
Stich, Stephen P. 1991. "Narrow Content Meets Fat Syntax." In *Meaning in Mind: Fodor and His Critics*, edited by B. M. Loewer and G. Rey, pp. 239–254. Oxford: Blackwell.
Suárez, Mauricio. 2010. "Scientific Representation." *Philosophy Compass* 5: pp. 91–101.
Swoyer, Chris. 1991. "Structural Representation and Surrogative Reasoning." *Synthese* 87: pp. 449–508.
Syropoulos, Apostolos. 2008. *Hypercomputation: Computing Beyond the Church-Turing Barrier*. Berlin: Springer.
Teller, Paul. 1980. "Computer Proof." *The Journal of Philosophy* 77: pp. 797–803.
Tucker, Chris. 2018. "How to Explain Miscomputation." *Philosophers' Imprint* 18: pp. 1–17.
Tucker, John V., and Jeffery I. Zucker. 2004. "Abstract Versus Concrete Computation on Metric Partial Algebras." *ACM Transactions on Computational Logic* 5: pp. 611–668.
Turing, Alan M. 1936. "On Computable Numbers, with an Application to the Entscheidungsproblem." *Proceedings of the London Mathematical Society, Series 2*, 42: pp. 230–265. Reprinted in *The Essential Turing*, edited by B. Jack Copeland (2004a), pp. 58–90. Page references are to Copeland.
Turing, Alan M. 1937. "On Computable Numbers, with an Application to the 'Entscheidungsproblem.' A Correction." *Proceedings of the London Mathematical Society, series 2*, 43: pp. 544–546.
Turing, Alan M. 1939. "Systems of Logic Based on Ordinals." *Proceedings of the London Mathematical Society, series 2*, 45: pp. 161–228.
Turing, Alan M. 1948. "Rounding-Off Errors in Matrix Processes." *The Quarterly Journal of Mechanics and Applied Mathematics* 1: pp. 287–308.
Turing, Alan M. 1950. "Computing Machinery and Intelligence." *Mind* 59: pp. 433–460.
Turner, Raymond. 2013. "The Philosophy of Computer Science." In *Stanford Encyclopedia of Philosophy*, edited by E. N. Zalta. http://plato.stanford.edu/entries/computer-science/.

Tymoczko, Thomas. 1979. "The Four-Color Problem and Its Philosophical Significance." *The Journal of Philosophy* 76: pp. 57–83.

Ullman, Shimon. 1979. *The Interpretation of Visual Motion.* Cambridge, MA: MIT Press.

Ulmann, Bernd. 2013. *Analog Computing.* Munich: Oldenbourg Verlag.

Urquhart, Alasdair. 2010. "Von Neumann, Gödel and Complexity Theory." *Bulletin of Symbolic Logic* 16: pp. 516–530.

Van Gelder, Tim. 1995. "What Might Cognition Be if Not Computation?" *The Journal of Philosophy* 92: pp. 345–381.

Van Rooij, Iris, Mark Blokpoel, Johan Kwisthout, and Todd Wareham. 2019. *Cognition and Intractability: A Guide to Classical and Parameterized Complexity Analysis.* Cambridge: Cambridge University Press.

Vardi, Moshe Y. 2012. "What Is an Algorithm?" *Communications of the ACM* 55: p. 5.

Von Neumann, John. 1927. "Zur Hilbertschen Beweistheorie." *Mathematische Zeitschrift* 26: pp. 1–46.

Von Neumann, John. 1958. *The Computer and the Brain.* London: Yale University Press.

Wang, Hao. 1974. *From Mathematics to Philosophy.* London: Routledge & Kegan Paul.

Weihrauch, Klaus. 2000. *Computable Analysis: An Introduction.* Berlin, Heidelberg: Springer.

Weihrauch, Klaus, and Ning Zhong. 2002. "Is Wave Propagation Computable or Can Wave Computers Beat the Turing Machine?" *Proceedings of the London Mathematical Society* 85: pp. 312–332.

Weisberg, Michael. 2013. *Simulation and Similarity: Using Models to Understand the World.* New York: Oxford University Press.

Weiskopf, Daniel A. 2011. "Models and Mechanisms in Psychological Explanation." *Synthese* 183: pp. 313–338.

Welch, Philip D. 2008. "The Extent of Computation in Malament-Hogarth Spacetimes." *The British Journal for the Philosophy of Science* 59: pp. 659–674.

Weyl, Hermann. 1949. *Philosophy of Mathematics and Natural Science.* Princeton: Princeton University Press.

Wheeler, John A. 1990. "Information, Physics, Quantum: The Search for Links." In *Complexity, Entropy, and the Physics of Information*, edited by W. H. Zurek, pp. 3–28. Redwood City, CA: Addison-Wesley.

Wiener, Norbert. 1948. *Cybernetics or Control and Communication in the Animal and the Machine.* Cambridge, MA: MIT Press.

Wilson, Robert A. 1994. "Wide Computationalism." *Mind* 103: pp. 351–372.

Wilson, Robert A. 2004. *Boundaries of the Mind: The Individual in the Fragile Sciences: Cognition.* Cambridge: Cambridge University Press.

Winsberg, Eric. 2010. *Science in the Age of Computer Simulation.* Chicago: University of Chicago Press.

Wolfram, Stephen. 1985. "Undecidability and Intractability in Theoretical Physics." *Physical Review Letters* 54: pp. 735–738.

Wolfram, Stephen. 2002. *A New Kind of Science.* Champaign, IL: Wolfram Media.

Woodward, James. 2003. *Making Things Happen: A Theory of Causal Explanation.* New York: Oxford University Press.

Yanofsky, Noson S. 2011. "Towards a Definition of an Algorithm." *Journal of Logic and Computation* 21: pp. 253–286.

Yao, Andrew Chi-Chih. 2003. "Classical Physics and the Church-Turing Thesis." *Journal of the ACM (JACM)* 50: pp. 100–105.

Zach, Richard. 2019. "Hilbert's Program." In *The Stanford Encyclopedia of Philosophy*, edited by E. N. Zalta. http://plato.stanford.edu/entries/hilbert-program/.

Zipser, David, and Richard A. Andersen. 1988. "A Back-Propagation Programmed Network That Simulates Response Properties of a Subset of Posterior Parietal Neurons." *Nature 331*: pp. 679–684.

Zuse, Konrad. 1967. "Rechnender Raum." *Elektronische Datenverarbeitung 8*: pp. 336–344.

Name Index

Aaronson, Scott, 42n39, 65n32
Abbott, Laurence, 8n4, 164, 261
Abrahamsen, Adele, 110n27
Ackermann, Wilhelm, 29, 29n13, 33
Aharonov, Dorit, 66, 67, 74
Aitken, Wayne, 81
Amit, Daniel, 106, 107–108n25, 142, 236
Andersen, Holly, 146n3
Andersen, Richard, 106n20, 246, 249, 253
Anderson, Charles, 108n25, 233, 236n9, 245
Anderson, James, 103n15, 110n27
Anderson, Neal, 4, 7n3, 16n16, 126n9
Andréka, Hajnal, 81n55, 83, 84n57
Astrachan, Owen, 55
Avigad, Jeremy, 76

Bahar, Sonya, 14n13, 75n42, 117n44, 118n45
Barrett, David, 157
Barrett, Jeffrey, 81
Bartels, Andreas, 229n2
Barto, Andrew, 161, 162
Bassett, Joshua, 251
Bechtel, William, 110n27, 146n3, 157, 158, 166, 247, 247n20, 257
Becker, Wolfgang, 236, 237
Beggs, Edwin, 80n53
Bengio, Yoshua, 106n22
Bernays, Paul, 34
Bernstein, Ethan, 55, 67, 72
Bishop, John Mark, 129n20, 207n4
Blackmon, James, 130n26
Blake, Ralph, 80
Blass, Andreas, 63n24, 64n29
Block, Ned, 101n11, 127n16, 182, 189, 199n35, 205–206, 219
Blum, Lenore, 76n47
Boden, Margaret, 22n22
Boghossian, Paul, 43n40
Boker, Udi, 28n4, 28n8, 64n29
Bontly, Thomas, 186n22
Boolos, George, 33n20, 33n21, 80n51, 191n28
Boone, Worth, 145, 158n15
Botvinick, Matthew, 161, 161n19, 162, 162n21
Bournez, Oliver, 75n43

Brabazon, Anthony, 61n14
Brattka, Vasco, 76n46
Braverman, Mark, 76n46
Bringsjord, Selmer, 77n49
Brown, Curtis, 132n29, 135n33, 138n43
Buckner, Cameron, 113
Burge, Tyler, 177, 179, 199n35, 213, 260n29
Button, Tim, 81
Buzaglo, Meir, 62n15

Campbell, Douglas Ian, 143n46
Cannon, Stephen, 233
Cao, Rosa, 21n21, 24n25
Carandini, Matteo, 152n10, 165
Chaitin, Gregory, 183n17
Chalmers, David, 119, 119n1, 123, 126, 127n10, 129n20, 129n21, 130–132, 132n29, 133, 133n30, 134, 136, 136n38, 136n39, 136n40, 137, 137n42, 138–139, 139n44, 140–144, 149n4, 150n6, 193, 201, 203, 207, 215n11
Chater, Nick, 99n7
Chemero, Anthony, 9n7
Chirimuuta, Mazviita, 153n11, 157, 158, 164, 164n23, 165, 165n25–26, 254, 261–263
Chomsky, Noam, 42
Chrisley, Ronald, 123n5, 129n20, 134n32
Church, Alonzo, 26, 26n1, 27, 27n3, 28, 28n6, 29n12, 30, 31n17, 32, 33, 33n22, 34, 34n25, 35n28, 36n31, 39, 40, 45, 55, 62, 65
Churchland, Patricia, 8n4, 11, 13, 14, 16, 102, 110n27, 184, 186, 197, 245
Churchland, Paul, 110n27, 184, 245, 255
Clark, Andy, 99n7, 245n18
Cleland, Carol, 65, 65n30, 120n2
Cobham, Alan, 66
Coelho Mollo, Dimitri, 3n2, 5, 21n19, 143, 145, 163, 168–171, 194, 195, 214, 215–216, 216n13, 217–221
Cohen, Rina, 56n8, 57
Collett, Matthew, 250n22
Collett, Thomas, 250n22
Colombo, Matteo, 17
Conklin, John, 250n22
Cook, Stephen, 76n46

NAME INDEX

Copeland, Jack, 2, 4, 7n3, 11, 11n9, 12, 26n2, 27, 28n4, 28n8, 35, 35n30, 39n33, 43, 43n41, 43n42, 45n46, 46n48, 49n2, 55–56, 56n5–7, 58, 60, 62, 63, 65, 65n31, 68n35, 70n36, 73, 78, 79, 80n52, 81n54, 85n60, 93, 95, 98n4, 99n8, 116n40, 122–124, 129n20, 129n21, 129n24, 135n36, 143, 143n46, 150, 150n6, 201n1, 206n3
Crane, Tim, 2, 93, 178, 188n24, 191n27, 196n33
Craver, Carl, 145, 146n3, 151, 152–153, 156, 158n15, 160, 165n25–26, 171
Cucker, Felipe, 76n47
Cummins, Robert, 2, 4–5, 88, 88n1, 89, 90–95, 95n3, 96, 99, 101, 103, 110, 111, 111n32, 112, 112n33, 112n36, 113, 115, 116, 118, 152, 152n9, 163, 167, 180n8, 180n9, 180n11, 181, 182, 234, 239, 240, 240n14, 241, 243

Daniels, Norman, 15n15
Darden, Lindley, 146n3
Dasgupta, Dipanker, 173, 173n32
Davies, Brian, 80n53, 81n54
Davies, Martin, 209n8, 260n29
Davis, Martin, 31, 31n17, 32n18, 34n24, 35n29, 36n31
Dayan, Peter, 8n4, 164, 261
De Mol, Liesbeth, 45n45
Dean, Walter, 63n21, 65n32
Demopoulos, William, 101n11
Dennett, Daniel, 55, 184, 190
Dershowitz, Nachum, 28n4, 28n8, 62, 63n24, 64n29
Descartes, René, 29
Deutsch, David, 70, 77
Dewhurst, Joe, 5, 17, 145, 171, 172n29, 176, 215, 216n13, 217–218, 219, 220, 221, 226n18
Dodig-Crnkovic, Gordana, 7n3
Dresner, Eli, 41n36, 129n25, 239n13
Dretske, Fred, 19, 20n18, 180n10, 181, 181n12, 182
Dreyfus, Hubert, 9n7

Earman, John, 70n37, 76n48, 77, 81n54, 83–84
Edmonds, Jack, 66
Egan, Frances, 129n25, 157, 157n13, 166n28, 175n2, 176, 198, 199n34, 213, 213n10, 219, 258–261
Einstein, Albert, 81
Elber-Dorozko, Lotem, 160n16, 161n18, 162n21, 254
Eliasmith, Chris, 108n25, 116n43, 223, 236n9, 245, 250n22
Essick, Greg, 246
Etesi, Gábor, 81, 84n57

Etienne, Ariane, 250n22, 251–252
Euclid, 29

Feferman, Solomon, 76n47
Fernau, Henning, 108n26
Figdor, Carrie, 199n35
Floridi, Luciano, 182
Fodor, Jerry, 2, 3, 9, 93, 101n11, 110n28, 112n35, 122n3, 127n15, 149, 150n6, 152, 152n9, 163, 178, 178n6, 179, 180n8, 182, 184, 199n34, 243–244, 255
Folina, Janet, 28n8
Fortnow, Lance, 65
Frege, Gottlob, 29
French, Steven, 229n2
Fresco, Nir, 6n1, 7, 17, 18, 58, 100n10, 129n25, 145, 150n5, 150n6, 177, 183n16, 201n1, 206–207n3, 216n12, 216n14
Frigg, Roman, 9n6, 74n41, 105n18, 106n19, 229n1

Gallistel, Charles, 110n28, 232n3, 243n16, 250n22
Gandy, Robin, 27, 29n13, 32n19, 33n23, 35, 38, 39–41, 42, 42n38, 45, 46, 47, 49–53, 54, 58–59, 62, 63n22, 64, 68–69, 73, 79, 80–81, 85, 87, 100n10
Garson, James, 113
Gherardi, Guido, 75n44–45, 76n47, 78n50
Gibbon, John, 110n28
Giere, Ronald, 229n2
Glennan, Stuart, 146n3
Glimcher, Paul, 233
Gödel, Kurt, 11–12, 11n9, 12nn11–12, 27, 29, 29n11, 29n14, 30–32, 30n16, 31n17, 32n18, 33, 34–35, 34n24, 35n27, 36n31, 38, 38n32, 43n42, 44n44, 45, 62, 62n16, 65n32
Godfrey-Smith, Peter, 129, 129n24, 132n28, 135–136, 135n34–35, 136n37, 189
Goel, Vinod, 101n11
Goff, Philip, 127n12
Gold, Arie, 56n8, 57
Goldman, Mark, 233n6
Goldreich, Oded, 66n33, 66
Goodman, Nelson, 101n11
Grier, David, 40n34
Griffiths, Thomas, 99n7, 245n18
Grush, Rick, 245–247, 245n18, 253n26
Grzegorczyk, Andrzej, 75n46, 76–78
Gu, Jun, 108n26
Gurari, Eitan, 99n5
Gurevich, Yuri, 61, 62, 62n17–19, 63, 63n23, 63n24, 64–65, 64n26, 64n28, 64n29, 68n35
Gutfreund, Hanoch, 106

NAME INDEX

Hafting, Torkel, 251n23
Haimovici, Sabrina, 157, 163
Hamkins, Joel, 56, 56n8, 57, 57n10
Harbecke, Jens, 160n16, 223n16
Hardcastle, Valerie Gray, 149, 171
Harel, David, 64, 72
Hartmann, Stephan, 74n41, 105n18, 106n19, 229n1
Haugeland, John, 2, 93, 101n11, 149n4
Heeger, David, 152n10, 165
Hemmo, Meir, 195n30, 207n4, 225n17
Herbrand, Jacques, 27, 32
Hertz, John, 106, 106n23
Hess, Robert, 236, 237
Hilbert, David, 29, 29n10, 29n11, 29n13, 30, 33, 34
Hildreth, Ellen, 252n24, 258
Hinckfuss, Ian, 120
Hinton, Geoffrey, 106n21, 106n22, 107–108n25, 110n27, 114n39
Hoffmann, Geoffrey, 173
Hogarth, Mark, 81, 81n56, 82
Homer, Steve, 65
Hopcroft, John, 60, 64
Hopfield, John, 106, 107, 107n24, 107n25, 109, 116, 117, 242
Horowitz, Amir, 186n22, 221
Horsman, Clare, 240n14
Horsman, Dominic, 240n14
Horst, Steven, 9n8
Hubel, David, 165, 260
Humphreys, Paul, 9n6
Huneman, Philippe, 155
Hutto, Daniel, 9n7

Illari, Phyllis McKay, 146n3
Impagliazzo, John, 60

Jeffrey, Richard, 33n20–21, 80n51, 191n28
Jerne, Hiels, 173

Kaplan, David, 145, 145n1, 158n15, 163, 165n24, 165n25, 165n26
Kemp, Charles, 245n18
King, Adam Philip, 232n3, 243n16, 250n22
Kitcher, Patricia, 260n29
Kleene, Stephen, 27, 27n3, 28, 28n5–6, 31n17, 33, 34, 36n31, 38, 39n33, 62
Klein, Colin, 129n22, 138n43
Klein, Horst-Manfred, 236, 237
Knuth, Donald, 62n18, 63n20
Koch, Christof, 8n4, 11, 13, 14, 15n14, 16, 111n29, 197, 245
Koepke, Peter, 58

Kolmogorov, Andrei, 62n18, 183n17
Komar, Arthur, 77
Kreisel, Georg, 43n42, 77
Kripke, Saul, 33n23, 43n40, 45
Krizhevsky, Alex, 106n22
Krogh, Anders, 106, 106n23

Lacombe, Daniel, 75n46
Ladyman, James, 186, 229n2
Lappi, Otto, 145n1, 153n11, 166n27–28, 254
LeCun, Yann, 106n22
Lee, Jonny, 3, 4, 17, 177n4, 178n5
Legendre, Géraldine, 114, 114n38
Leibniz, Gottfried Wilhelm, 29
Leigh, R. John, 233, 233n5
Levin, Janet, 127n14
Levy, Arnon, 152n7, 155, 158
Lewis, Andy, 56, 57n10
Lewis, David, 101n11
Lewis, Harry, 60, 61n12–13, 191, 192
Löwe, Benedikt, 57–58
Lycan, William, 120n2

Machamer, Peter, 146n3
Malament, David, 81
Maley, Corey, 14n13, 101n12, 116n42, 170, 172, 244, 244n17
Manchak, John Byron, 80
Mancosu, Paolo, 29n13
Marr, David, 153, 154, 160–161, 166, 166n27, 166n28, 171, 180n10, 219, 247–248, 252, 252n25, 257–258, 259, 260, 260n29, 261, 262, 265
Matthews, Robert, 129n25
Maudlin, Tim, 207n4
Mazur, Stanislaw, 75n46
McCarthy, John, 34n26
McClelland, James, 103n17, 107–108n25, 111n30, 113, 114n39
McCulloch, Warren, 34n26, 103, 103n14, 201n2
McGarraghy, Seán, 61n14
McNaughton, Bruce, 250n22
Melnyk, Andrew, 129n20
Mendelson, Elliot, 28n8
Miłkowski, Marcin, 22n22, 58, 129n23, 135, 138n43, 140n45, 145, 145n2, 150, 150n5, 152, 177, 216n12, 216n14, 216n15
Millhouse, Tyler, 129n25
Millikan, Ruth Garrett, 171, 181, 182
Milner, Robin, 63
Minsky, Marvin, 34n26, 103n16, 111n30, 116, 201n2
Mittelstaedt, Hort, 250n22
Mittelstaedt, Marie-Luise, 250n22

Miyata, Yoshiro, 114
Morgan, Alex, 173n31
Morton, Peter, 260n29
Moschovakis, Yiannis, 63–64, 87
Müller, Vincent, 7n3
Munakata, Yuko, 8n4, 111n29

Nadel, Lynn, 150, 245
Nagin, Paul, 60
Neander, Karen, 171
Németi, István, 81, 81n55, 83, 84n57, 84n59
Németi, Peter, 83, 84n57
Newell, Allen, 2, 55, 93
Niv, Yael, 161, 161n19, 162, 162n21
Norton, John, 42n38, 80n53, 83–84

O'Brien, Gerard, 110n27, 199n35, 245
Odifreddi, Piergiorgio, 63n20
O'Keefe, John, 150, 245
O'Neill, Mark, 61n14
Opie, Jon, 110n27, 199n35, 245
O'Reilly, Randall, 8n4
Orlandi, Nico, 9n7

Palmer, Stephen, 106, 106n23, 243n16
Papadimitriou, Christos, 60, 61n12, 191, 192
Papayannopoulos, Philippos, 75n43, 76n47, 102n13, 244, 244n17
Paschalis, Vasilis, 63–64, 87
Pavlov, Boris, 78, 99n6
Peacocke, Christopher, 199n35, 260n29
Pellionisz, Andras, 103n15
Penrose, Roger, 77
Piccinini, Gualtiero, 2, 2n1, 4–6, 6n1, 7n3, 9n5, 10, 11, 14n13, 15–16, 16n16, 17, 18, 19–20, 21, 22–23, 43n42, 55, 58, 70, 70n38–39, 71–73, 73n40, 75n42, 78, 100n10, 103n14, 111, 114, 114n37–38, 116n41, 117n44, 118n45, 126n9, 127n13, 129n23, 137n42, 139, 140n45, 143, 145–148, 149, 150n5, 150n6, 151–153, 158, 158n15, 159, 160, 161n20, 163, 164n22, 165n25, 167–169, 170–173, 173n31, 175n1, 175n2, 176, 180–182, 183n16, 186n21, 186n22, 191, 192n29, 194–195, 196, 196n32, 197, 199, 216, 218, 219, 221, 222, 225–228
Pitowsky, Itamar, 70n37, 78, 79, 81, 81n55, 84n58, 85, 96–98, 99, 118
Pitts, Walter, 34n26, 103, 103n14, 111n30, 201n2
Plotkin, Gordon, 183n18
Pnueli, Amir, 63n25
Polger, Thomas, 122n3, 157n13
Post, Emil, 26n1, 28, 45, 45n45
Potgieter, Petrus, 80n52

Pour-El, Marian, 70n37, 75n46, 77–78, 147
Primiero, Giuseppe, 17
Proudfoot, Diane, 28n4
Putnam, Hillary, 119–120, 122, 122n3, 122n4, 123–124, 124n6, 125n8, 127, 127n15, 128–129n17, 130–132, 130n27, 135, 136n37, 136, 202–203
Pylyshyn, Zenon, 2, 3, 9, 93, 101n11, 110n28, 111n31, 112n34, 112n35, 197, 243

Quine, Willard Van Orman, 184
Quinon, Paula, 33n22

Ramsey, William, 23–24, 23n23, 24n24, 24n26, 95, 180n8, 180n9, 180n11, 181, 182, 184–185, 230
Rapaport, William, 130n26
Rescorla, Michael, 28n4, 164n22, 175n1, 176, 177, 186n21, 189–190, 190n26, 196n32, 199n35, 200, 201, 208
Richards, Ian, 70n37, 75n46, 77–78
Richardson, Robert, 146n3
Roberts, Bryan, 80
Robinson, David, 219, 233–234, 233n5, 236n9, 250
Rogers, Hartley Jr., 63n20
Rosen, Gideon, 41n35
Rosenfeld, Edward, 103n15
Rosinger, Elemér, 80n52
Rosser, Barkley, 27
Roth, Martin, 113, 114–115
Rumelhart, David, 103n17, 106n21, 107n25, 113, 114n39
Rusanen, Anna-Mari, 145n2, 153n11, 166n27–28, 254
Russell, Bertrand, 80
Ryder, Dan, 245

Scarantino, Andrea, 180, 181n12, 182
Scarpellini, Bruno, 77
Scheutz, Matthias, 129n21, 129n24, 135, 135n33, 136n38, 215n11
Schiller, Henry Ian, 216n14
Schmidhuber, Jürgen, 7n3, 106n22
Schneider, Susan, 9–10n8
Schwarz, Georg, 111, 112n33, 112n35, 115–116
Schweizer, Paul, 21n20, 24n25, 129n19, 129n20, 129n21
Scott, Dana, 183n19
Seager, William, 127n12
Searle, John, 9n7, 18n17, 120, 121–122, 123, 124, 125n8, 127, 128–129, 129n18, 129n21, 130n27, 135, 136, 202–203
Segal, Gabriel, 260n29

NAME INDEX

Sejnowski, Terrence, 8n4, 11, 13, 14, 16, 102, 110n27, 186, 197, 245
Seung, Sebastian, 107n25, 233, 236, 236n9, 237, 238n10, 250, 255
Shadmehr, Reza, 111n29
Shannon, Claude, 34n26, 183, 191
Shapiro, Lawrence, 122n3, 157, 157n13, 158, 260n29
Shapiro, Stewart, 28n4, 28n7, 28n8, 46n47, 62n15
Sharp, Patricia, 251
Shea, Nicholas, 186n22, 208, 221
Shenker, Orly, 195n30, 207n4, 225n17
Shepard, Roger, 243n16
Sher, Gila, 184n20
Shor, Peter, 67
Shub, Michael, 76n47
Sieg, Wilfried, 27, 29n9, 29n10, 30n15, 33n23, 34, 35, 35n30, 36n31, 38, 39n33, 41, 42, 43n42, 49n1, 68n34, 100n10
Siegel, Michael, 63n25
Siegel, Ralph, 246
Siegelmann, Hava, 116
Silverberg, Arnold, 260n29
Simon, Herbert, 2, 93, 154
Singerman, Eli, 63n25
Smale, Steve, 76n47
Smith, Brian, 6–7, 6n1, 6n2, 9, 71, 74, 265
Smolensky, Paul, 107n25, 114, 114n38
Soare, Robert, 35
Sompolinsky, Haim, 106
Sosic, Rok, 108n26
Sprevak, Mark, 3, 7n3, 17, 58, 79, 120n2, 130n26, 135n33–34, 136n39, 137n42, 138n43, 144, 175n1, 176, 178n5, 180–181, 182n14, 186, 186n21, 188n23, 189n25, 191, 196, 196n31, 199–200, 201, 216n14, 218, 260n29
Staiger, Ludwig, 80n52
Sterelny, Kim, 260n29
Stern, Peter, 8, 15n14
Stewart, Ian, 80
Stich, Stephen, 2, 93, 150n6, 177, 194
Strachey, Christopher, 183n19
Suárez, Mauricio, 232n4
Sutskever, Ilya, 106n22
Swoyer, Chris, 121, 229, 232, 253n26
Sylvan, Richard, 56n5
Syropoulos, Apostolos, 56n5

Tank, David, 107n25, 117
Taube, Jeffrey, 251
Teller, Paul, 42n39

Tenenbaum, Joshua, 99n7, 245n18
Travis, John, 8, 15n14
Tucker, Chris, 17
Tucker, John, 63n25, 80n53
Turing, Alan, 4, 12n10, 26, 26n1, 26n2, 27, 28, 28n6, 29n12, 30, 32, 33n22, 33n23, 34, 34n26, 35, 35n27, 35n29, 36, 36n31, 37–40, 42, 43, 43n42, 44–49, 52–55, 61, 62, 64, 64n28, 65, 75, 75n44, 75n45, 76n47, 243
Turner, Raymond, 191
Tymoczko, Thomas, 42n39

Ullman, Jeffrey, 60, 64
Ullman, Shimon, 252, 252n24
Ulmann, Bernd, 102n13, 244
Urquhart, Alasdair, 65n32
Uspensky, Vladimir, 62n18

Vardi, Moshe, 63
Vazirani, Umesh, 66, 67, 72, 74
Von Neumann, John, 30n15

Wang, Hao, 34–35, 44n44, 62n16
Weaver, Warren, 183
Weihrauch, Klaus, 75n46, 78n50
Weisberg, Michael, 9n6, 105n18, 157, 229n1
Weiskopf, Daniel, 155, 157, 158
Welch, Philip, 81
Weyl, Hermann, 80
Wheeler, John, 7n3
Wiener, Norbert, 183
Wiesel, Torsten, 165, 260
Williams, Ronald, 106n21
Williamson, Jon, 146n3
Wilson, Robert, 186n22
Winsberg, Eric, 9n6
Wise, Steven, 111n29
Wolf, Marty, 100n10, 201n1, 206n3
Wolfram, Stephen, 7n3, 70, 77
Woodward, James, 254

Yang, Yi, 143n46
Yanofsky, Noson, 63
Yao, Andrew Chi-Chih, 66
Yuille, Alan, 99n7

Zach, Richard, 29n10
Zee, David, 233, 233n5
Zenzen, Michael John, 77n49
Zhong, Ning, 78n50
Zipser, David, 106n20, 246, 249, 253
Zucker, Jeffrey, 63n25
Zuse, Konrad, 7

Subject Index

abstract
  causal organization, 139
  causal structure, 2, 139, 142
  as distinct from concrete computation, 6, 41n35, 63n25, 71, 72
  entities/object, 63, 65, 72, 91, 158n14
  See also explanation: abstract; property: abstract; machine: abstract
abstraction, 41n36, 157–158
  as contrasted with idealization, 42n38, 51
accelerating/accelerated machine.
  See machine: accelerating/accelerated
algorithm
  analog, 61
  definability, 63–65
  distributed, 61
  effective procedures and, 2, 26, 29, 41, 64, 65, 65n31
  hybrid, 61
  interactive, 61
  level (algorithmic), 159, 231, 247, 248, 258, 265
  non-effective, 64n26, 65
  parallel, 64, 64n29
  physically implementable, 64
  probabilistic, 167
  quantum, 61
  real-time, 61
  sequential, 61, 64, 64n28, 64n29
  symbolic, 64n28
algorithmic (machine) computation, 26, 41, 47, 49, 60, 61–69, 79, 83, 84, 86, 87
analog, 89, 107, 116n42, 117–118, 153, 159, 234, 236, 244n17
  computation, 17, 75n43, 96, 100, 100n10, 102–103, 137, 148, 217, 244
  computer, 14, 75n42, 87, 101n12, 118, 147, 244
  digital as distinct from, 17, 100–101, 101n11, 116, 117–118
  machine, 86
  real-valued neural network, 116
  representation, 244, 255

architectural accounts of computation, 2–3, 115, 118, 138, 139, 147, 149
architectural dogma, 88, 92, 119
artificial intelligence (AI), 9, 103, 106
attractor neural network (ANN). See neural network: attractor
automata (automaton), 60, 127n11, 133, 134, 135, 138, 143, 148, 201, 205, 206, 207, 208, 209–210, 214, 218, 226
  abstract, 1, 14, 71, 130
  cellular, 7, 138
  combinatorial state (CSA), 132, 133, 138, 205
  finite-state (FSA), 10, 20, 98, 111, 111nn30–31, 120, 123, 124, 128, 137
  implementation of, 2, 125, 126, 134, 136, 201, 208, 210, 211, 212, 214, 216–217, 218, 221, 224
  maximal, 214, 215, 215n11, 216
automata theory, 2, 26, 33, 34n26, 60, 116, 137, 191–192, 265
automatic formal system, 2, 149n4

broad conception of computation (BCC), The, 58, 83

Church-Turing thesis (CTT), 12, 28, 28n6, 31n17, 33–35, 34n26, 55, 57, 60, 64n29, 65, 65n31, 66, 70, 70n39
  CTT-Algorithm (CTT-A), 61, 66, 67, 84, 85
  CTT-Bold (CTT-B), 55–56, 57, 70
  CTT- Extended (CTT-E), 66–67
  CTT-Original (CTT-O), 61, 65n31
  See also Physical Church-Turing thesis
classical theories of cognition, 9, 103, 180, 185, 197, 243, 245
classification criteria (desiderata), 17, 24, 88, 92, 241
Cobham-Edmonds thesis, 66
cognitive criterion (desideratum), 6, 7, 9, 266
cognitive science, 3n2, 9, 10, 14, 21n21, 94, 99, 103, 128, 128n17, 135, 145, 152, 155, 158, 164, 177, 180, 196, 245, 267
communication, theory of (Shannon), 183.
  See also information theory
competence/performance distinction, 42–43

SUBJECT INDEX

computability
  axioms of, 36, 44, 52
  bit, 76n46
  effective, 26, 27, 30, 31, 32, 34, 35, 40, 45, 46–47, 75
  founders of, 26, 62, 65, 268
  human, 4, 44, 52
computability theory, 2, 25, 60, 74, 119, 146, 167, 174, 191, 265
computable function, 27, 28, 33, 35–36, 39, 41, 46, 61, 62, 66, 75–76, 77
computation
  effective, 26, 27, 28, 34, 37, 38, 40, 41, 43n42, 44, 46, 64, 65, 268
  human, 27, 47, 53, 61, 62, 62n15, 65
  indeterminacy of, 201n1, 212, 216n14, 222, 225
  machine, 4, 26, 27, 40–41, 46, 47, 49, 54, 55, 59, 61–65, 67, 69, 87, 100n10
  neural, 9, 21n21, 99, 103–117, 253
  parallel, 46–47, 49, 52
  probabilistic, 99
  quantum, 66, 67, 70
  real-valued, 75–76, 77, 78
  relativistic, 79–84
  theory of, 6, 7, 9, 133, 191
  See also abstract: as distinct from concrete computation; algorithmic (machine) computation; analog: computation; digital: computation; generic (machine) computation
computational complexity, 60, 65–67, 125n7, 266
computational description, 8–9, 9n5, 10, 13–14, 19, 21, 21n20, 22, 25, 94, 186, 187, 194
computational equivalence, 126, 134, 178, 189, 216, 217, 219–220, 221
computational externalism, 186, 186n22, 221
computational functionalism, 3, 127–128, 127n16, 196, 267
computational model, 8, 66, 74, 138, 145, 145n1, 162, 164n23, 261
computational perspectivalism, 194
computational sufficiency thesis (CST), 126–128, 126n10, 127n15, 136n41, 207n5
computational taxonomy, 17, 178, 207–211, 213, 214, 217, 222, 226, 227, 264
computational theories of cognition (CTC), 88, 198
computational theory of the mind, 3, 6, 9–10, 129n19, 196, 197, 267
computational vehicle. *See* vehicle
computationalism, 3, 122, 123, 127, 196
computational-level theories (CL).
  See explanation: computational; level: computational

computer model, 7, 8, 9, 9n6, 230
computer science, 1, 2, 8, 49, 62, 63, 63n25, 72, 92, 96, 114, 118, 125, 148, 177, 191, 192, 214
  textbooks, 34, 34n26, 60, 61, 61n12
  theoretical, 65, 67, 74
conceptual criterion (desideratum), 6, 7, 265
concrete computation. *See* computation: concrete; abstract: as distinct from concrete computation
connectionism, 103, 112
  conservative, 111, 111n32
  machine, 111n32
  PDP (parallel distributed processing), 103, 114n39
  system, 111, 112, 112n33
content, 12, 23, 95n3, 126n10, 133n30, 144, 150, 175n1, 177, 179–180, 180n8, 181, 182, 183, 187, 188, 189–190, 192, 193, 194, 199n34, 211–212, 213n10, 214, 216, 218, 219, 221–228, 259–260, 260n29, 264, 266
  adaptive-based, 182
  causal-based, 182
  change in, 198–199, 200, 213
  computational, 182, 193, 197–198
  formal, 213n10
  as functional role, 182
  informational, 11, 14, 177, 179, 180n8, 181, 182, 183, 191
  intentional, 177, 184
  isomorphism-based, 184–185
  mathematical, 213n10
  mental, 126n10, 127n16, 181, 182, 183n15, 197
  naturalistic accounts of, 182, 185
  theories of, 182, 185, 195, 196, 197, 200, 243
  See also pluralism: about computational content; representation; semantics

decidability, 29, 30, 31, 40, 79. *See also* undecidability
digital, 2, 64n28, 89, 138, 153, 159, 244
  analog as distinct from, 17, 100–101, 101n11, 116, 117–118
  architectural profile, 3
  computation, 2n1, 3, 100, 100n10, 145, 148, 177, 217
  computer, 9, 10, 14, 26, 50, 78, 104–105, 120, 121, 161, 201, 243
  mechanism, 147

edge detection, 22, 159, 166, 172, 173, 219, 227, 248, 252, 257, 259, 262
effective computation. *See* computation: effective

SUBJECT INDEX

eliminativist theories, 184–185
empirical criterion (desideratum), 6–7, 265
energy function, 106, 107, 108, 109, 110, 142, 156–157, 237
Entscheidungsproblem (Hilbert and Ackermann), 29, 29n12, 32–33
explanation
  abstract, 157–158, 164n23
  algorithmic, 258
  causal, 11, 16
  computational, 8, 9n5, 11, 16, 93, 146, 148, 149, 151–157, 157n14, 158–159, 160, 162, 163, 164, 165n25, 166, 174, 188, 216n13, 218, 219, 226–229, 241, 253, 258, 260, 261, 263
  computational-level (Marr), 257–258, 265 (*see also* level: computational)
  decompositional, 152, 155, 156–157, 163
  desideratum, 22–23, 265
  function-theoretic (FT) (Egan), 258–259
  implementational, 160, 258
  mechanistic, 24n26, 102, 146, 147, 148, 149, 151–154, 154n12, 157, 157n13, 158, 159, 160, 162–164, 166, 174, 260, 268
  optimality, 261, 263

functional analysis, 93–94, 93n2, 95, 102, 152–153, 152n8–9, 157, 157n13, 159
functionalism. *See* computational functionalism

Game of Life (Life), 47, 49, 50, 51, 52–53, 54, 61, 68, 69, 99
Gandy machine, 52, 53–54, 58, 59, 60, 68–69, 72–73, 85–86, 87
Gandy's thesis, 50, 52, 54, 55
Generic (machine) computation, 49, 55–60, 62, 67, 69, 86, 87, 148

halting problem/function, 56, 57, 59, 81, 83, 84, 85, 99, 192
human computability, Turing's analysis of, 4, 26, 27, 33–39, 40, 41, 42, 43, 44, 46, 52, 61
human computation (calculation). *See* computation: human
human computer ("computer"), 7, 8, 10, 12, 14, 18, 21, 26, 26n2, 27, 36, 38, 39–45, 46, 47, 52, 53, 54, 60, 61, 65n31
HUMAN computer, 53–54, 68, 86
hypercomputation (hypercomputing), 44, 46, 56n5, 83, 116
hypercomputer, 55–56, 62

idealization, 42, 42n39, 43n40, 71, 153, 230, 237
  as contrasted with abstraction, 42n38, 51

implementation
  as an account of computation, 4, 119–144 (chap. 5)
  Chalmers's account of, 119, 130–133, 135–137, 138, 140, 141, 143, 144
incompleteness results, 12, 12n11, 29, 31–32, 33. *See also* undecidability
indeterminacy of computation. *See* computation: indeterminacy of
information theory, 165, 262
  algorithmic, 183, 183n17
  Shannon's, 191
  See also communication, theory of
instantiation, 88n1, 89, 90, 93, 94–96, 99, 99n8, 100
instrumentalist theories, 184, 185
intentionality, 18n17, 177
interpretation, 6, 7, 18–19, 20, 21, 28n4, 90, 95, 130, 153, 153n11, 180, 183–184, 191, 192, 199, 245, 246, 247, 259, 261
interpretative model, 164, 261
intuition, 12, 13, 14, 15n15, 35, 198

lambda-definability, 27, 31n17, 32
level
  algorithmic, 153, 159, 247, 248, 258, 265
  of analysis, 247
  computational (Marr), 166, 166n28, 247–248, 257–258, 259, 260, 261
  functional, 216n13
  hierarchy of, 162, 164
  implementational, 160, 161, 216n13
  mechanistic, 161, 163
  tri-level framework (Marr), 160–161
local causation, Gandy's principle of, 51, 53, 58, 69, 73, 80
logical dogma, 2, 88, 92, 119
London-Tower Bridge diagram, 90, 91, 234, 243
lookup table, 11, 102–103, 242

machine, 19, 26, 40, 41, 46, 49, 50, 53, 55, 57, 63, 71–72, 132, 180, 210, 242, 244, 245n18, 248
  abstract, 63, 72, 74
  accelerating/accelerated, 74–75, 80, 80n52, 81n54, 86, 267
  calculating (computing), 40, 60, 54, 57, 59, 61, 72, 73, 74, 90, 91, 100n10, 178, 188n24, 241
  non-physical, 73
  notional, 55–56, 73, 74
  physical, 71, 72, 73, 87, 91, 208, 241
  physically realizable, 72, 73
  Pitowsky's (averaging), 96–98, 101, 118
  real-RAM, 75–76

machine (*cont.*)
  relativistic (RM), 79–80, 81, 82, 83, 84, 85, 86
    (*see also* computation: relativistic)
  shrinking, 80–81, 81n54
  supertask, 80
  vending (VEND), 189–190
  See also algorithmic (machine) computation; generic (machine) computation; oracle machine (*o*-machine)
mechanism, 17, 89, 97, 112, 153, 221, 226, 242, 244, 250, 257, 258, 259, 260
  biological, 106, 165
  causal, 136, 164
  computing, 10–11, 13, 16, 17, 19, 23, 197, 215, 216, 219, 222, 223
  concrete, 19
  external, 222–223
  inner, 253–254
  level of, 135, 160–161, 162
  model-to-mechanism mapping (3M) framework, 165, 165n26
  neural, 16, 252n25, 253, 254
  non-computing, 11, 149–150, 151
  sketch of, 153, 162, 164
mechanistic account of computation, the, 1, 4, 58, 129, 145–174 (chap. 6), 194, 219
medium-independence, 5, 20, 139–140, 147, 148, 149, 149n4, 151, 167, 168–169, 170–171, 173, 174
mirroring, input-output, 231, 232, 240
miscomputation, 10, 17, 148, 266
model, 8, 74, 114n39, 154, 155, 158, 161, 162n21, 164n23, 229, 230, 240–241, 243, 245, 245n18, 253, 261, 268
modeling account/notion of computation, the, 1, 5, 238–243, 263
modeling, input-output, 31, 232, 234, 239, 241, 243, 247, 248, 249, 250, 252, 253, 254, 255, 256, 258, 261, 263
  See also interpretive model; computational model; computer model; neural networks: model
multiple realization. *See* realization: multiple

naturalistic constraint, 3–4, 195–197
naturalistic theories (of content). *See* content: theories of
neither semantic nor non-semantic (NSNNS) view, 177, 177n4, 179, 190n26, 212
neural computation. *See* computation: neural
neural integrator, the (in the oculomotor system), 232–235, 249, 253, 254
  as an internal model, 235–238

neural network, 99, 103–110, 138
  abstract, 104, 137
  analog (real-valued), 116, 117
  architecture, 110, 111
  artificial, 245
  attractor (ANN), 103, 106, 107, 110–111, 118, 142, 236
  computation and, 110–111, 112, 115, 116, 118, 138, 142, 147–148
  digital, 117
  feed-forward, 106
  Hopfield, 107, 116–117, 118
  line attractor, 117, 118, 236, 237, 238
  model, 105–106, 106n20
  physical, 104, 105
  theory of, 173
  topology of, 173
  See also n-queens problem (network); computation: neural
neuroscience (brain sciences; neural sciences), 1, 4, 5, 8, 10, 15, 24n25, 118, 145, 146, 152, 160, 164, 165, 165n25, 195, 197, 235n8, 261n30
  cognitive, 111, 160, 180, 227, 229, 241, 249, 250, 253, 263, 266
  computational, 8, 103, 106, 111, 160, 161n18, 232, 245
non-semantic views of computation, 4–5, 130, 133, 137, 150n6, 175, 181–185
  See also semantic view of computation
normalization equation/function, 164, 165, 165n24, 261, 262
normativity, 42, 43n40, 181
notional machine. *See* machine: notional
n-queens problem (network), 68, 103, 105, 107–110, 112–113, 115, 117, 125, 125, 142, 144, 155

objectivity, 3, 10, 18, 19, 20–21, 24–25, 148, 194, 265
oracle machine (o-machine), 56
organizational invariance/ invariant, 5, 139, 140, 149n4

pancomputationalism, 3n2, 4, 15, 126, 150n6, 192–193, 192n29
  limited, 126, 133n31, 140–141, 143, 192–193
  unlimited, 93, 119, 125, 126, 133n31, 134
  very limited, 4, 193, 264
parallel computation. *See* algorithm: parallel; computation: parallel
path integration, 250–251, 251n23
PDP (parallel distributed processing). *See* connectionism: PDP

Physical Church-Turing thesis, 70, 75
  PCTT-B (bold physical), 70, 77–78
  PCTT-M (modest physical), 70, 76, 78, 79, 83–84
  super bold, 79
  See also Church-Turing thesis (CTT)
physical symbol system, 2
pluralism
  about computation, 4, 177n4
  about computational content, 180, 181, 182, 197, 232
probabilistic computation. *See* algorithm: probabilistic; computation: probabilistic
procedure
  decision, 33, 44
  effective, 2, 26, 29, 34n26, 40, 41, 60, 64, 65, 65n31, 148
  finite, 30–31, 31n17, 30n16, 35n27, 38, 45, 46
  infinite, 44
  mechanical, 11, 12, 30, 30n16, 32, 38n32, 43n42, 46, 62n16
  mundane, 65, 65n30
  recursive, 31n17
program, execution of, 2, 23, 88, 88n1, 91, 92, 94, 98n4, 106, 111, 112–115, 112n36, 114n37, 118
proof theory, 2, 265
property
  abstract, 41n35, 72, 122n4
  extrinsic, 211
  intrinsic, 210
  medium-independent, 41n35, 153, 158–159, 160, 161, 168, 169, 187
  non-abstract, 89
  non-semantic, 176, 185, 191, 192, 195, 212
  organizationally invariant, 140, 142
  semantic, 1, 5, 17, 39, 149, 150, 174, 175–180, 180n8, 181–186, 188, 189, 190, 191, 192, 195, 196, 199–200, 208, 212, 214, 218, 226, 264, 265, 266
  ·sensitivity to formal, 186, 199
  structural, 152, 154, 157, 159, 160, 166
  syntactic, 149, 177, 179, 191, 199, 243

quantum computation. *See* computation: quantum

randomness, 78, 99, 167
realization (realizable, realizability), 72, 80, 104, 122n4, 129n18, 161, 168
  multiple, 122, 122n3, 123, 139, 157, 157n13, 163, 168, 169, 216, 224

physical, 67, 70, 72–73, 74, 78, 79, 83, 86–87
universal, 123, 129n18, 132, 134n32
real-RAM machine model. *See* machine: real-RAM
reals, computability over the, 75–77, 78
recursive function, 27, 55, 62, 62n16
reinforcement learning, 161, 161n18, 162
relativistic computation. *See* computation: relativistic; machine: relativistic
representation
  admissible, 75n44
  analog, 244, 245
  cognitive, 177, 243
  input-output (simulation), 5, 95, 232
  mental, 23, 88, 168, 236, 243, 246, 250
  neural, 21
  structural (s-representation), 232, 243
rule, 2, 58, 67, 85, 92, 93, 132, 143, 146, 147–148, 149, 150, 167–168, 170–171, 173

semantic view of computation, the, 1, 4, 5, 144, 175–200 (chap. 7), 208n6, 212, 213, 219, 225, 226, 228
  C-semantic, 178–179, 188, 189, 212
  E-semantic, 178, 179, 189, 208, 212
  master argument for, 189, 192, 207–228
  objections to, 175, 189–201
  standard argument for, 188, 189, 212
  See also non-semantic views of computation
semantics
  denotational, 184
  interpretational, 88, 243
  operational, 183
  non-semantic theories of, 179–185
  See also content: theories of
simple mapping account, the, 129, 129n19
simulation, 7, 8, 9, 9n6, 57, 58, 66, 230
simulation representation (input-output representation). *See* representation: input-output (simulation)
simultaneous implementation, 201–206, 207–208, 207n4, 208, 215n11, 216, 225, 228. *See also* computation: indeterminacy of
spin-glass system, 106, 107, 142, 143–144
step-satisfaction, 2, 139, 149
  as an account of computation (Cummins's), 4, 5, 88–118 (chap. 4), 243
structural representation (S-representation). *See* representation: structural
substantivity constraint (premise), 3, 3n2, 3, 24
supertask, 44, 56, 80, 80n53, 81, 84–85, 85n60
surrogative reasoning, 253, 253n26

SUBJECT INDEX

syntactic accounts of computation, 150, 150n6
syntactic engine, 2
syntactic structure, 10n8, 91, 198. *See also* property: syntactic
syntax, 39, 199n34

taxonomy. *See* computational taxonomy
teleological function, 5, 20, 143–144, 146, 148, 149, 150, 151, 167, 169, 170–174, 194, 195, 197
the-right-things-compute (desideratum), 14, 17, 148, 188, 241, 264
the-wrong-things-don't-compute (desideratum), 10, 11, 14, 15–16, 17, 148, 149, 188, 242, 264
triviality results, 21, 119n1, 120, 144, 150n6, 268
    avoiding, 129–130, 133, 203
    implications of, 124–129, 133
    Putnam's, 119, 123–124, 129, 202
    Searle's, 119, 120–122, 129, 130, 202
    strong, 119, 133–134, 135–137, 144, 207n4
    weak, 133–135, 140, 141, 144, 207n4

Turing machine, 14, 28, 41, 53, 53n4, 54, 60, 61n13, 62n16, 65, 69, 71, 72, 73, 76, 81, 82, 84, 85, 86, 112, 137, 138, 191, 206
    abstract, 104
    accelerating, 73, 80
    infinite-time, 56–60, 57n10, 67, 69, 79, 83
    physical, 104
    probabilistic, 66
    universal, 10, 26, 55, 57, 77, 79
type 2 theory of effectivity (TTE), 75–76n46

undecidability, 30n15, 31, 33, 79. *See also* decidability

vehicle, 3, 146, 147–148, 150–151, 153, 159, 168, 170–171, 186n22, 187, 198, 208, 213, 213n10, 265

"what-if-things-had-been-different questions," 254

Zipser and Andersen's model, 106n20, 246, 249, 253